Laboratory Manual of Cell Biology

Edited for the
British Society for Cell Biology by

D.O.Hall and Shirley E.Hawkins
School of Biological Sciences,
University of London King's College

 THE ENGLISH UNIVERSITIES PRESS LIMITED

ISBN 0 340 18746 8 Boards
ISBN 0 340 16814 5 Paperback

First printed 1975

The English Universities Press Ltd
St Paul's House, Warwick Lane, London EC4P 4AH

Printed and bound in Great Britain by Clarke, Doble & Brendan Ltd

Computer Typesetting by Print Origination, Bootle, Merseyside, L20 6NS

Contents

Foreword

In many schools, and even some colleges and universities, cell biology is treated as a theoretical rather than an experimental subject. Indeed, the idea of setting up class experiments in cell biology seems to fill some teachers with panic. This manual of experiments should be especially welcome for all those who are properly concerned with the experimental basis of such an important discipline. Particular credit goes to the editors, Prof. David Hall and Dr. Shirley Hawkins, as well as to the individual contributors. In introducing the Laboratory Manual of Cell Biology it is a pleasure to thank them all on behalf of the British Society for Cell Biology.

M. G. P. STOKER, M.D., F.R.S.
Hon. President, British Society for Cell Biology,
Imperial Cancer Research Fund Laboratories
Lincoln's Inn Fields
London WC2A 3PX

Introduction

Cell biology has become a subject taught in its own right to undergraduates both in the biological and medical sciences over the last five to eight years. However, in order to teach it successfully, the teacher must rely heavily on practical techniques both at the simpler and more complex levels. The student should be exposed to the methodology, which is often distinct from biochemistry and which requires different manipulative techniques for full understanding of cellular functioning.

At the present time there is a dearth of published laboratory experiments in cell biology, and the teacher has to be inventive in order to present a varied practical course. With this in mind, the British Society for Cell Biology has commissioned this manual in the hope that it will help both to broaden existing courses and to encourage the setting up of new courses for undergraduates and for advanced school students.

All members of the Society were invited to submit experiments which they had used successfully, written to a standard format in order to provide the maximum amount of information, even though some may appear excessive. We felt that it was better to have too much rather than too little information. We have edited each manuscript with this in mind, so that teacher, technician and student can readily find their way around each experiment.

The overall nature of the manual was determined by the experiments which were submitted. We have arranged these into fourteen broad sections and, within each section, have attempted to start with the simpler experiments followed by those of increasing complexity. This is purely arbitrary, but it is more difficult to set up simple experiments which are both instructive and interesting to the student.

We would like to receive comments, favourable or not, from people who may use these experiments, since this will help us, both in our own teaching and in any future publications by the Society. The Society has set up an Education Sub-Committee concerned with the overall aspects of teaching Cell Biology. Any income from this manual will be used by the Sub-Committee for the furtherance of this purpose.

D.O. HALL
SHIRLEY E. HAWKINS
University of London King's College
May 1973

Section 1

Cell and Tissue Culture Techniques

1 Subculture of a cell line: initiation of monolayer cell cultures in bottles

Time required
20 min.

Assumed knowledge
Standard aseptic techniques as applied to virology (in addition to measures to avoid bacterial contamination, pipettes must be operated with rubber bulbs to protect both the operator and the cultures from infection with viruses or mycoplasmas).

Theoretical background
The first stage in any subcultivation procedure is to obtain the cells in suspension. The nature of the cell-to-matrix bond is not fully understood, but it is susceptible to attack both by proteolytic enzymes and by chelating agents, which have an affinity for the calcium and magnesium ions responsible for preserving the integrity of the bond. The method to be described uses a mixture of trypsin and ethylene diaminetetraacetate (EDTA, Versene). This combination has the advantage of acting some fifty times faster than either component alone. Hence exposure to agents which potentially damage cellular membranes is kept to a minimum. Further effect of the enzyme and chelating agent is inhibited by suspending the cells in growth medium which contains serum. At this stage it is important to produce a suspension comprising single cells, since clumps may adversely affect the quality of subsequent cultures.

The number of new cultures seeded from one original culture depends on the cell line, the frequency of subcultivation and the time available for incubation. A seeding rate of 5×10^4 cells/cm^2 is generally found to be suitable. The subcultivation regime to be described should produce healthy, confluent cultures within three days.

Necessary equipment

	Equipment/media, etc.	Amount required per student
1	Growth medium: Eagles'[1] Minimum essential medium	50 ml

[1] Eagles, H.
1959 Science 130, 432

1

containing 10% calf serum,
0·1% sodium bicarbonate
(final concentration),
200 iu/ml penicillin,
100 iu/ml streptomycin
Source: Wellcome Reagents Ltd.

2 Trypsinisation medium: 10 ml
Versene (1:5000) Containing
trypsin (5%), giving
a final concentration of 0·125%
Source: Wellcome Reagents Ltd.

3 HEp-2 cells (human   1 x 4oz medical
carcinoma of larynx)[2]   flat culture
Source: Biocult

4 4 oz medical flat bottles   3
(sterile)
Source: Universal Glass Container Ltd.

5 500 ml beaker   1

6 10 ml pipette (sterile)   2

7 2 ml pipette (sterile)   1

8 10 ml pipette bulb   1

9 2 ml pipette bulb   1
Source: Esco Rubber Co.

10 Microscope (x 5 objective,   1 per 3
x 8 eyepieces)   students

11 37°C incubator   1 per class

[2]Moore, A.E. Sabachewsky, L. and Toolan, H.W.
1955 Cancer Res. 15, 598
N.B. Other established cell lines may be used in place of
HEp-2. Additional pipettes will be required if the students
are instructed to make their medium from its components.

Table 1   *(reproduced by permission, from Whitaker, 1972)*

Dimensions of vessels commonly used for cell culture

Vessel	Type of glass	Dimensions (cm)	Culture area (cm^2)	Total volume (ml)	Volume of medium (ml)
4in x ½in test tube	Pyrex soda	4·7 x 1·2	2	8	1
6in x ⅝in test tube	Pyrex soda	7·0 x 1·4	10	20	2
2 oz medical bottle	Soda	3·0 x 7·5	22·5	60	10
Baby feeding bottle	Pyrex	3·0 x 11·0	33	250	20
4 oz medical bottle	Soda	4·0 x 9·0	36	120	10–20
12 oz medical bottle	Soda	6·5 x 14·0	91	360	40
20 oz medical bottle	Soda	7·0 x 16·0	112	600	75
Roux bottle	Pyrex	10·0 x 20·0	200	1 200	100
Thompson bottle	Pyrex	15·0 x 25·0	375	4 000	250
Povitski bottle	Pyrex	16·0 x 37·0	600	5 000	500

Cleaning of Glassware

1 1% Pyroneg (Diversey Ltd) free-steamed at 100°C for one hour, cell debris removed by
   brushing whilst still hot, and then rinsed in hot tap water.
2 Decon 75 at 1:100 or 1:200 overnight.
The washing process by both methods is completed by several rinses in cold tap water
followed by at least six rinses in distilled water.

Table 2 Data for the use of Antibiotics in Cell Cultures (*reproduced by permission, from Whitaker, 1972*)

Antibiotic	Trade Name	Manufacturer	Reported Antibacterial Spectrum	Reported Inhibitory Concentration* for most Contaminants	Maximum Non-cytotoxic Concentration*+	Concentration*+ of Antibiotic Suitable for use in cell cultures
Kanamycin-sulphateϕ	Kanasig	Sigma	Broad spectrum	0·6 - 50	1,000	10
Amphotericin Bϕ	Fungi Zone	Squibb	Yeasts, moulds	0·2 - 2	4 - 8	2 - 4
Nystatin	Mycostatin	Squibb	Yeasts, moulds	10 - 100	75 - 100	25 - 50
Chloramphenicolϕ	Chloromycetin	Parke-Davis	Broad spectrum	0·2 - 20	10 - 15	5 - 10
Oxytetracycline HCL with ascorbic acid	Terramycin	Pfizer	Gram −ve bacilli	0·4 - 50	10 - 20	10
Sodium Penicillin G		Beecham	Gram +ve organisms	0.005 - 20	900 - 1000	100 - 200
Streptomycin sulphate		Glaxo	Gram −ve organisms	1 - 50	900 - 1000	50 - 100
Dehydrostreptomycin sulphate		Glaxo	Gram −ve organisms	1 - 50	1000	50 - 100
Neomycinϕ		Andrews	Broad spectrum	0·1 - 25	800 - 1000	10 - 100
Polymyxin Bϕ		Burroughs Wellcome & Co	Gram −ve bacilli	0·5 - 50	50 - 75	20 - 50
Bacitracin		Andrews	Gram +ve cocci Gram +ve bacilli Gram −ve cocci		750 - 1000	100
Viomycin		Parke-Davis	Broad spectrum & mycobacteria	0·7 - 100	700 - 1000	100
Chlortetracycline	Aureomycin	Lederle	Broad spectrum	0·1 - 10	30 - 50	20 - 40

* Concentration expressed as μg/ml except penicillin and nystatin which are expressed as international units/ml
+ Cells tests: Human embryo skin; Primary monkey kidney; HEp2
ϕ Recommended for the elimination of most micro-organisms

Type of Group
This experiment is suitable for individual students working in large classes.

Instructions to the student
1 Take HEp-2 culture, pour off the spent growth medium and overlay the cell sheet with 10 ml prewarmed (37°C) trypsinisation medium. Leave for 30 s.
2 Pour off the trypsinisation medium until it starts to drip (this leaves about 0·5 ml behind in the bottle). Leave the bottle on the bench for 1 minute.
3 Arrange three new culture vessels on the bench vertically with the caps loosened.
4 Add 3·5 ml growth medium to the original culture vessel, wash the cells off the glass with a pipette and vigorously suck up and down several times to break up clumps of cells.
5 Distribute 1 ml aliquots of this suspension to the new culture vessels, leaving 1 ml behind in the original vessel.
6 Add a further 9 ml growth medium to each vessel. Tighten cap securely.
7 Label each bottle (add one to the indicated subculture level on the cell seed bottle to give the new subculture number).
8 Incubate at 37°C for 2–3 days.

Recording results
1 Observe the colour of the phenol red indicator in the medium of each culture and roughly estimate its pH. A high pH (red−purple) may be due to poor cell growth or to escape of CO_2 from the atmosphere from inadequately sealed vessels.
2 Note the degree of confluency.
3 Describe the predominant morphology of the cells.

Extensions and applications
Subcultivation is the basic process in handling a cell line. It enables large quantities of cells or many cultures to be produced for experimental purposes or for storage.

Further reading
Whitaker, A.M.
1972 'Tissue and Cell Culture', Baillière Tindall, London

Hints for the Teacher
It is important to ensure that the trypsinisation medium is warmed to 37°C before the start of the experiment. It is useful to prepare two batches of cultures for the class on separate occasions in case one batch is lost through accidental contamination.

A.M. WHITAKER
Wellcome Research

2 Subculture of a cell line: initiation of monolayer cell cultures in test tubes

Time required
45 min.

Assumed knowledge
As for previous experiment, together with some knowledge of the use of an improved Neubauer cell counting chamber.

Theoretical background
Many of the processes used in this experiment are similar to those described in the one previously. Cell suspensions tend to settle out rapidly and thus must be shaken frequently when used to fill out large numbers of vessels. Since most of the manipulations are carried out with the test-tubes in a vertical position it is also important to shake the tubes carefully to ensure an even cell suspension before incubation.

Necessary equipment

	Equipment/media, etc,	Amount required per student
1	Growth medium: Eagles' Minimum essential medium containing 10% calf serum, 0·1% sodium bicarbonate (final concentration), 200 iu/ml penicillin, 100 iu/ml streptomycin *Source:* Wellcome Reagents Ltd.,	150 ml
2	Trypsinisation medium: Versene (1:5000) containing trypsin, (5%) giving a final concentration of 0·125% *Source:* Wellcome Reagents Ltd.	50 ml
3	HE-p2 cells (human carcinoma of larynx) *Source:* Biocult	3 x 4 oz medical flat cultures

4	4 oz medical flat bottle (sterile) Source: Universal Glass Container Ltd.	1
5	¼ oz Bijou bottles (sterile) Source: Universal Glass Container Ltd.	1
6	6in x 5/8in test tubes in racks each covered with a loose-fitting oxoid metal cap (sterile) or 5in x ½in test tubes—if these are used, seed tubes with 1 ml. Source: Universal Glass Container Ltd. (Racks by Luckham Ltd.)	50
7	500 ml beaker	1
8	10 ml pipette (sterile)	5
9	2 ml pipette (sterile)	5
10	10 ml pipette bulb	1
11	2 ml pipette bulb	1
12	Rubber stoppers for test tubes (sterile) Source: Esco Rubber Co.,	50
13	Trypan blue stain (0·5% aq) BOH	2 ml
14	Improved Neubauer Counting Chamber with cover slip Source: A.R. Horwell.	1 per 3 students
15	Microscope (x 5 x 10 objectives; x 8 eyepiece)	1 per 3 students
16	37°C incubator Source: A.R. Horwell.	1 per class

Type of group

This experiment is suitable for individual students working in large classes.

Instructions to the student

1 Take one bottle HEp-2 culture, pour off the spent growth medium and overlay the cell sheet with 10 ml prewarmed (37°C) trypsinisation medium. Leave for 30 s.
2 Pour off the trypsinisation medium until it starts to drip (this leaves about 0·5 ml behind in the bottle). Leave the bottle on the bench for a further 1 minute.
3 Treat the other two cultures in the same way.
4 Wash the cells off the glass with 2 ml growth medium from a pipette. Aspirate thoroughly to obtain a single cell suspension.
5 Pool the cell suspension and adjust the volume to 10·5 ml.
6 Transfer 0·5 ml of this suspension to a ¼ oz bottle. Add 1·5 ml trypan blue stain. Mix.
7 Introduce one drop of the stained suspension into an improved Neubauer Counting Chamber and count the total number of cells in each of the four sets of 16 squares (magnification x 80).

N.B. This stain only penetrates those cells whose membranes are damaged. Stain is excluded by healthy cells. Although the correlation is not exact, all stained cells are counted as dead and all unstained cells as viable.

8 Calculate the total number of viable and non-viable cells by the application of the following formula.

$$N/4 \times D \times 10\ 000 = \text{cells/ml}$$

when N = total number of cells in all 4 sets of 16 squares

D = dilution factor caused by the addition of stain.

9 Resuspend 10×10^6 viable cells in 100 ml growth medium in a 4 oz medical flat bottle.
10 Dispense 2 ml aliquots of this suspension into each of the test tubes.
11 Remove the metal caps and replace with rubber stoppers. Press these home firmly, but sensibly (warning—if the mouth of the test tube is cracked the tube may break and cause serious cuts).
12 Shake the test tube rack to obtain an even cell distribution and incubate at an angle of 20°.

Recording results

1 Observe the colour of the indicator in each tube—see previous experiment.

2 Note area of cell sheet.
3 Describe cell morphology.

Extensions and applications
Test tubes are the smallest convenient individual culture vessels available. Thus they are particularly useful when large numbers of replicate cultures are required, for example for titration of viruses.

Further reading
Whitaker, A.M.
'Tissue and Cell Culture', Baillière Tindall,
London
1972

Hints for the teacher
Rubber bungs may be rendered non-toxic for cell cultures by boiling in soap powder, followed by thorough rinsing with tap and distilled water. It is useful to prepare a stock suspension of cells to keep in reserve in case students fail to obtain a sufficient number of cells from their bottles.

A.M. WHITAKER
Wellcome Research Labs.

3 Subculture of a cell line: preparation of coverslip cultures

Time required
20 min.

Assumed knowledge
See previous experiment.

Theoretical background
In addition to those factors covered by the previous two experiments, the process to be described makes use of an alternative buffer system. Most cell culture media include a buffer system modelled on the naturally occurring CO_2-bicarbonate system present in blood plasma. However, this has a number of disadvantages. Sodium bicarbonate provides sub-optimal buffering over the physiological range, CO_2 is readily lost from the media causing a rise in pH; and moreover sodium bicarbonate is quite toxic for some cells. A range of Zwitterionic Biological Buffers has recently become available which are non-toxic and show optimal buffering over the physiological range. 4—(2-hydroxyethyl)-1-piperazine ethane sulphonic acid (HEPES) is one such buffer. It will control the pH of a culture medium close to neutrality in free gas exchange with air.

Necessary equipment

	Equipment/media, etc.	Amount required per student
1	Growth medium: Eagles' Minimum essential medium containing 10% calf serum, 3% HEPES buffer (1M) 200 iu/ml penicillin, 100 iu/ml streptomycin *Source:* Wellcome Reagents Ltd.	50 ml
2	Trypsinisation medium: Versene (1:5000) containing trypsin, (5% solution) giving a	20 ml

final concentration of
0·125%
Source: Wellcome Reagents Ltd.

3	HEp-2 (human carcinoma of larynx) *Source:* Biocult	2 x 4 oz medical flat cultures
4	4 oz medical flat bottle (sterile) *Source:* Universal Glass Container Ltd.	1
5	¼ oz Bijou bottles (sterile)	1
6	500 ml beaker	1
7	10 ml pipette (sterile)	5
8	2 ml pipette (sterile)	5
9	10 ml pipette bulb *Source:* Esco Rubber Co.,	1
10	2 ml pipette bulb	1
11	Glass cover slips *Source:* Chance & Propper	6
12	4in Petri dishes (sterile)	2
13	Trypan blue stain (0·5%aq) BDH	2 ml
14	Forceps (sterile)	1 pair
15	Improved Neubauer Counting Chamber with cover slip *Source:* A.R. Horwell	1 per 3 students
16	Microscope (x 5 and x 10 objectives; x 8 eyepieces)	1 per 3 students
17	37°C incubator	1 per class

Type of group
This experiment is suitable for individual students working in large classes.

Instructions to the student
1 Take one bottle HEp-2 culture, pour off the spent growth mediums and overlay the cell sheet with 10 ml prewarmed (37°C) trypsinisation medium. Leave for 30 s.

2 Pour off the trypsinisation medium until it starts to drip (this leaves about 0·5 ml behind in the bottle). Leave the bottle on the bench for a further 1 minute.
3 Treat the other bottle of HEp-2 in the same way.
4 Wash the cells off the glass with 2 ml growth medium from a pipette. Aspirate thoroughly to obtain a single cell suspension.
5 Pool the cell suspensions and adjust the volume to 10·5 ml.
6 Transfer 0·5 ml of this suspension to a ¼ oz bottle. Add 1·5 ml trypan blue stain. Mix.
7 Estimate number of cells/ml in a Neubauer Counting Chamber (see previous experiment).
8 Resuspend 3×10^6 cells in 40 ml growth medium.
9 Transfer 3 clean sterile cover slips with sterile forceps into each of two Petri dishes.
10 Add 20 ml of the resuspended cells to each petri dish.
11 Incubate. Make sure that there is adequate humidity in the incubator by the provision of an open dish of distilled water.

Extensions and applications
Cover slip cultures make it possible to prepare permanent stained records of cultured cells or of cytopathic effects occuring in them.

Further reading
Whitaker, A.M.
1972 'Tissue and Cell Culture'. Baillière Tindall, London,

Hints for the teacher
Cover slips should be sterilised individually, for example each in a stoppered tube. It is often difficult to separate cover-slips if they are sterilised together.

A.M. WHITAKER
Wellcome Research Labs.

4 Comparison of protective affects of glycerol and dimethylsulphoxide on cell freezing

Time required

Session	I	1½ hr.
Session	II	15 min.
Session	III	1¼ hr.
Session	IV	½ hr.

A gap of 1½ hr. should be left between sessions I and II, a gap of whatever time is convenient between II and III and a gap of 24–48 hr. between III and IV.

Assumed knowledge

Basic cell structure. Superficial knowledge of molecular diffusion. Principle of sequestration of water molecules by hydrophilic compounds. Sterile technique. Manipulation and trypsinisation of monolayer cell cultures. Counting of cells by either Coulter counter or haemocytometer.

Theoretical background

Animal cells from tissue cultures may be stored indefinitely in liquid N_2, in the presence of a cryoprotective agent. The agent (e.g. glycerol or dimethyl sulphoxide) protects the cell from physical injury due to ice-crystal formation and may also protect against injury from osmotic damage. The cells should be frozen slowly (1°C/min) and thawed rapidly to obtain maximum survival. This rate is a compromise between the rapid rate of freezing desirable to minimise ice-crystal formation and the slow rate of freezing desirable to permit exit of water from the cell again to prevent ice-crystal formation within cellular organelles.

Necessary equipment

	Sessions I and II	Amount required (per student unless otherwise stated).
1	0·25% trypsin. Dilute from	

	2·5% (Flow Labs) in balanced salt solution or isotonic saline citrate	10 ml
2	50 cm² flask of L-929 cells (or alternative) subconfluent (in log phase of growth)	1
3	Minimal Essential Medium (Eagle) with Hanks' salts containing 0·35 g/l sodium bicarbonate, 10% calf serum, 50 μg/ml Gentamicin (Flow Labs)	20 ml
4	Dimethyl sulphoxide (autoclaved) (DMSO) *Source:* BDH	2 ml
5	Glycerol (autoclaved)	2 ml
6	1 ml ampoules *Source:* Gallenkamp	3
7	Ice bath	1
8	Ampoule sealer (O_2-Gas flame)	1 (shared)
9	Insulated cooling box (expanded polystyrene, approximately 1·5 cm wall thickness, 20–30 cm long x 10 cm wide x 10 cm deep. These sizes are approximate and need not be too closely adhered to; however, the wall thickness should not be altered more than 20%.)	1 (shared)
10	2 ml syringes *Source:* Gillette Surgical	3
11	Bijou bottles *Source:* Universal Glass Container Ltd.	3
12	Universal container *Source:* Gallenkamp	1
13	Liquid N_2 freezer (vapour phase)	

8

Source: (Union Carbide,
or BOC)   1 (shared)

14 Coulter or Celloscope cell
    counter or haemocytometer
    slide and microscopes   1 shared/each

15 −70°C freezer chest
    or 'Drikold' chest   1 (shared)

16 Pipettes: 2 ml, Pasteur
    (autoclaved in cans)
    Source: Gallenkamp   1 tin

Session III

1   25 cm² culture flasks
    Source: Falcon, Greiner,
    Flow Labs.   3

2   Water bath with lid   2 (shared)

3   Gloves and goggles   3 sets

4   Medium   20 ml

5   1 ml syringes   3

6   Ice tray   1

7   1 ml pipettes   1 tin
    10 ml pipettes   1 tin
    Pasteur pipettes   1 tin

8   Viability stain (1%
    Naphthalene Black (BDH) in
    balanced salt solution or isotonic saline 5 ml

9   Slides and cover slips

10 Microscopes (10 x objective)

Session IV

1   Isotonic citrate (counting
    fluid; for Coulter counter
    only)   100 ml

2   Universal containers   2

3   Microscopes

4   Leishman's stain (or May
    Grunwald & Giemsa)

Source: Gurr   5 ml

5   Trypsin
    Source: Flow   10 ml

6   Haemocytometers
    Source: Gallenkamp   1

Type of Group
6-12 students working individually but divided into three groups.

Instruction to the student
A suspension of cultured cells, prepared by trypsinisation, is frozen slowly in the presence of a preservative and held at −196°C for a short time before thawing and examining cell viability by dye exclusion and growth potential. The relative values of glycerol and dimethylsulphoxide as preservatives are compared.

Session I

1 Trypsinise the 50 cm² bottle of monolayer-growing cells. These should be subconfluent and in the log phase of growth.
2 Resuspend the cells in 2 ml medium and remove 0·2 ml of the cell suspension for counting on Coulter counter or haemocytometer.
3 Dilute the suspension if necessary in medium to obtain a cell concentration of $2·5 \times 10^6$ cells per ml.
4 GROUP A. Add dimethylsulphoxide to final concentration of 10% v/v.
   GROUP B. Add glycerol to final concentration of 10% v/v.
   GROUP C. Add both dimethylsulphoxide and glycerol to final concentration of 10% each.
5 Place the DMSO-containing samples on ice.
6 Inoculate ampoules at 1 ml each.
7 Seal the ampoules in gas/oxygen flame.

N.B. Ampoules must be perfectly sealed for liquid phase storage or inhaled nitrogen will cause an explosion on thawing.

8 Place in cooling box (keep DMSO samples on ice till all ampoules ready) and place at -70°C for 1½–2 hours. This will cool the cells at approximately 1°C/min. to below the critical level (-50°C).

Session II

9 Transfer the ampoules from the cooling box to the liquid-nitrogen freezer. (For reasons of safety

and ease of handling, vapour-phase storage should be used.)

N.B. Gloves and goggles *MUST* be worn when handling material in liquid nitrogen.

Session III

10 Thaw the frozen cells rapidly by transferring the ampoules to a water bath at 37°C.
    If the ampoules are thawed from the liquid nitrogen phase, it is important to transfer the ampoules to a container with a lid to give protection if the ampoules explode.

11 Inoculate two 25 cm² flasks, one ampoule per flask (with 0·5 ml cell suspension) and 4·5 ml medium. Inoculate two more flasks with 0·1 ml of cell suspension and 4·9 ml medium. (For some cell strains it may be an advantage to remove the dimethyl sulphoxide before inoculation of flasks, by centrifugation at approximately 100 g; this may be added as an extra experimental variable if desired.

12 Add 1 drop naphthalene black to 2 drops of the cell suspension remaining in the ampoule and estimate the viability by determining the number of unstained cells (viable) as a proportion of the total number of cells.

Session IV

13 Stain one flask with Leishman and compare the degree of confluence in the flasks from Groups A, B and C.

14 Remove the medium from the remaining flasks, add 5 ml of trypsin solution, leave for 15 min at 37.°C. Resuspend the cells and count 1 ml in 50 ml of counting fluid using the Coulter counter or count directly on haemocytometer.

Discussion

1 Which cryoprotective agent is best? Are their effects additive? Suggest a possible mechanism of action.

2 Is naphthalene black exclusion a good criterion of viability?

3 Is cell survival dependent on inoculum density at thawing? If so, why should it be?

Extensions and applications

Cultured cell lines are prone both to contamination by microorganisms and to genotypic or phenotypic instability. Freeze storage in liquid nitrogen has proved invaluable to guard against loss of a culture or its characteristics. A common regime is:

1 Grow up a standard cell population, check that it possesses the required characteristics and it is free of contamination from (a) other cell lines, or (b) microorganisms including mycoplasma and, if possible, viruses.

2 Freeze a large enough stock to provide enough seed material for the predicted programme of work.

3 Thaw a seed ampoule every three months, grow up to required stock, check characteristics and discard previous stock. In this way the culture is never retained longer than three months and retains some degree of consistency.

Other parameters of variation which may be investigated are (a) other cryoprotectives such as sucrose or polyvinylpyrolidone, (b) the rate of

Recording

Protective	Dye exclusion on thawing	Estimated % confluence after 24—48 hours	Counted after 24 hours	
			High-concentration inoculum	Low-concentration inoculum
Glycerol				
DMSO				
Both				

Prepare composite table with means and variation and compare different conditions.

ooling in a controlled-rate cooler or in cooling boxes
f different insulation, (c) serum concentration, (d)
ne use of conditioned medium to inoculate cells into
fter thawing.

Care should be taken to avoid contact of DMSO to
rubber liners, etc., as it may dissolve toxic components out of the rubber. DMSO is a powerful skin
penetrant so make sure your hands are clean before
you use it!

R.I. FRESHNEY
Glasgow.

urther reading

1 Doebbler, G.F.
       1966 Cryobiology, 3, 2
2 Leibo, S.P., Farrant, J., Mazur, P., Hanna, M.G.
       and Smith, L.H.
       1970 Cryobiology, 6, 315
3 Lovelock, J.E. and Bishop, M.W.H.
       1959 Nature, 183, 394
4 Mazur, P.
       1962 Nat. Cancer Inst. Mono., 7, 13
5 Merryman, H.T.
       1966 Cryobiology, Academic Press, London
6 Paul, J.
       1970 Cell and Tissue Culture, Livingstone,
       Edinburgh
7 Porterfield, J.S., Ashwood-Smith, M.J.
       1962 Nature, 193, 548
8 Stulberg, C.C., Peterson, W.D. and Berman, L.
       1962 Nat. Cancer Inst. Mono., 7, 17.

Hints for the teacher

Sealing of ampoules and any manipulation of liquid
nitrogen are best performed by the teacher. Some
practice in sealing ampoules should be gained before
the class takes place.

Labelling ampoules is best done with transparent
polythene Scotch Tape (Sellotape)—the kind you can
write on—as magic marker or wax pencil tends to
come off.

Make sure you know where each set of ampoules is
located in the liquid-nitrogen freezer or recovery may
take too long and slow thawing may commence.
-70°C to liquid nitrogen—this step must be very rapid,
as the cells should not rise above -50°C.

If haemocytometer or Coulter counters are to be
used make sure the students know beforehand how to
use them.

Each student will need a 50 cm^2 flask of cells
which should be set up at 10^5 cells per ml in 10 ml
five days before the class and should be just ap-
proaching confluence on the day of trypsinisation.
Any cell strain available may be used. Differences
between strains may be examined if desired.

All pipettes, glassware and reagents down to item
7, Session III, on the materials list should be sterile.
Dimethyl sulphoxide and glycerol may be autoclaved.

11

5 Features of cell behaviour and morphology in culture

Time required
2 hr. to set up cultures. 2 hr. on following day to examine cultures.

Assumed knowledge
It is assumed that the student is moderately familiar with sterile technique and that he has received a small amount of formal teaching on the purpose, problems and technique of cell culture. In normal practice, second-year university classes find little trouble with this practical.

Theoretical background
The purposes of this practical are (a) to introduce you to a general technique of cell biology, which can be used for more advanced experiments (see 'Extensions and applications'), (b) to demonstrate features of cell behaviour in culture. There is no complex theory behind the technique, but the observations made should provoke a large number of questions, some of which are outlined in the 'Discussion' section.

Necessary equipment
Two different techniques can be used. A. Petri-dish culture, normally requiring inspection with an inverted phase contrast microscope. B. Hanging-drop culture requiring cavity slides and inspection by normal phase contrast microscopy. Method A requires more capital equipment but is cheaper to use repetitively than method B. Method B provides the best view of the cells. Method A is, with some variation, used extensively in research, while B is not.

Method A
1 Hot room or incubator at 37°C (gassed with humidified air + 5% CO_2).
2 Equipment for heat sterilising glassware and instrument storage tins.
3 Sterile filtered media (Flow Labs., Biocult, etc), the simpler media can be prepared in the laboratory if equipment for sterile filtration is available.
4 Cylinder of Air + 5% CO_2 and pressure-reduction valve (for gassing cultures)

5 7–9 day incubated chick embryos
6 Sterile calf (alternatively horse or chicken) serum (Flow Labs)
7 Equipment and materials for actual culture (per student)
   (a) Clamp stand plus clamp to hold glass plate (approx. 6in x 4in) (binocular shield)
   (b) 1 pair fine, 1 pair coarse forceps, 2 cataract knives or Swann-Morton scalpels with no.15 blades. 2 x 60mm dia. glass Petri dishes. Heat sterilised and packed preferably in individual tins or alternatively wrapped in sets in paper tissue
   (c) Sterile 60mm plastic Petri dishes; these should be of a grade suitable for cell culture (Falcon, Biocult, Esco, Sterilin etc.) (Allow between 2 and 6 per student.)
   (d) Sterile Pasteur pipettes (with cotton wool plugged ends). (Allow 2–4 per student.) Teats
   (e) 70% ethanol. Cotton wool swabs. 1 Bunsen burner per 4 students
   (f) Magic markers or wax pencils. Containers for egg shells, waste embryos, etc.
8 Media
   (a) Hank's BSS medium or Eagle's saline (Flow or Biocult)
   (b) Culture medium: 10% serum, 10% tryptose phosphate broth (Difco), 80% Eagle's MEM medium (Flow or Biocult)
9 Optical equipment: low-power binocular microscope, inverted phase-contrast microscope (Eyepiece and stage micrometers and grids, per student or two students)

Method B
Basic facilities and biological materials as for Method A (1–6), except that gassing equipment is not needed.
Equipment and materials for actual culture is as follows: (per student unless otherwise stated.)
1 Clamp stand plus clamp to hold glass plate (approx. 6in x 4in) (binocular shield)
2 1 pair fine, 1 pair coarse forceps, 2 cataract knives or Swann-Morton scalpels with no.15 blades (Gallenkamp)
   2 x 60 mm dia. glass Petri dishes. Heat sterilised and packed preferably in individual tins or alternatively wrapped in sets in paper tissue
3 3–6 cavity slides, preferably flat-bottomed cavities. 3–8 sterile No.1 cover slips large enough to just cover the cavity. It is desirable to clean these very thoroughly and sterilise by autoclaving

or UV light.

4 1 cloth-covered board (approx. 6in x 4in)

5 500 ml flask filled with distilled water and gauze plug so that water can be boiled and kept sterile. 1 per 4 students.

6 Wax kettle containing 50% 56°C melting point paraffin wax (BDH) (for wax and vaseline) and 50% yellow vaseline. 1 paintbrush. 1 per 6 students. (*Avoid setting the wax on fire*).

7 70% ethanol. Cotton wool swabs. 1 Bunsen burner per 4 students.

8 Magic markers or wax pencils. Containers for egg shells, waste embryos etc.

9 Media
   (a) Hank's BSS medium or Eagle's saline
   (b) Culture medium:
        10% Serum
        10% tryptose broth (Difco)
        80% Eagle's MEM medium

10 Optical equipment Low-power binocular microscope. Phase-contrast microscope with 10 x 40 x objectives. (Eyepiece and stage micrometers and grids, per student or two students)

Type of group

Large classes can manage this practical with ease, provided that a fairly large number of trained demonstrators can be on hand, say 1 demonstrator to 8 students. The students may find it advantageous to work in pairs.

Instructions to the student

1 Sterile precautions.
   (a) When instruments are not in use, keep their ends in a sterile container or folded in a sterile tissue.
   (b) If you let the end of an instrument touch your hands or the bench, re-sterilise it by dipping the end in 70% ethanol.
   (c) When opening bottles of medium, hold them tilted, flame the necks before and after pouring. Replace cap immediately after flaming.
   (d) Keep a binocular shield between you and the work in hand.

2 Embryonic chick cell culture. Method A.
   (a) Swab a 7-day egg with 70% ethanol. Break open at blunt end and transfer embryo by its neck to a sterile glass Petri dish.
   (b) Pour some Hank's saline into the Petri dish over the chick embryo. Remove heart with fine forceps and transfer to a second glass Petri dish containing Hank's. It may be a help

to use the binocular low-power microscope for this dissection.
   (c) Using a pair of fine forceps and a fine scalpel, cut the ventricle into pieces about 1 mm square. These are called explants.
   (d) Pipette about 10 of these explants into a Falcon plastic Petri dish in so little medium that when the drop is spread the pieces of tissue are stranded on the bottom of the dish. Move explants so that they are about 2 mm apart in pairs.
   (e) Leave the Petri dish with its cover on for 10–20 minutes to allow the explants to adhere to the dish.
   (f) Then carefully pipette enough culture medium into the dish so that the explants are just covered. Take care not to disturb and detach the explants. (If you do, remove nearly all medium and go back to (e).) Replace cover.
   (g) Label outside of upper Petri dish, near the edge, and transfer to tray for incubation at 37–38°C. If a gassed humidified incubator is not available, place Petris in a vacuum 'desiccator' with water in bottom below tray and arrangements for bubbling air + CO_2 into it.
   (h) Cultures available for inspection next day.

Examine the cultures, preferably using an inverted phase-contrast microscope. If they are not available, a fairly satisfactory result can be obtained by draining off nearly all the medium from each dish and then floating a cover slip over the culture. The culture is then examined with a normal phase-contrast microscope. This latter technique does not need a hot room.

3 Method B
   (a) Follow Method A as far as 2 (c). Then cover glass Petri containing explants.
   (b) Take cloth board and carefully pour a small quantity of boiling distilled water over the cloth so that it is thoroughly dampened. Mop off excess water with a sterile tissue. The purpose of preparing this moist board is to ensure that the cultures do not dry out while being prepared in the next stage. Place your binocular shield over the board. Lay 4 to 6 cover slips on the board using sterile forceps and pipette a small drop of medium on to the centre of each cover slip.
   (c) Transfer one explant to the drop of medium on each cover slip. Most people find it is easiest to fish for them using one scalpel

blade point to push an explant on to the blade of another scalpel, and then to reverse this action to put the explant into the drop. Make sure that the explant is in contact with the glass and that it is not held by surface tension against the top of the drop.

(d) Take a cavity slide in one hand and pick up a cover slip complete with explant using a pair of forceps in your other hand. Quickly but carefully invert the cover slip so that it covers the cavity. Make sure that the drop does not run off the cover slip while you do this. (Small spread drops are least likely to run). As soon as the cover slip is in place hold it lightly there at a corner and immediately go to e.

(e) Using the paintbrush, put a drop of molten wax on one corner of the cover slip so that as soon as the wax hardens the coverslip is held to the slide. Then quickly wax round the edge of the cover slip to ensure that there is no air gap through which the culture drop might dry by evaporation.

(f) Label the cavity slides, transfer to incubation trays. (Cardboard slide trays are suitable for this: the slide can be placed upside down in the tray).

(g) Incubate at 37–38°C and inspect after 18–24 hours. Examine by phase-contrast microscopy, preferably in a hot room where cell movement and action will continue and where there should be no obscuration of the optical image by condensation on the other side of the cavity.

4 Variations. Other chick tissues can be used to give other cell types in culture, e.g. epithelia. Other species can be used, provided that the culture temperature is appropriate for them. Cell suspensions can be used to establish the cultures, either bought commercially or prepared in your laboratory. This last method allows you to grow cell lines. If you wish to do this use Method A, and add your suspension plus culture medium to the Petri dish. Then incubate.

Recording results
The results recorded depend on the students approach in the Discussion section: standard recording methods are advised.

Discussion
Though the student can gain a certain satisfaction from the contemplation of his success in culturing cells, the practical can be much more profitable if a number of matters are considered. Some of these are outlined here:

Describe the general morphology of the outgrowth of cells. How does this outgrowth arise? Make suggestions.

What is the population density of cells near the edge of the outgrowth and close to the explant: measure using an eyepiece grid.

How fast do the cells move? Measure the diameter of the outgrowth achieved in the culture time. This gives an upper limit.

Try to measure the rate of cell movement by direct observation of a culture at 37°C.

Is there a discrepancy between the two measurements? If so, give your explanation.

Does the explant change in size appreciably during the culture period?

Where do the cells that give rise to the outgrowth come from?

Does mitosis in the outgrowth account for much of the increase in number in these cells? Measure mitotic frequency and try to observe mitotic duration on cultures maintained in the hot room.

Observe the general shape of individual cells and their relation to neighbouring cells. Do the cells overlap frequently? If you can make measurement of this, give your explanation of cell shape and of the general packing of the cells.

Are the cells aligned in relation to the explant? If they are, how do you account for this?

If population density is fairly constant in all parts of the outgrowth from the explant, how might you explain this?

How would you test the hypotheses you have set up to explain cell packing, shape and population density?

It is recommended that the students should be shown a time-lapse film of cell behaviour in culture which should include a sequence on fibro-blast movement.

Extensions and applications
Cell culture methods are widely used for growth of cells for biochemical and virological studies, for the investigation of cell nutrition and cell interactions. In practice, methods based on the Petri-dish technique are more widely used than those based on the hanging-drop method.

If techniques for preparing cell suspensions are available, this practical can easily be developed, using Method A, into a means of estimating cell viability and for investigation of critical inoculum size effects (plating-out efficiency tests).

Further reading
Paul, J.
  1970 'Cell and Tissue Culture', E. & S. Livingstone, Edinburgh
Harris, M.
  1964 'Cell Culture and Somatic Variation'. Holt, Rinehart & Winston. New York

Hints for the teacher
If your laboratory does not carry out routine cell culture, it is advisable to carry out trial runs with the media you propose to use before the practical. Do not carry out practicals involving use of bacteria or fungi in the same part of the building at the same time, or for a few days before the cell culture practical.

A.S.G. CURTIS
Glasgow.

6 *Xenopus* cells in culture

Time required
Primary cultures—embryonic cells: 1 hr; adult cells: 4-5 hrs. Observation at intervals over 7-10 days.

Assumed knowledge
Sterile technique

Theoretical background
Amphibian cells are ideal for elementary experiments in cell culture, because they proliferate, divide and move at room temperature; they are larger than avian or mammalian cells (so can be observed with dry objectives); they tend to be epithelioid rather than fibroblastic *in vitro;* they grow well in Leibovitz L-15 medium in free gaseous exchange with the atmosphere (thus cover glasses in simple Petri dishes can be used). In two previous papers we described simple methods for observing *Xenopus laevis* (the South African clawed toad) cell morphology and mitosis and for studying DNA and RNA synthesis by autoradiography. In this note, we give two methods for establishing Xenopus cell cultures: from embryos and from young adults.

Necessary equipment
1 Primary cultures from embryos
  (a) *Xenopus laevis* embryos (almost hatched)
  (b) 3 sterile 5 cm Petri dishes containing amphibian phosphate buffered saline (APBS).
  (c) Tissue culture dishes (Esco Rubber Ltd.) if an inverted microscope is available. If not, use 5 cm or 10 cm Petri dishes + flamed 18 x 18 mm cover glasses stuck to the bottom with vaseline (autoclaved) for cell culture, and a cavity slide for observation
  (d) Sterile, plugged Pasteur pipettes
  (e) Microscope, preferably with phase contrast optics; incubator at 25°C
  (f) Alcohol, cotton wool, bunsen burner, clean scalpel
2 Primary cultures from young adults
  (a) young, metamorphosed *Xenopus*
  (b) potassium permanganate (BDH.)
  (c) sterile 1 in 500 diluted MS222 (Sandoz)
  (d) haemocytometer
  (e) sterile, plugged 10 ml pipettes
  (f) bench centrifuge; capped, sterile centrifuge tubes

(g) sterile 50 ml bottles
(h) stirring bar (sterile) and magnetic stirrer
(i) 25 ml flasks (capped and sterile)
(j) fine scissors and forceps
(k) alcohol, cotton wool, bunsen burner
(l) microscope, incubator, culture dishes, as above

3 Media
(a) Growth medium—60% Leibovitz L-15 medium, 10% foetal calf serum, 30% double distilled water. The water should be autoclaved before use. The serum and L-15 medium can be obtained from Flow Laboratories.
(b) APBS (amphibian phosphate-buffered saline)—6·0 g NaCl; 0·2 g KCl; 0·1 g Na_2HPO_4; 0·02 g KH_2PO_4; 2·0 g glucose; 0·002 g phenol red in 1 litre double distilled water. Sterilised by autoclaving.
(c) Versene (EDTA) trypsin solution—20 g trypsin + 0·2 g EDTA (BDH) per litre of APBS; neutralized with sodium bicarbonate (to give pH 7·5—8·0); sterilised by filtration; 2·5% trypsin solution can be obtained from Flow Laboratories.

All media should contain 100 iu/ml benzylpenicillin and 100 μg/ml streptomycin sulphate both obtainable from Glaxo and 2 μg/ml Fungizone (E.R. Squibb). Add antibiotics to APBS *after* autoclaving it.

Type of group
Students working singly or in pairs; no limit to total size of group.

Instructions to the student
1 *Primary cultures from embryonic material*
Swab down the working area with alcohol. Take *Xenopus* embryos and place in the first dish of APBS, transferring as little water as possible. Gently draw the embryos up and down in a sterile Pasteur pipette to remove the surrounding jelly membranes. Wash the embryos in the APBS of the first dish by pipetting it gently up and down, then transfer with as little liquid as possible to the second dish. Wash as before, then transfer to the third dish of APBS and wash again. Thus, the embryos are washed in three changes of APBS containing antibiotics. Use a fresh pipette for each transfer. Finally transfer the embryos to the empty culture dish, again using a fresh pipette and carrying over as little liquid as possible. Three embryos are required for each 5 cm tissue culture dish, 12 for each 10 cm Petri dish. Flame a clean scalpel blade and chop the embryos into small pieces.

Cover the fragments with a little versene/trypsin solution and leave at 25°C, mixing occasionally by gently tipping the dish. After 30 min, add 4 ml (5 cm dish) or 12 ml (10 cm dish) of growth medium. Gently swirl the medium until any remaining fragments have broken up and the cells have been spread over the whole dish surface. Replace in the 25°C incubator and observe about every two days.

2 *Primary cultures from adult material*
Sterilise the body surface of the animals by immersion in very weak potassium permanganate solution for 30 min, then anaesthetise them in the MS222. Swab down the working area with alcohol. Place an animal on its back and, using sterile instruments, remove the ventral skin. Re-sterilise the instruments (by flaming) and remove the body wall muscle, exposing as much of the viscera as possible. Re-sterilise the instruments and, taking great care not to break the gut or bladder, remove the kidneys and place them in APBS in a sterile Petri dish (some workers might find a stereomicroscope helpful for the dissection; all should practise the dissection before trying to establish primary cultures). Mince the tissues with sterile scissors to give small fragments, transfer to the flask containing the stirring bar and stir for about 5 min to remove surface debris. Decant off the APBS and replace with 10 ml of versene/trypsin solution. Stir for 15 min, then decant off the versene/trypsin solution and discard it—this first digest consists mainly of red blood cells. Add a further 10 ml versene/trypsin solution and stir for 30 min. Decant this solution into a sterile centrifuge tube and add a further 10 ml to the flask. Collect the cells from the decanted solution by centrifuging at low speed for 5 min. Pour off the supernatant fluid and re-suspend the cells by gently pipetting with 10 ml growth medium, then transfer the cell suspension to a sterile 50 ml bottle. Repeat the collection procedure at 30 min intervals until 4 digests have been re-suspended and added to the bottle. Using a haemocytometer, determine the final cell density (do not count cell fragments or red blood cells). Mix sufficient cell suspension with growth medium in a sterile bottle to give 2×10^5 cells in 4 ml for each 5 cm dish and 1×10^6 cells in 12 ml for each 10 cm dish. Incubate at 25°C and observe at intervals of about two days. One toadlet will give enough kidney cells for two 10 cm dishes.

Improved results are obtained if collagenase (Sigma, type I) or pancreatin (Sigma, grade II) is added to the versene/trypsin solution immediately before use—to give a final concentration of 0·1%.

Pancreatin is much cheaper than collagenase.

Extension and applications:
Instead of, or in addition to, culturing kidney cells, culture cells from the body wall muscle or the heart ventricle (minus pericardium). Kidney and heart give epithelial cell cultures, muscle gives mainly fibroblastic cells. However, 5 hearts are required to give sufficient cells for one 10 cm dish.

These cultures can be used for observing the stages of mitosis, cell morphology and cell movement, for growth studies, for studying the effects of different temperatures on cell growth and behaviour, for investigating DNA and RNA synthesis by autoradiography, or for most other general studies on the behaviour of vertebrate cells *in vitro*.

Further reading:
1 Balls, M. and Godsell P.M.
   1972 'Animal Cells in Culture—Methods for Use in Schools', Journal of Biological Education, 6, 17—22
2 Freed, J.J. and Mezger-Freed, L.
   1970 'Culture Methods for Anuran Cells', Methods in Cell Physiology, 4, 19—47
3 Godsell, P.M. and Balls, M.
   1973 'DNA and RNA Synthesis in Animal Cells in Culture—Methods for Use in Schools'. Journal of Biological Education, 7(4), 19—24

Hints for the teacher
The main problem in setting up these cultures is contamination due to poor sterile technique. The results from embryonic cultures vary from one group of embryos to another, but a range of cell types usually emerges. Each student should prepare several 5 cm dishes of embryo cultures. Permanent cell lines are readily obtainable from young *Xenopus* kidneys. Harris Biological Supplies, now produce a *Xenopus* cell culture kit based on our 1972 paper. This would provide a useful background to attempts at primary culture. Harris Biological supplies also supply amphibian cell culture medium, APBS, and EDTA/trypsin solution in small quantities.

P.M. GODSELL
M.BALLS
Norwich.

7 The pattern of growth of a plant (sycamore) cell suspension culture and its modification by plant growth hormones

Time required
Two 3 hr. periods, one to prepare media and one to initiate the cultures. Three subsequent periods of about 1 hour for sampling the culture, these periods at 7 day intervals. A final period of 2—3 hrs for a sampling with cell counts and estimation of cell protein. Access in between these times so that dried cells can be weighed. Total duration of experiment 5 weeks. If students are provided with the culture media the duration is reduced to 4 weeks and the initial 3 hr. session omitted.

Assumed knowledge
Basic training in aseptic technique and in the use of the microscope.

Theoretical background
The experiment provides growth curves for batch cultures of higher plant cells. A knowledge of the kinetics of growth of liquid bacterial cultures (lag, exponential, linear, progressive deceleration and stationary phases) will enable comparison to be made between these kinetics and those of the plant cell culture. The student will have to decide whether such a comparison is possible with data from only 4 sampling times during 28 days' incubation. The experiment assumes that the growth effects and effects on cell composition which develop during a single culture passage are indicative of the 'essentiality' and 'specificity' of the growth hormones used. The student should calculate the minimum 'dilution' factor of 'essential' hormones caused by the observed growth if the cells are assumed to be unable to synthesise the hormone in question or a natural substitute for it.

Necessary equipment
Sterilising facilities. Horizontal platform shaker(s) in a temperature-controlled room or sufficient incubator

shakers (Gallenkamp). A flask shaker (Baird and Tatlock Microid shaker, or similar); a bench-mounted centrifuge; a drying oven at 60°C; a pH meter (with N/10 and N/100 NaOH and small pipettes), one for each 9 students. Facilities for aseptic transfer, preferably either laminar air-flow cabinet(s) or sterile transfer room(s).

For the students: First period

Ample supply of glass-distilled water. Heller's inorganic salt solution x 10; vitamin solution x 10; solution of 2, 4-dichlorophenoxyacetic acid (2, 4-D) 10 mg/l (BDH); solution of kinetin (K) 100 mg/l (BDH). *Students in groups of 3:* for each group, the following are needed.

1 Graduated measuring cylinders 3 x 25 ml, 1 x 100 ml, 1 x 250 ml, 1 x 500 ml
2 10 ml graduated pipette
3 4 x 1000 ml flat-bottomed flasks
4 2 weighing bottles
5 1 x 50 ml beaker
6 Wash-bottle of glass-distilled water
7 36 x 100 ml wide mouth Erlenmeyer flasks
8 Glass writing pencil
9 Supply of non-absorbent cotton wool, gauze and aluminium foil.

Second and subsequent periods

Each student requires:

1 A 24-28 day old stock suspension culture of sycamore cells (volume of culture of ca. 60 ml. Cell density not less than 1.5×10^6 cells ml^{-1})
2 A sterile solution of urea—400 mg/100 ml prepared by aseptic dilution of 40% filter-sterilised urea solution of Oxoid Ltd.
3 A sterile 2 ml graduated automatic pipette (A.R. Horwell) for addition of the urea solution to the flasks of medium prepared during the first period
4 A sterile 5 ml graduated automatic pipette fitted with a long stainless steel cannula (internal diameter 3 mm) for transfer of cell suspension from the stock culture to the flasks of new medium
5 A home-made counting chamber and cover-slip (see Fig. 1)
6 Three boiling tubes
7 Two acid-resistant plastic vials with screw caps (capactiy approx. 50 ml, Sterilin)
8 A 3 cm Hartley funnel (Gallenkamp) and filter flask and 2.5 cm discs (50) of glass fibre paper (Whatman GF/C)
9 Water pump or vacuum point
10 3 x 9 cm Petri dishes

Figure 1 Counting chamber prepared from a standard microscope slide (0.8–0.9mm thickness). The glass strips are cut from a similar slide and mounted with araldite. The broken outline indicates the positioning of the cover-slip.

11 Desiccator
12 4 x 15 ml graduated centrifuge tubes
13 Compound microscope
14 Tally counter
15 Wide bore Pasteur pipettes
16 50 ml measuring cylinder
17 Packet of sterile aluminium foil squares and forceps to handle these
18 Reagents and apparatus to perform the Lowry, Rosenborough, Farr and Randall (1951) protein estimation (for details see Layne, E. 1957; *'Methods in Enzymology'* vol. 3). Spectrophotometric and turbidometric methods for measuring proteins: protein estimation with Folin—Ciocalteu Reagent (BDH) in *'Methods in Enzymology'* Vol. 3, p. 448. Academic Press, New York 1963.

For each 12 or 15 students one bath fitted with test tube racks at 70°C and a similar bath at 85°C.

The technical staff should produce the stock solutions: Heller's inorganic solution x 10, vitamin solution x 100 (see below) and the solutions of 2, 4-D and kinetin (as detailed above) in adequate amount for the number of students involved (see under (7) below). The composition of the basic culture medium is as follows:

Basic culture medium

Heller's inorganic solution (supplied to students as a x 10 concentrate, i.e. 100 ml per litre):

	mg/litre in *final* medium
KCl	750
MgSO$_4$.7H$_2$O	250
NaNO$_3$	600

	mg/litre in *final* medium
$NaH_2PO_4.2H_2O$	130
$CaCl_2.6H_2O$	110
$ZnSO_4.7H_2O$	1·0
H_3BO_3	1·0
$MnSO_4.4H_2O$	0·1
$CuSO_4.5H_2O$	0·03
KI	0·01
$FeCl_3.6H_2O$	1·0

Vitamin solution
(supplied to students as a x 100 concentrate)

thiamine HC1	(BDH)	1·0
pantothenic acid	(BDH)	2·5
choline chloride	(BDH)	0·5

Other constituents to be weighed:

meso-inositol	(BDH)	100
cysteine HC1	(BDH)	10
sucrose		20 000

Adjust pH to 5·6—5·8 before autoclaving

| urea (Oxoid sterile sol.) | 200 |

This is added aseptically to the cold sterile medium.

Where available Analar reagent used. All glassware should be Pyrex, very carefully cleaned and given a final rinse with distilled water before use. Solutions should be prepared with glass distilled water (double glass distilled if available).

The flasks of medium prepared by the students are autoclaved by technical staff at 15 1b/in^2 for 15 min. Sterile pipettes should be prepared by autoclaving in sealed tins or wrapped carefully in aluminium foil or autoclavable transparent plastic. The aluminium foil squares (2·5 in square) should be dry sterilised (150°C for 1 hr) in tins or Petri dishes. If domestic aluminium foil is used a double layer is needed— pieces 2·5 x 5 in folded to 2·5 in square. The dilution of the purchased urea solution is done aseptically with sterile water and dispensed into plugged bottles or flasks for the students.

The stock callus of sycamore *Acer pseudoplatanus* is obtainable from the Botanical Laboratories, University of Leicester. It is received as a tube culture on solidified medium. The technical staff maintain this callus and prepare suspension cultures from it as required. The medium for growth of the culture as a callus or as a cell suspension is the basic medium detailed above plus 1·0 mg/l 2,4-D and 0.25 mg/l kinetin. For callus it is solidified by incorporating 0·8% Difco Ionagar. With solid medium add the urea with medium cooled but before solidification. For callus propagation and multiplication transfer on each

occasion at least 200 mg fresh weight; subculture every 4—5 weeks. To initiate a suspension transfer approximately 400 mg to each 250 ml wide mouth Erlenmeyer culture flask containing 60 ml medium. After 24—28 days incubation, subculture by transferring 10 ml to new medium (1 in 7 dilution). The suspension culture can be serially subcultured indefinitely, but maintain callus as a safeguard against contamination of suspensions—when the culture is not being used for teaching, the propagation only of callus demands less time. If the suspensions are growing satisfactorily they should at the time of subculture contain not less than 2·0 x 10^6 cells/ml (for technique of cell counting see 'Instructions to students').

Type of group
Note suggestion of students working in threes to prepare media; individually to establish and monitor the cultures. Only the limitation of facilities (e.g., facilities for working aseptically) limit the size of the group.

Instructions to the student
You are going first to prepare three different culture media and then to study the growth of sycamore cells in these media.

First period
This is devoted to media preparation, and you work in groups of three. The 36 wide-mouth 100 ml Erlenmeyer flasks have first to be fitted with properly made cotton wool plugs tied in gauze. Have a specimen plug checked by the demonstrator before preparing the remaining 35. First prepare 500 ml of *double strength* medium (formula given under 'Necessary equipment'). To do this, weigh and dissolve in turn the sucrose, *meso*-inositol and cysteine HC1 in approximately 250 ml distilled water. Now add 100 ml of the stock solution 'Heller's inorganic' and 10 ml 'vitamin solution'. Adjust the total volume to 500 ml. Now use this to add 150 ml to each of the 3 remaining litre flasks. Label these 'Omitted', 'Auxin', and 'Cytokinin'. To the 'Omitted' flask add 150 ml distilled water; to the 'Auxin' flask sufficient of the solution of 2, 4-dichlorophenoxyacetic acid (2, 4-D) to give a final concentration of 1 mg/litre and to the 'Cytokinin' flask sufficient of the solution of kinetin (K) to give a final concentration of 10 mg/l. Add sufficient water to the 'Auxin' and 'Cytokinin' flasks to bring the volumes to 300 ml. Adjust the pH of all three solutions to pH 5·6—5·8. Now fill 12 of the 100 ml flasks each with 25 ml of the medium

'Omitted'—label these flasks O. Similarly 12 flasks with 'Auxin' medium—label these flasks A, and 12 flasks with 'Cytokinin' medium and label these C. Cover the cotton wool plugs of these flasks with 2·5 in squares of aluminium foil pressing the foil to the flask necks (this foil is to protect the cotton plugs from 'drip' in the autoclave and to keep the rims of the flasks free from dust). Additionally label all the 100 ml flasks with your bench or class number. These flasks are then loaded into the autoclave baskets and will be autoclaved for you by the technical staff ready for you to use in the next class period. They are autoclaved at 15 lb/in² steam pressure for 15 minutes.

Second period

The first task is to complete the culture media by adding the urea. You now have your share of the culture media—4 flasks of each of the three media. You now work on your own. The urea solution provided is sterile (it has been sterilised by filtration) and contains 400 mg urea per 100 ml. You are to add 1·5 ml of this to each flask of medium using aseptic technique. Use for this purpose the sterile 2 ml graduated pipetting unit.

You are also provided with a 250 ml Erlenmeyer flask containing approximately 60 ml of a sterile suspension culture of sycamore (*Acer pseudoplatanus*) cells. This is to be used to initiate suspension cultures in the media you have prepared. To do this, transfer aseptically to each flask 2·5 ml of suspension (shake the suspension each time before withdrawing the inoculum) using the sterile 5·0 ml graduated pipetting unit provided. The demonstrator will outline the procedure to be followed to achieve successful aseptic transfer. Now aseptically pick up one of the sterile 2·5 in square of aluminium foil using the sterile forceps provided. Discard one at a time the cotton wool plugs from your culture flasks and replace with a square of aluminium foil placed centrally over the flask opening and then pressed to the side of the flask neck by an appropriate movement downwards of the loop made between your thumb and first finger of the right hand (or left hand if left handed).

Your cultures will now be incubated on a rotary platform shaker (speed 120 rev/min) at 25°C.

You have some of the stock cell suspension left over. You are now to determine (a) its packed cell volume, (b) its cell dry weight and (c) the number of cells it contains per ml. You will then calculate these parameters for the cultures you have just initiated, bearing in mind that 2·5 ml of the stock suspension is

now present in 29·0 ml of culture (25 + 1·5 + 2·5 ml). After 7 days, 14 days and 21 days incubation you will withdraw one culture from each of the three media to measure packed cell volume and cell dry weight. After 28 days incubation, the final culture for each medium will be harvested and used to determine packed cell volume, cell dry weight, cell number per ml, and cellular protein. At each harvest a specimen of the cultures should be examined in the microscope (both with standard optics and if available, phase contrast) and drawings made to illustrate degree of aggregation (lower power) and to show the range of shape and size of individual cells (high power). Phase contrast should enable you to record the frequency of cells in which protoplasmic streaming is taking place and also frequency of cells in which protoplasts appear to be shrunken (dead cells?). Packed cell volume: transfer 15 ml culture to the graduated centrifuge tube. Centrifuge at 2000g for 5 min. Express cell volume as ml cell pellet per ml culture. The pellet can be redistributed and returned to the culture vessel (you will not have too much culture when sampling the 100 ml culture flasks). Cell number: the cell-suspension culture not only contains free cells but aggregates of cells. It is necessary to separate the cells of the aggregates for the cell count. For the stock culture and for your 28 day cultures take 2 ml and for your 7, 14 and 21 day cultures 4 ml of suspension. Transfer the sample to a boiling tube, add 10 ml (for 4 ml samples) and 5 ml (for 2 ml samples) of 8% aqueous solution of chromium trioxide (BDH). Heat in the bath at 75°C for 45 min, cool, wash out with water into (acid-resistant) plastic screw capped vials. Shake at maximum speed in the flask shaker for 10 min. Measure volume of suspension in vial. Now transfer suspension to the counting slide (Fig. 1) to fill the 3 channels. Put on the cover-slip, press to obtain Newtonian rings (concentric interference fringes), allow cells to settle. Examine under the microscope at a magnification of x 100. The depth of the channels is almost exactly 1·0 mm. Calculate the field area of the microscope, multiply this by the depth of the suspension in the channel to give the volume of suspension scanned per field (this will be close to 0·8 μl and, if desired, this volume may be assumed without detracting in any way from the experiment). Now, using your tally counter, count the number of cells in ten fields in each channel until 10 channels have been scanned i.e., 4 slides are examined. The accuracy of the counting procedure can be assessed by comparing the means for each 10 fields. Cells in all have been counted in 80 μl of the suspension in the vial and this suspension was derived from 2 (or 4) ml of the original culture. Calculate the number of cells (in millions per ml e.g., 0.9×10^6 cells/ml).

Dry weight: measure the volume of the remaining culture and collect on a glass fibre disc in the Hartley funnel. Wash cells with water. Transfer disc to Petri dish. Dry at 60°C overnight. Cool in desiccator. Weigh. Check average weight of a disc dried under the same conditions. Calculate cell dry weight (mg) per ml culture.

Protein content: the disc of dried cells from the day 28 harvest is returned to the Hartley funnel, and the cells washed first with boiling 70% ethanol (2 x 5 ml) and then with cold acetone (2 x 5 ml) and the disc and cell residue air dried. Transfer the disc with cells to 1·0 ml N NaOH and heat in the 85°C bath for 30 minutes. This solution is then filtered and then further processed according to the procedure outlined in the key reference given above (Layne, 1957 in Meth. Enz. Vol. 3). *N.B.* the OD should be measured in a spectrophotometer at 750 nm or in a colorimeter using the nearest appropriate filter. It is recommended that the calibration curve be prepared beforehand by the technical staff using bovine serum albumin, Fraction V (BDH) and that a separate protocol be available to the student for this estimation.

Recording results

Plot graphs of data for packed cell volume and cell dry weight for day 0, 7, 14, 21, and 28 for each test medium. In a table, show data for day 0 (stock culture) and day 28 (experimental cultures) for packed cell volume, cell dry weight and cell number, all per ml culture, for packed cell volume and cell dry weight per 10^6 cells and for the day 28 experimental cultures protein per ml and per 10^6 cells.

Discussion

What can you say about the general growth pattern of the cultures? What would have to be done to delimit properly changes in the growth rate of the cultures with time? What can you say from your results regarding the requirement of the cells for an exogenous supply of plant growth hormones? Can you say from your observations on the 'Omitted' medium cultures that such a supply is essential? If not, what further work would be necessary to establish this? Do your results suggest that *either* an auxin *or* a cytokinin will meet the growth requirement of the cells for a plant growth hormone? If so what do you make of such a finding? What is the mean generation time of the cultures—does this change as incubation proceeds? Are there differences in the degree of aggregation of cells, in the morphology of the cells, or in the colour of the cultures in the different media? Why is the urea filter sterilised? Why does chromic acid separate the cells in the cellular aggregates? Why are the cultures continuously shaken during incubation? How would you try to find out what limits the final cell concentration achieved in such cultures?

Further reading

Simpkins, I., Collin, H.A. and Street, H.E.
    1970 'The Growth of *Acer pseudoplatanus* Cells in a Synthetic Liquid Medium, Response to the Carbohydrate, Nitrogenous and Growth Hormone Constituents, Physiologia Plantarum, 23, 385

Witham, F.H.
    1968 'Effect of 2, 4-Dichlorophenoxyacetic Acid on the Requirement of Soybean Cotyledon and Tobacco Stem Pith Tissues,' Plant Physiology, 43, 455.

Street, H.E. (ed.)
    1973 'Plant Tissue and Cell Culture,' Blackwell's Scientific Publ. Ltd., Oxford.

Hints for the teacher

Students should have the procedure of aseptic transfer demonstrated to them. They may need some help in satisfactorily carrying out the cell count. How consistent is the whole class in counting the stock culture? Suspensions should not be allowed to stand about but should be quickly placed back on the shakers.

H.E. STREET
Leicester.

8 Amphibian hearts in organ culture

Time required

½–1 hr for setting up cultures; observation at intervals up to 21 days, according to the type of experiment.

Assumed knowledge

Sterile technique; basic vertebrate anatomy.

Theoretical background

Heartbeats are myogenic in origin, and whole hearts isolated from adult amphibians continue to beat for up to 6 months in organ culture.[2] The beating rate is greatly affected by temperature[4] and by the addition of drugs, such as adrenalin and acetylcholine.

If the prospective heart-forming region is removed from an amphibian embryo, the heart will develop and commence beating *in vitro*.[1]

The experiments outlined provide a simple, but useful, introduction to organ culture studies of development and maintenance of function *in vitro*.

Necessary equipment

1 Animals

   Newly-metamorphosed, adult *Xenopus laevis laevis* (the South African clawed toad)

   Adult *Triturus vulgaris* (the common newt) or *Triturus cristatus* (the crested newt)

   Stage 24 to 27 *Xenopus* embryos (Nieuwkoop and Faber, 1967)[3]

2 Adult heart culture

   Weak potassium permanganate solution (BDH)

   Sterile 1:500 diluted MS222 (Sandoz)—sterilise by autoclaving (reuseable)

   Fine forceps and fine scissors

   5 cm Petri-dishes (sterile)

   5 ml pipettes (plugged, sterile)

   Alcohol, cotton wool

   Dissecting microscope

3 Developing heart culture

   Fine glass needles (made by drawing out 1 mm thin-walled capillary tubing)

   Sterile Pasteur pipettes

   5 cm Petri dishes (sterile)

   5 ml pipettes (plugged, sterile)

   Dissecting microscope

4 Media and solutions

   (a) Growth medium—60% Leibovitz L-15 medium, 10% foetal calf serum, 30% double distilled water. The L-15 and serum can be obtained sterile from Flow Laboratories. The water should be autoclaved.

   (b) APBS (amphibian phosphate-buffered saline —6·0 g NaCl; 0·2 g KCl; 0·1 g Na_2HPO_4 0·02 g KH_2PO_4; 2·0 g glucose; 0·002 g phenol red in 1 litre double distilled water Sterilised by autoclaving. (Any sterile vertebrate physiological solution will do.)

   (c) Solutions 1 and 2 should contain 100 iu/ml benzylpenicillin and 100 μg/ml streptomycin sulphate (both obtainable from Glaxo), with 2 μg/ml Fungizone (E. R. Squibb). Add antibiotics to APBS *after* autoclaving.

   (d) 1% adrenalin chloride solution (BDH); 1% acetylcholine chloride (BDH) solution.

Harris Biological Supplies supply complete culture medium and APBS.

Type of group

Students working singly or in pairs; no limit to total size of group.

Instructions to the student

1 Adult heart culture

Sterilise the body surface of the toadlets (or newts) by immersing them in weak potassium permanganate solution. Swab down the working area with alcohol. Anaesthetise an animal in MS222, then swab its ventral surface with alcohol. Remove the ventral skin (not possible with newts) with flamed dissecting instruments. Reflame the instruments and open the body cavity, folding back the muscle (muscle plus skin in the case of newts). Reflame the instruments then remove the heart by gently lifting the ventricle and cutting through the large blood vessels attached to the heart. Place the heart in a Petri dish containing APBS, then remove extraneous tissue and the pericardium. Transfer the heart carefully to a Petri dish containing 5 ml culture medium. Place the dish in an incubator at 20 to 25°C. Sterility is vital; the medium should be changed about every 5 days.

   (a) How long can the hearts be kept beating in culture? How does the heartbeat rate change during the culture period? Do all parts of the heart beat together or in sequence? (*Xenopus* hearts will not continue to beat for as long as will *Triturus* hearts).

   (b) How does temperature affect the rate of

beating?

(c) How do drugs affect the rate of beating? Add 0·1 ml of the adrenalin chloride solution, or 0·1 ml of 1%, 0·1% or 0·01% acetylcholine chloride solution. (Adrenalin or shock treatment—a prick with a sterile needle or a bang on the bench near the culture dish—will often restart beating in a heart which has not beaten for a few days.)

Heart development in vitro

emove the extra-embryonic membranes from the nbryos (if not yet hatched); with sterile Pasteur pettes, pass the embryos through 3 or 4 changes of PBS containing antibiotics; use a fresh pipette for ch transfer, transferring as little liquid as possible ith the embryos. Remove the presumptive heart gion (see diagram) from each embryo with fine glass edles, then place the fragments in Petri dishes ntaining culture medium. Observe the explants ily to see development of the heart and onset of artbeat. This experiment shows that, once deterined, differentiation of an organ will follow—even if e part concerned is isolated from the rest of the nbryo.

Hints for the teacher

These simple experiments provide a good introduction to organ culture—'the maintenance or growth of tissues, organ primordia, or the whole or part of an organ *in vitro,* in such a way as to allow differentiation and preservation of the architecture and/or function.'

Xenopus toadlets are preferable in some ways, as they are more easily reared in the laboratory and are therefore available all the year round (Gerrard & Haig). *Triturus* hearts survive longer in culture, but newts are usually obtained from the wild (through animal suppliers), and users should be conscious of the need to conserve wild populations.

M. BALLS
R.S. WORLEY
Norwich.

Stage 23   Stage 27

 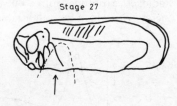

gure 1 Removal of heart anlagen from Xenopus embryos.

urther reading

Copenhaver, W.M.
    1955 In 'Analysis of Development' edited by B.H. Willier, P.A. Weiss and V. Hamburger. W.B. Saunders, Philadelphia

Millhouse, E.W., Chiakulas, J.J. and Scheving, L.E.
    1971 'Long-term Organ Culture of the Salamander Heart. Journal of Cell Biology, 48, 1–14

Nieuwkoop, P.D. and Faber, J.
    1967 'Normal Tables of Xenopus laevis (Daudin),' 2nd edn., North-Holland, Amsterdam.

Stephenson, E.M.
    1968 'Temperature Tolerance of Cultured Amphibian Cells in Relation to Latitudinal Distribution of Donors,' Australian Journal of Biological Sciences, 21, 741–57

Section 2

Tracer Techniques

1 Comparison of methods for detecting radio-activity. Oxidation of 1-(^{14}C) and 2-(^{14}C) acetate by *E. coli*

Time required
4–5 hr.

Assumed knowledge
For second-year students familiar with basic microbial techniques. We use this experiment to introduc
students to the use of radioactive isotopes an
radioactive techniques.

Theoretical background
Radioactive isotopes differ one from another in th
energy with which they emit radiation.

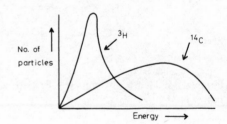

Figure 1

Different methods of detecting radioactivity hav
been developed to deal with radiation of differe
energies. Geiger–Müller counters and gas-flo
counters can be used for high-energy radiation, suc
as ^{14}C, but scintillation counters must be used f
low-energy radiation, such as ^{3}H.

In the first part of this experiment you wi
compare the percentage efficiency of these thre
methods of detecting radioactivity using ^{14}C an
^{3}H-labelled materials.

Necessary equipment
Each group of students will require the following.

Item	Amount required

Type of group
4 or 5 pairs of students. It is convenient to run this experiment as a class experiment in which each pair of students takes one time point. The class results can then be combined to obtain sufficient data for analysis.

Part I
E. coli suspension grown on
-[^{14}C] acetate as source
f carbon

Millipore filter apparatus	2
Millipore filters, 0·45 μm ore size	1 box
Glass counting vials (Koch-Light)	200
Bray's solution	2 litres
Planchet containing filter with uitable (^{3}H)-labelled compound	
Counting equipment	

Part II

Wash bottles with spargers aerators) (Gallenkamp as-distribution tubes)	4
Water baths at 30°C	2
0·1N NaOH	2 litres
20 mM acetate growth nedium (sterile)	500 ml
E. coli suspension grown overnight in 20 mM acetate growth medium	200 ml
1 ml disposable syringes and ong (3in) needles	
Hyamine hydroxide in methanol M (BDH)	100 ml
Fluted filter paper circles approx. 7 cm diameter)	
Hard glass test tubes	24
Serum caps to fit test tubes Oxoid)	
5N H$_2$SO$_4$	500 ml
Forceps	(1 per pair)
Toluene counting fluid Koch-Light)	21
1-[^{14}C] acetate 100μCi/ml Radiochemical Centre)	1·5 ml
2-[^{14}C] acetate 100μCi/ml Radiochemical Centre)	1·5 ml
Aluminium planchets and glue	
Automatic pipettes (e.g. Pumpettes) or syringes Gallenkamp)	
Infra-red lamp	
Radiation monitor.	

Instructions to the student
Part I   Comparison of Methods of Detecting Radioactivity

You are provided with a suspension of *E. coli* W grown on 2-[^{14}C] acetate as sole source of carbon. Filter 0·2 ml of this suspension through a 0·45 μm Millipore filter, wash well with water, and glue the filter to a planchet. Filter another 2 x 0·2 ml portions of the same suspension as before, but dry it and place it in a glass counting vial and add 10 ml of Bray's fluid to one filter and 10 ml of toluene fluid to the other. Determine the number of counts/min registered by

1   the end window Geiger–Müller counter
2   the gas-flow counter, and
3   the scintillation counter.

A standard solution containing ^{14}C is provided with the scintillation counter.

Repeat this experiment with the vials provided, which contain a ^{3}H-labelled compound or a ^{14}C-labelled compound. What can you deduce about the nature of the ^{14}C and ^{3}H radioactive emissions?

Part II   Oxidation of 1-[^{14}C] acetate and 2-[^{14}C] acetate by E. coli W

You are provided with two wash bottles containing acetate-grown *E. coli* W cells actively growing in medium (150 ml) containing either 1-[^{14}C] acetate or 2-[^{14}C] acetate. At 15 minute intervals transfer 0·5 ml portions of the 0·1N NaOH (100ml) [which will trap any ^{14}CO$_2$ liberated as Na$_2$ ^{14}CO$_3$] with the given syringe to a test tube. Pipette 0·2 ml of 1·0 M hyamine hydroxide in methanol onto a fluted filter paper and place this filter paper in the top of the test tube. Stopper the test tubes with a serum cap. Inject 0·5 ml 5N H$_2$SO$_4$ through the rubber cap and into the sample with the given syringe, mix, and leave for at least 30 min. Any ^{14}CO$_2$ liberated will be trapped on the filter paper. At the end of this time transfer the filter paper to a counting vial with forceps, fill the vial with Toluene containing fluid (15 ml) and count in the scintillation counter. Do not forget to run a blank.

Recording results
Part I
Calculate the percentage efficiency of the various

instruments and hence calculate the number of radioactive disintegrations per minute in your samples.

Part II

Plot the rate of liberation of $^{14}CO_2$ as a function of time after addition of the isotope. Account for any variations in the rate of oxidation of 1-$[^{14}C]$ acetate and 2-$[^{14}C]$ acetate in terms of what you know about the operation of the TCA–Krebs cycle and of the Glyoxylate cycle.

Discussion

The results of this experiment vary from one day to another in a way which seems to depend on the rate at which the cells grow. Would you expect such differences? How might they be accounted for? It helps when considering this kind of experiment to work out the fate of labelled carbon atoms by tracing their detailed fate through the relevant metabolic reactions.

Extension and applications

The analysis of the fate of isotopically labelled carbon compounds has proved to be one of the most powerful techniques available to biochemists. This kind of experiment in which the end products of metabolism are looked at is the most convenient initial experiment. Any conclusions drawn from such experiments, however, must remain tentative and need to be complemented by further detailed enzymological analysis.

Hints for the teacher

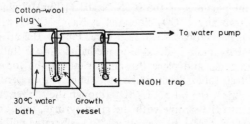

Figure 2

The wash bottles are most conveniently and safely set up as shown in Fig. 2. Using acetate of the suggested specific activity, this experiment is, in fact, quite safe to use with students who have never handled radioactivity before. The following safety precautions should, however, be followed as in all other radioactive experiments. Disposable syringes are more convenient than pipettes as well as being safe

Safety precautions for the use of radionuclides

The following rules are general guide lines for minimising any risks involved in manipulation of radionuclides and *must be observed.*

1 Never pipette by mouth. Radioactive substances are most dangerous when taken internally.

2 Label all containers with the radioactive labels so that there will be no question as to the contents

3 Do not smoke or eat anything while working in the laboratory with radioactive materials.

4 Perform all manipulations with radioactive substances over a protective absorbent mat which can be discarded appropriately after use.

5 If spillage occurs, notify the demonstrator immediately. Any spilled material must be cleaned up as soon as possible.

6 Be sure to dispose of all waste material according to instructions, making certain that they are put in appropriate and clearly marked containers.

7 At the end of the laboratory period be sure to check hands, clothing, furniture, etc., for contamination, using a portable monitor. After such a check, be sure to wash your hands before leaving the laboratory.

Growth of *E. coli.* (National Collection of Industrial Bacteria, Aberdeen)

50 ml of overnight culture of *E. coli* are added to 100 ml of fresh sterile $20mM$ acetate medium and placed in the first wash bottle 3 hours before addition of isotope in order to obtain actively growing cell suspension. At zero time, 0·1 ml (10 μCi) of stock ^{14} acetate solutions are to be added and samples taken at 15 minute intervals from the alkaline trap.

J.M. ASHWORTH
Esse

2 Studies on the precursors of DNA.

Time required
Part time over 2 successive days.

Assumed knowledge
Aseptic technique and the problems involved in the use of isotopes. Facility in chromatography.

Theoretical background
In many experiments a knowledge of the rate of synthesis of DNA (and hence the rate of cell growth) is required. This is often measured by incubating cells with tritiated thymidine and measuring the incorporation of radioactivity into acid insoluble material. These two short experiments illustrate some of the intermediates involved between thymidine and DNA and give a time course for their formation.

Necessary equipment
Requirements for Experiment 1
Roux bottle L929 (2×10^7 cells)–exponentially growing

H–thymidine ($30 \mu Ci/100$ nmoles/ml) 0·5 ml (Radiochemical Centre)

BSS (Earles, pH 7·4) 200 ml

5% Trichloroacetic acid

Thymidine triphosphate ⎫ marker solutions
DP   ⎬ for chromatogram.
MP   (1 mg/ml)
Thymidine ⎭ (All from Sigma)

Whatman 3 MM paper

Solvent for chromatography (as below)

Chromatography tank containing
    isobutyric acid 100 parts
    water 55·8
    ammonia (0·880) 4·2
    EDTA (0·1 M) 1·6
    final pH = 4·6

Toluene scintillator (5 g PPO/litre Analar toluene)

Hyamine hydroxide (Koch-Light)

Ultra-violet lamp, goggles and dark room

Tubes, centrifuge

Pipettes and automatic unit for radioactive sample

37°C incubator or hot room

Scraper (rubber policeman on bent glass rod)

Centrifuge (refrigerated e.g. MSE 6L) and tubes

Ice and ice buckets

Requirements for Experiment 2
1 Roux bottle L929 cells (2×10^7 cells) (Flow Labs, Biocult)

20 Vials (sterile). Trident glass containers which double as scintillation vials (Johnson and Jorgensen Cat. No. 3/11/3905)

Trypsin/versene* 5 ml

EC_{10}† 100 ml

^{3}H-thymidine 30 μCi/100 nmoles/ml

Eppendorf or similar hand controlled pipettes (100 μl) with sterile tips (V.A. Howe)

Stop clock

Pipettes (sterile)

BSS (Earles, pH 7·4) 500 ml

5% Trichloracetic acid

Dioxan scintillator (7 g PPO and 100 g naphthalene/litre Analar dioxan) (Koch-Light)

Toluene scintillator (5 g PPO/litre Analar toluene) (Koch-Light)

Ethanol

Hyamine hydroxide (Koch-Light)

Hot (37°C) room containing an incubator and a bench

Cell counter (Coulter or haemocytometer)

Type of group
These experiments can be carried out by pairs of students or by individuals. The total number would be limited by space.

Instructions to the students
1. To determine the nature of the intermediates between ^{3}H-thymidine and DNA
Label a Roux bottle containing L929 cells for 1 hr at 37°C with $15\mu Ci$ ^{3}H-thymidine at a final concentration of 10^{-6} M. Remove excess thymidine by washing several times with *cold* BSS and prepare a concentrated cold acid extract for chromatography. To do this scrape the cells into *cold* BSS and remove into a

---

*The trypsin/versene contains per 200 ml

0·1 g trypsin (1:250) (BDH)	0·18 g Na_2HPO_4
0·12 g sodium citrate	0·03 g KH_2PO_4
1·52 g sodium chloride	0·03 g versene (EDTA)
0·03 g potassium chloride	

adjusted to pH 7·8 and sterilised by filtration.

†EC_{10} is Eagles minimum essential medium containing 10% calf serum and penicillin and streptomycin. It is buffered with Hepes buffer. This is important as students will be constantly opening the incubator door and it is impossible to maintain the correct pCO_2 under these conditions. It is for a similar reason that the incubator should be sited in a hot room i.e. to maintain a constant temperature.

centrifuge tube held in ice. Centrifuge at 2–4°C (800 g for 5 min). Resuspend the cells in a small volume (0·5 ml) cold 5% TCA. Recentrifuge.

Set up a chromatrogram applying about 50 μl samples of the acid extract and 10 μl samples of the tracers provided. The chromatogram should run overnight (descending). (Overrunning is not advisable as the thymidine spot is easily lost.) Dry the chromatogram, visualise spots using uv lamp in dark room (*remember* to wear goggles), cut out the spots and count the radioactivity by shredding the paper into 10 ml of toluene scintillator containing 0·5 ml hyamine hydroxide solution.

2. To compare the rate of equilibration of ^{3}H-thymidine into the acid-soluble pool with its rate of incorporation into DNA

Trypsinise a Roux bottle containing L929 cells as follows:

Carefully remove the medium and replace with 2–3 ml trypsin/versene solution. Incubate (37°C) for 3–5 min until the cells are loosened from the glass. Transfer the cells to a sterile 4 oz bottle and make up to 25 ml with EC_{10}. Pipette to obtain a single cell suspension and count the cells.

Set up 20 vials containing 2×10^5 cells in 1 ml EC_{10} and incubate them overnight at 37°C.

Label vials with 3 μCi ^{3}H-thymidine at a final concentration of 10^{-5}M for various times, e.g. 0, 1, 2, 5, 15, . . . ,120 min. Remove excess thymidine by washing several times with cold BSS and then add 1·0 ml cold 5% TCA. Remove 0·5 ml of this TCA extract to a vial and add 10 ml dioxan scintillator. Wash the cells a further four times with cold 5% TCA and then twice with ethanol. Air dry. Add 0·5 ml hyamine hydroxide and incubate at 60°C for 10 min. Cool and add 5 ml toluene scintillator.

Count the acid-soluble and acid-insoluble radioactivity.

Recording results

For Experiment 1 the results can be expressed as percent of total counts in the appropriate fractions and in Experiment 2 the results should be expressed graphically.

Discussion

The student should try to fit his results into a scheme of events and try to interpret the reason for the curves obtained. Does the rate of DNA synthesis found in Experiment 2 correspond to that which is expected. Students may also like to attempt the following problem:

Dishes were established containing $0·30 \times 10^6$ L929 cells in 3 ml medium. The cells were incubated for 60 min with ^{3}H-thymidine (5 μCi) at the concentration given below and incorporation of radioactivity into DNA and the acid-soluble pool measured. Autoradiography showed 33% of the cells to be making DNA.

Thymidine per dish (nmoles)	Radioactivity in DNA, acid soluble pool ($10^3 \times$ dpm per dish)	
3	50	68
12	28	46
30	17	32

Interpret the results.

Given that the DNA content of a cell in G_1 is 10 pg calculate the length of S-phase assuming a constant rate of DNA synthesis.

Also calculate the approximate concentration of endogenously synthesised dTTP within an S-phase nucleus.

Extensions and application

These experiments illustrate the techniques used and some of the problems encountered in the use of isotopes in general and particularly in the use of tritiated thymidine for the study of DNA synthesis.

Further reading
1 Cleaver, J.E.
        1967 'Thymidine Metabolism and Cell Kinetics' North Holland Publishing Co. Amsterdam.
2 Adams R.L.P.
        1969 Exp. Cell Res., 56, 49–58.

Hints for the teacher

Precautions in the use of radioisotopes must be stressed and adequate facilities provided for the disposal of waste.

The Trident containers are very suitable for use in Experiment 2 and they can be covered with loose fitting disposable caps for autoclaving and throughout the experiment. They are virtually the same size as scintillation vials and should wear their own clip on caps for the final stage of counting.

R.L.P. ADAMS
Glasgow

28

3 The correlation between DNA synthesis and and tritiated thymidine uptake in organ culture

Time required

1 To set up cultures: 2 hr
2 Termination of cultures: 5 periods ½ hr each
3 DNA extraction, isotope counting and DNA assay: 8 hr
4 Histological analysis: 2 hr.

Assumed knowledge

1 Sterile technique as practised in tissue culture or microbiology
2 Use of liquid scintillation counter and spectrophotometer
3 Use of microscope
4 Histological section cutting and radioautographic technique (technician).

Theoretical background

Thymidine is a specific precursor for the synthesis of DNA and the rate of incorporation of radioactively labelled thymidine is therefore used as an index of the rate of DNA synthesis in living tissues.[1] Two methods for measuring labelled thymidine incorporation are commonly employed: to extract the DNA from the tissue and to measure its specific activity using liquid scintillation counting and secondly to employ radioautography as a means of measuring the proportion of nuclei which have incorporated the label and to use the number of radioautographic grains per nucleus as an index of the relative amount of label retained by the nucleus. The present experiment is intended to show the extent to which these two methods give comparable results.

Tissues maintained in organ culture are used, since reproducible time-related changes in the rates of DNA synthesis are obtained in this system. The *rotating-bottle* culture method is chosen because it is simpler[8] than conventional *raft* or *grid* organ culture methods[7,9] and has been found applicable to a variety of mature mammalian tissues.

Necessary equipment

General

Sterilizing oven
Cylinder of oxygen plus 5% CO_2 (Medical Quality, British Oxygen Co.) with constant-pressure valve (BOC. M. 30-09) and flexible polythene tube branching to at least 4 outlets
1 reciprocal flask shaker in thermostat-regulated water bath for every 5 groups of students
Refrigerator
Water bath (100°C max)
Vortex stirrer.
Low-speed centrifuge (MSE super minor) with provision for multiple 15 ml tubes
Automatic liquid scintillation counter
Spectrophotometer (e.g. Pye Unicam SP500) with 1 ml capacity cuvettes
Photographic dark room
Facilities for histological section cutting, staining etc.
Research microscope with oil-immersion objective
Calculating machine.

Equipment per group

1 Cork board approx. 4in x 6in. Chart pins. Bunsen burner. Glass jar (5 lb) with screw top and containing cotton-wool pad
2 Tissue-culture apparatus chemically clean, sterilised in dry heat and packed in suitable metal containers: beakers 250 ml (2), dissecting scissors (2 pairs), forceps, blunt (1 pair), forceps, pointed (1 pair), spatula, 8in long with ¼in wide blade (2), Swann-Morton type metal scalpel handles with No. 11 blades (Gallenkamp) (2), measuring cylinder 50 ml (2), 2 oz (approx 50 ml) flat medical bottles with *metal* screw caps and rubber liners previously de-toxified by boiling (12), graduated pipettes 10 ml plugged with cotton wool (6), Petri dishes with covers 3½–4in (6), dissecting slides consisting of 3½ x 4½in photographic plates with a strip of autoclave tape attached to one side (6), gassing attachments consisting of 3 x 3in L-shaped glass tube with one limb plugged with cotton wool (6), sheet of aluminium foil 12 x 12in (1), pad of cotton wool
3 Glassware, chemically clean, for fixation, storage and processing of cultures. Quantities are given per group, though in a small class individual packaging would not be necessary. Pasteur pipettes, tips 1½–2 mm diam. (12), specimen tubes with plastic caps (24), conical centrifuge tubes 15 ml (18), pipettes graduated 1 ml (6), pipettes graduated 5 ml (6), liquid scintillation vials (18).

Materials

1   Animals: *per group* Mature rat 2—4 *months* old (1), newborn rat 2—4 *days* old (2)

2   Preparation of animals: anaesthetic ether, industrial alcohol

3   Culture medium: *per group* sterile bottle with 100 ml of Trowell's T8 medium, containing 20% foetal bovine serum and 2% antibiotic solution (5000 iu/ml of Penicillin and Streptomycin) available from Biocult).

4   Isotope: thymidine 6-^{3}H (Radiochemical Centre), 1 mCi made up to 40 ml with sterile 0·9% NaC1, stored in bottle with injection cap 2·4 ml per group, supply of disposable syringes 1 ml with needles

5   Fixation and storage: Carnoy's Fluid (acetic acid: chloroform: alcohol: 10:30:60), 0·9% NaC1

6   Histological processing and radioautography. In addition to the normal reagents: Kodak AR10 stripping film plates, aqueous pyronin/methyl green stain (as used in cytochemical test for RNA)

7   DNA extraction: 0·3N KOH, 1·2N perchloric acid (PCA), 0·2N PCA, 2N PCA

8   Scintillation fluid: 5 g PPO, 0·3 g POPOP dissolved in 1000 ml Analar toluene, after which add 500 ml Triton X-100 (BDH, Scintillation grade)

9   Diphenylamine test: 2% diphenylamine in glacial acetic acid (Analar), 160 mg acetaldehyde in 100 ml water, 2N PCA, calf thymus DNA 16 mg in 10 ml 1N PCA

N.B. For sterility purposes each group should have its own bottle of culture medium. All other solutions, including isotope, can be made available for the whole class.

Type of group

Divide the class up into pairs. The experiment involves a comparison of new-born and mature cultures and the schedules for equipment and method assume that each group of 2 people will culture material of both ages. However, depending on the availability of animals, the technical expertise of the student, and the time available, groups could be divided between those using new-born and those using mature material.

Instructions to the student

Preparation of animals and cultures

Take glass jar, soak cotton pad with ether and place new-born rats inside. Allow 20 min for complete anaesthesia and death. Swab bench with alcohol. Using full aseptic precautions pipette 5 ml of medium into each of 12 screw-top medicine bottles (culture bottles), pipette 5 ml into each of 2 Petri dishes, pipette a small pool (approx. 0·1—0·2 ml) onto each of 2 dissecting slides (*not* onto tape strip) and cover with Petri dish lid. Arrange scissors and forceps in beaker of alcohol. Place spatulas and scalpels in measuring cylinder laid on its side. Fill a second measuring cylinder with alcohol. When animals are dead remove and pin ventral side uppermost to cork board. Swab skin with alcohol. Put mature rat in ether jar. Using scissors and forceps make a ventral incision in the abdominal skin of newborn rats and dissect skin from upper abdomen and thorax. Swab instruments, dip in alcohol, flame and use again for opening thorax and exposing lungs. Swab and return to beaker. Flame second set of instruments, carefully dissect out lung lobes taking care not to crush them with forceps. Place in Petri dish with medium. Swab instruments and repeat for mature rat, placing lungs in a separate Petri dish. Using flamed scissors and forceps divide both sets of lungs into separate lobes. Transfer the large left lobe of the newborn rat lung to a dry part of the dissecting slide, blotting off excess medium, then transfer it to the tape strip. Using the pair of scalpels slice the lung into 1½mm wide strips, starting at the broad edge of the lung. Divide the strips into 1½ mm cubes and transfer to the pool of medium on the same slide. Repeat for the second left lobe and then for other lobes until a total of 80 cubes (explants) have been made. Using the pair of spatulae, flamed and cooled, place 12 explants in each of 6 culture bottles. Fit gassing attachments into bottles, connect to multiple outlets from gas cylinder, cover with aluminium foil and flush for 10 min with oxygen/CO_2 at 10 lb/in^2 (69 N/m^2). Fit screw tops. Repeat for mature lung material to give a total of 12 culture bottles. The water bath is set at 37°C with the flask shaker set at an amplitude of approx. 4 cm and a speed of 120 oscillations per minute. Culture bottles are placed upright in the shaker and incubated continuously.

Design of time series

Three pairs of bottles (new-born and mature) should be taken down at 3 conveniently spaced intervals up to 12 hours. The remaining three pairs are taken down at 24, 48 and 72 hours. Two hours before termination add 0·2 ml of tritiated thymidine solution (final concn. 1 μCi/ml) to the appropriate bottles and return to incubate. (Provided the screw tops are replaced without delay, the loss of oxygen is negligible and re-gassing is unnecessary.) At the end of the incubation period, wash the explants in cold NaCl solution, fix half (6) in Carnoy's fluid and place the other 6 in a specimen tube of NaCl which is then stored in the freezing compartment of the refrigerator.

Radioautography (requiring technical assistance) Section explants at 6μm, coat with AR10, expose for 1 week. If the technician has experience in making permanent stained radioautographs follow the technique normally used. Otherwise, when the preparations are required for microscopic analysis, put one drop of pyronin/methyl green stain onto the emulsion and fit a cover glass. Count labelled nuclei under an oil-immersion objective. An eyepiece graticule and tally counter are useful though not essential. Record the number of labels in 1000 nuclei in at least one section from each group of 6 explants. Calculate the mean and standard deviation for labelling index (LI) in each group.

DNA extraction
Thaw samples (which may be stored frozen for several days), transfer to 15 ml centrifuge tubes, add 1 ml 0·3N KOH and incubate overnight at 37°C. Suspend on a mixer (e.g. 'Whirlimixer', Fisons). If the explants have not been digested, incubate at 70°C for a maximum of 2 hours more. Cool on ice, add 0·6 ml 1·2N cold PCA and keep on ice for 20 min. Spin at 2000 rev/min for 5 min. Reject supernatant and wash precipitate twice with 0·7 ml 0·2N cold PCA, rejecting supernatant each time. To the washed precipitate add 0·5 ml 2N PCA and incubate 30 min at 70°C. Cool and spin. Retain supernatant. Wash precipitate twice with 0·5 ml 0·2N PCA to give a total of 1·5 ml supernatant (DNA solution).

Scintillation counting
Add 0·5 ml of each sample of DNA solution to 5 ml of scintillation fluid contained in scintillation vials. Include a blank containing 1 N PCA in place of DNA solution. Load vials into counter, set for tritium emission, and count for at least 10 min.

Diphenylamine reaction for DNA
Starting with calf thymus DNA solution make a series of 6 standard dilutions (at least 1 ml each) from 0 to 80 μg/ml using 1 N PCA as a dilutant. To 0·8 ml of DNA solution (standard dilutions and test samples from experiment) add 0·8 ml diphenylamine reagent and 0·04 ml acetaldehyde solution. Incubate overnight at 30°C or 30 min at 70°C. A blue colour

reaction should develop. Put 1 ml of solution into spectrophotometer cuvettes and measure the optical density at 600 nm. Rinse cuvettes in 1 N PCA between successive samples.

Recording results
Make separate graphs for new-born and adult material. On both graphs compare LI with thymidine incorporation measured by scintillation counting.

Labelling index (LI)
Plot labelling index (left hand vertical axis) against time (horizontal axis).

Rate of incorporation of thymidine (RIT)
This is proportional to the specific activity of extracted DNA—expressed as counts per min (cpm) per μg DNA. This is calculated as follows. Taking the data for optical density of the calf thymus DNA serial dilutions, plot a standard curve of the optical density (vertical axis) against DNA concentration (0–80 μg/ml). Estimate the DNA concentration of each test sample by comparing its optical density against the standard curve. Calculate the absolute amount of DNA in the 0·5 ml sample used for scintillation counting. From the cpm reading for the test samples, subtract the cpm obtained in the blank vial (1 N PCA only). For each sample, express the specific activity of the DNA as cpm per μg DNA. Using these values as indices of the rate of incorporation of thymidine (RIT), incorporate them in the graph, using the right-hand vertical axis as a measure of RIT.

Discussion
Are there any serious discrepancies in the time response using these two classic methods for measuring the rate of 'DNA synthesis'? LI measures the proportion of nuclei incorporating [3]H TdR, while RIT is dependent not only on the proportion of nuclei but on the amount of [3]H TdR incorporated per nucleus. Some measure of the latter can be obtained from radioautographs by means of grain counting: under oil immersion, count the number of grains per nucleus, taking 10 readings from each group of 6 explants. How do changes in grain count relate to RIT?

Assuming a constant rate of DNA replication, what factors could produce changes in RIT? Consider enzymic degradation of the label, competition with TdR synthesized *de novo*, and changes in metabolic pathways. How could these, or any other hypotheses of your own, be tested experimentally? What confidence can be placed in these 2 techniques as reliable methods for measuring DNA replication?

Extensions and applications

Organ culture is an ideal way of examining the direct effect of chemical and physical stimuli on the growth, differentiation and metabolism of differentiated tissues. DNA replication is an essential prerequisite of cell division, and the experiment described above was developed as part of a study designed to examine the physiological factors which regulate cell proliferation during normal growth of the organism. Much information on the biological action of hormones was derived from organ culture experiments, while changes in the composition of the culture medium provide valuable data concerning tissue metabolism—reference is given in the next section to a test which can be applied to measuring the rate of glucose consumption in the culture system described[3]. Apart from the lung, many other organs may be cultured successfully using the same technique. In general, organs from newborn animals are more successful than those from older animals.

Further reading

1 Cleaver, J.E.
  1967 'Thymidine Metabolism and Cell Kinetics', North Holland, Amsterdam.
2 Hell, E., Berry, R.J., and Lajtha, L.G.
  1960 A Pitfall in High Specific Activity Tracer Studies, Nature, 185, 47.
3 Huggett, A.St.G. and Nixon, D.A.
  1957 Enzymic determination of blood glucose, Biochem. J., 66, 12.
4 Littlefield, J.W.
  1965 Studies on Thymidine Kinase in Cultured Mouse Fibroblasts. Biochem. Biophys. Acta, 95, 14–22.
5 Munro, H.H. and Fleck, A.
  1966 The Determination of Nucleic Acids, Methods Biochem. Anal., 14, 113–76.
6 Painter, R.B. and Rasmussen, R.E.
  1964 A Pitfall of Low Specific Activity Radioactive Thymidine, Nature, 201, 409.
7 Paul, J.
  1970 'Cell and Tissue Culture', 4th Edn. Livingstone, Edinburgh.
8 Simnett, J.D. and Fisher, J.M.
  1973 An Assay System for Humoral Growth Factors, Nat. Cancer Inst. Monogr., 38, 19–28.
9 Trowell, O.A.
  1959 The Culture of Mature Organs in a Synthetic Medium. Exp. Cell Res., 16 118–47.
10 Weissmann, R., Smellie, R.M.S. and Paul, J.
  1960 Studies on the Biosynthesis of DNA by Extracts of Mammalian Cells. Biochem Biophys. Acta, 45, 101–10.

Hints for the teacher

Safety precautions

From the point of view of excluding contamination, no special laboratory facilities are required for setting up organ cultures provided the benches are free from clutter and can be swabbed with alcohol. Dust is one of the chief sources of contamination—the doors and windows of the laboratory should be closed to exclude draughts. Within pairs, students should collaborate in such a way as to prevent bottles and dishes being open to contamination longer than necessary. Instruments require frequent flaming—take care not to put flaming instruments back into alcohol. A large tin can placed over the ignited beaker is the best way of extinguishing a fire so caused. Keep ether and killing jar on a separate bench.

Dispense ^{3}H TdR solution with disposable syringes. There are certain legal requirements to be met before carrying out experiments with radioactive substances. The person in charge should discuss the practical with the Radiation Protection Officer with particular regard to the cleaning or disposal of contaminated apparatus and the disposal of radioactive waste.

Sterile preparation

Sterilize equipment sufficiently in advance to give it time to cool. The T8/serum mixture should be prepared and dispensed in individual bottles beforehand. It may be stored in this way for several weeks in the deep freeze—but allow time for thawing. Include spares in case some bottles crack.

Animals

Select healthily growing new born rats—undersized ones give poor results. When dissecting organs into explants one of the pair should make note of the number—it is easy to lose track of this.

Labelling and recording

Label all culture bottles and make a note of the schedule of treatment each one is to receive in order to avoid confusion over when isotope is to be added and the cultures terminated.

- J.D. SIMNETT
Developmental Biology Laboratory
Newcastle upon Tyne

4 Incorporation of tritiated leucine and uridine into chicken red blood cells

Time required
Minimum of 5 hr on day of experiment and probably 3 hr for analysis of results about 2 weeks later.

Assumed knowledge
Familiarity with sterile technique required, and prior guidance on the handling of radioactive material advisable. Some understanding of the mechanism of protein synthesis.

Theoretical background
Almost all living cells are actively involved in the process of protein synthesis. This important biochemical pathway involves, first, the synthesis of RNA from a DNA template by RNA polymerase enzyme (transcription), followed by the use of this RNA message for the ordered assembly of amino acids into proteins on the cellular ribosomes (translation). By using RNA precursors (uridine) and protein precursors (leucine) labelled with the radioactive isotope tritium, we can follow the rates of these processes in chicken red blood cells in a variety of conditions. Use of the drug actinomycin D, which inhibits RNA synthesis, and the drugs puromycin and cycloheximide, which inhibit protein synthesis, provides additional information on the relationship between RNA and protein synthesis.

Moreover, by the autoradiographic technique, we are further able to determine how many cells are involved in these processes, and whether such synthesis is substantially nuclear, or cytoplasmic, or both.

Necessary equipment
Apparatus for a class of 20 students
5 Bench centrifuges (MSE) fitted to take 50 ml tubes
40 50 ml glass centrifuge tubes
10 sterile Bijou bottles (or McArtney bottles) (Gallenkamp)
40 pasteur pipettes
Sterile graduated pipettes—60 at 1 ml
Non-sterile graduated pipettes, 10 at 1 ml, & 20 at 10 ml

33

100 3 mm paper discs of 2·1 cm diam. (Whatman
    GF/C)
120 scintillation vials
10 Polystyrene blocks with 10 pins each
5 pencils
40 microscope slides
50 Rubber teats
10 tube racks for 50 ml centrifuge tubes
1 Shaking water bath at 37°C
5 vacuum pumps on sink taps
Tissues
10 bunsen burners
Good phase contrast/bright field microscopes for
    analysing cresyl blue (Gurr) stained cells and auto-
    radiographs
20 test tubes
10 Haemocytometer counting chambers
10 Time clocks

Chemicals for class of 20 students
50 g Ilford K5 Nuclear emulsion
1 g saponin (BDH)
1 mg Actinomycin D (Sigma)
100 mg Puromycin di HC1 (Sigma) *or*
1 g Cycloheximide (Sigma)
^{3}H Leucine and ^{3}H Uridine, 1 ml of each labelled as
    1 mCi/mM in aqueous solution (Radiochemical
    Centre, Amersham)

Solutions for class of 20 students
10 ml Brilliant Cresyl blue stain, freshly prepared &
    filtered
    (1% stain in Locke's solution)
8 litres non sterile chick Ringer's solution (i.e. Locke
    solution)
i.e. Sodium chloride   8·69g/litre
Potassium chloride   0·31g/litre
Calcium chloride   0·20g/litre
Magnesium chloride   0·10g/litre
Glucose   2·00g/litre
Potassium dihydrogen
    phosphate   0·20g/litre
di Potassium
    monohydrogen phosphate 0·80g/litre
600 ml of liquid scintillation fluid
2 litres sterile chick Ringer's solution (Locke)
2 litres 10% TCA with 0·125 mM Uridine (BDH)
2 litres 10% TCA with 0·012 mM Leucine (BDH)
2 litres 50% TCA
1 litre sterile chick Ringer's solution (Locke,) with
    8% bovine serum albumin (BDH)

Animals for class of 20 students
2 adult chickens, one normal and one rendered
anaemic by phenylhydrazine injection

Type of group
The group can vary, but the experiment is probably
ideal with a class of 20 students working as 10 pairs.
An experienced technician in attendance is to be
recommended.

Instructions to the student
Students will receive a sterile bottle containing 5 mls
of chicken red cells in culture medium. There are ten
different cultures, numbered as follows:
1 Normal bird, ^{3}H leucine
2 Normal bird, ^{3}H leucine + Actinomycin D
3 Normal bird, ^{3}H leucine + Puromycin or Cyclo-
    heximide
4 Normal bird, ^{3}H uridine
5 Normal bird, ^{3}H uridine + Actinomycin D
6 Anaemic bird, ^{3}H leucine
7 Anaemic bird, ^{3}H leucine + Actinomycin D
8 Anaemic bird, ^{3}H leucine + Puromycin or Cyclo-
    heximide
9 Anaemic bird, ^{3}H uridine
10 Anaemic bird, ^{3}H uridine + Actinomycin D.
Each student pair to be responsible for one bottle.
Between sampling they are incubated in a shaking
incubator at 37°C. Samples should be removed at
zero time (as quickly as possible after the blood cells
have been added to the bottle), 30 min, 1 hr, 2 hr, 3
hr and 5 hr, and treated as described in Hints for the
Teacher. Be sure to use the same sampling technique
each time, shaking the culture before sampling,
draining the pipette by blow out, etc. The lysate from
any one time sample goes on to two discs, which can
both carry the same code, i.e.
    1A indicates a disc from bottle (1) sampled at zero
time (A). Samples should be taken at convenient
predetermined times as was suggested above. It will
take a pair of students about 20 min to process an
aliquot. Samples of 0·2 ml should be removed by a
1 ml graduated pipette and teat (*DO NOT* pipette
radioactive solutions by mouth) at the stated time,
and dispensed into 50 ml of non-sterile Ringer for
washing by centrifugation. After two washes the pellet
should be resuspended in 0·2 ml of Ringer and lysed
by the addition of 1 mg of saponin. Centrifugation at
100 g for 5 minutes will bring down a white pellet,
leaving a clear red lysate. The lysate should be placed
on each of two numbered (in pencil) paper discs
supported on pins on the polystyrene board. The
counts from the two discs will be summated so careful

equal application to the two discs is not absolutely necessary. Paper discs should be air dried (a hot lamp will help) before being dropped into cold 10% TCA (plus leucine or uridine as appropriate).

Following removal of the last sample, about 0·5 ml of the remaining cell suspension should be well washed by centrifugation through Ringer's solution, resuspended in 0·5 ml of Ringer's solution with 8% bovine serum albumin, and the suspension smeared on 2 numbered slides per sample for the preparation of autoradiographs; process for autoradiography by the usual methods.

The scintillation counts, when available, should be plotted against time, the background count being subtracted from each. A zero time count much above 50 indicates either poor washing technique of cells or discs, or a delay between addition of blood cells and zero time sample.

Recording results
The scintillation counts can be recorded graphically against sample time, uridine curves on one sheet and leucine on another. The data from the analysis of autoradiographs will yield information about the percentage of the cells involved in the measured synthesis and provide an interesting comparison with the initial reticulocyte count from Brilliant Cresyl Blue.

Discussion
How many red cells are involved in making protein and RNA in the two birds? What is the effect of anaemia? What can one deduce from the use of Actinomycin D, Puromycin and Cycloheximide? Does Actinomycin D inhibit protein synthesis in your experiment? If so, why? To what extent has individual technique affected the results of any one pair of students? Differential nuclear and cytoplasmic grain counts on the autoradiographs is advisable. Why?

Extensions and applications
The techniques used in this experiment have been used very widely to probe the biochemical activity of cells and the relationship between RNA and protein synthesis.

Further reading
Harris, H.
   1974 'Nucleus & Cytoplasm' (3rd edn.) Clarendon Press, Oxford.
Maclean, N. and Jurd, R.
   1972 'The Control of Haemoglobin Synthesis' Biol Revs., 47, 393–437

Madgwick, W., Maclean, N. and Baynes, Y.
   1972 'RNA Synthesis in Chicken Erythrocytes', Nature (New Biology), 238, 137–8

Hints for the teacher
Home Office licence required for injection and bleeding of chickens. An experienced technician must begin preparation at least a week before and must assume 2 days' preparation before the day, and 1 day's follow-up afterwards. The anaemic chicken should be injected as follows: 8 days prior to experiment phenylhydrazine (BDH) at 10mg/kg body weight: 3 days prior, ditto: 2 days prior ditto but at 5mg/kg; 1 day prior ditto but 5mg/kg. The injections should be subcutaneous with a sterile syringe. The chickens should be bled from a wing vein by sterile syringe. This requires some experience and the help of a technician to hold the bird during the bleed. 2 ml of blood should be drawn from each bird and immediately mixed with 100 ml of sterile Ringer's solution to prevent clotting. Blood cells should be washed by two washings with sterile Ringer's solution and cells recovered by centrifugation at 100 g for 5 min. Two ml of blood can be washed in two 50 ml centrifuge tubes, 1 ml per tube, and the supernatant washing solution removed by a bench aspirator with a sterile Pasteur pipette. Following the final wash, the 2 pellets from each bird should be pooled after resuspension in 10 ml of sterile Ringer. At this stage, a sample should be taken for a cell count using a haemocytometer counting chamber. During the bleeding of the chicken, 1 drop of fresh blood should be mixed with 2 drops of the Brilliant Cresyl Blue solution, allowed to stand for 5 min, and then a reticulocyte count made by bright field microscopy of a blood smear on a slide.

Each pair of students should be provided with a sterile Bijou bottle (or McArtney bottle) numbered on top and side and already containing the following sterile mix, made up for a final volume of 5 ml: the twenty common amino acids except leucine, L-forms where relevant, 0·5 μmoles/ml; glucose, 1 mg/ml; ferrous ammonium sulphate, freshly prepared, 5 μg/ml; penicillin, 400 units/ml; streptomycin sulphate (Glaxo), 400 μg/ml; bovine serum albumin, 3 mg/ml, tritiated leucine or uridine, 0·1 mCi; leucine where appropriate, 0·5 μmoles/ml; Actinomycin D, where appropriate, 10 μg/ml; Puromycin, where appropriate, 200 μmoles/ml; Cycloheximide, where appropriate 100 μg/ml. The volume of liquid in each bottle should be adjusted to 3 ml with sterile Ringer. At zero time, 2 ml of washed blood cells in Ringer should be added to each bottle and the students alerted to take the first sample.

The paper discs after 1 hour in cold 10% TCA should be processed by heating to 90°C in 5% TCA to ensure solution of free amino acids from the leucine samples, and washing in room temperature 5% TCA to remove free nucleotide from the uridine samples. Following this extraction, they should be washed in ether, air dried, and counted in the scintillation counter.

The organisation of the practical is best done by having a technician prepare the initial blood cell suspension, the students remove, wash and lyse the timed samples and transfer the lysate to discs, and the technician washes the discs and carries out the scintillation counting. Individual students can be assigned to (1) brilliant cresyl blue reticulocyte count; (2) estimation of cell number by haemocytometer; (3) coating of slides for autoradiography with K5 emulsion.

All students should inspect and analyse their autoradiographs after about 10 days' exposure and appropriate development.

Note that the bulk of the radioactive material goes down the sink with the washings. A check should be made in each individual establishment to determine the regulations on disposal. If necessary they could be collected by aspirator into a large Buchner flask for later disposal. Do not allow students to mouth pipette radioactive solutions.

A haemocytometer count of each culture is recommended at the end of the experiment to check for cell lysis.

N. MACLEAN
Southampton

5 Measurement of ATP synthesis by a radio-isotope tracer technique: photosynthetic phosphorylation by chloroplasts

Time required
About 2½ hr, or 3½ hr if students are to prepare their own chloroplasts.

Assumed knowledge
Students should have had some instruction on the use of radioactive isotopes, such as handling and counting techniques.

The experiment provides a useful introduction to the use and handling of radioactive isotope tracers.

Theoretical background
The method of phosphorylation used in this experiment is applicable to a wide variety of phosphorylating systems, such as mitochondrial oxidative phosphorylation.

In the 'light' reactions of photosynthesis in chloroplasts, reducing equivalents (or electrons) produced by splitting of water are transferred through a series of electron carriers, ultimately to NADP. There are two light-dependent reactions involving chlorophyll in this sequence, photosystems I and II.

As in mitochondria, the sequence of electron transfer steps is linked to the synthesis of ATP. Freshly prepared, washed (P_1S_1) chloroplasts are capable of carrying out this process, although they have lost the ability to fix CO_2.

In this experiment, a section of the electron-transport chain will be investigated, which uses water as electron donor and the dye methyl viologen as an artificial electron acceptor:

In photosystem II: $H_2O = \frac{1}{2}O_2 + 2H^+ + 2e^-$
In photosystem I: $2e^- + 2MV_{ox} = 2MV_{red}$

Where e^- represents a reducing equivalent, and MV_{ox} and MV_{red} are the oxidized and reduced forms of methyl viologen. The reduced form reacts with oxygen:

$$2MV_{red} + O_2 + 2H^+ = 2MV_{ox} + H_2O_2$$

The chloroplasts contain catalase activity which breaks down H_2O_2:

$$H_2O_2 = H_2O + \tfrac{1}{2}O_2 .$$

So there is no resultant output or uptake of oxygen by this system. But during electron transport, phosphorylation takes place:

$$ADP + P_i = ATP.$$

The incorporation of ^{32}P-labelled orthophosphate into ATP can be measured by converting the un-reacted orthophosphate to a phosphomolybdate complex and extracting into a mixture of isobutanol and benzene (1:1). Organically bound ^{32}P remains in the aqueous layer and is measured by liquid scintillation counting.

No scintillator is required for counting of ^{32}P, since it can be detected by the Cerenkov radiation which it emits in solution. Alternatively other types of counter, such as a gas-flow counter, if available, may be used.

Necessary equipment
Apparatus and solutions for preparation of washed chloroplasts (see Section 9.7.):

Scintillation counter
Bench centrifuge (MSE)
Whirlimixer (Gallenkamp) or similar test-tube shaker
2 stoppered 10-ml centrifuge tubes, glass or (preferably) polycarbonate (Sterilin)
2 stoppered glass test-tubes (Gallenkamp)
4 scintillator vials (Koch-Light)
Illumination source and filter (see Section 9.8).
Filter pump with tube to suck off organic solvents, with a suitable trap (e.g. bottle, flask).
^{32}P-labelled orthophosphate, such as that supplied by the Radiochemical Centre. Dilute in 50 mM K_2HPO_4 to give 25 μ Ci/ml.

In addition, the use of a radioactive isotope will necessitate the following:

Covering for the bench, such as Benchkote (Whatman), or a layer of polythene sheet covered by a layer of blotting paper.
A portable radiation monitor, such as a small Geiger

counter..
Bin for radioactive waste.
Devices for dispensing solutions without the use of mouth pipetting, such as bulb pipettes, syringes etc. (Gallenkamp).
Methylviologen (BDH) 0·01M
Tricine (BDH) 0·2M
$MgCl_2$ 0·1M
ADP (Boehringer) 0·05M
Trichoroacetic acid (BDH) 10% (w/v)
Isobutanol/Benzene 1:1 (saturated by shaking with H_2O)
Water (saturated by shaking with 1:1 isobutanol/benzene)
Acid ammonium molybdate solution (5 g ammonium molybdate [BDH]in 40 ml 10 N H_2SO_4 made up to 100 ml with water)

K_2HPO_4 0·02M

Type of group
Individuals or pairs. Suitable for fairly large classes under adequate supervision. For large classes the time required for counting may be a limiting factor.

Instructions to the student
Precautions for handling radioactive solutions
The ^{32}P used as a tracer in this experiment is a strong emitter of beta-radiation and represents a potential hazard, if handled wrongly. The most serious risk is of ingestion, by mouth or through the skin. Therefore it is essential to follow certain rules:
1 Wear a laboratory coat to protect your clothes.
2 No smoking, eating or drinking in the laboratory, and no pipetting by mouth of any solution.
3 Label all tubes clearly, so that you and other people know where the radioactivity is.
4 Keep all radioactive solutions on a restricted part of the bench, with a suitable covering, and keep all unnecessary paraphernalia such as notebooks away from it.
5 If any radioactive solution is accidently spilt, clean it up immediately with an acid phosphate solution.

Method
Prepare washed chloroplasts as described in Experiment 9.7. These should be used as soon as possible, as their phosphorylating activity deteriorates with time.
Transfer into each of two centrifuge tubes:
1·0 ml 20mM tricine/5mM $MgCl_2$/0·2mM methylviologen, pH 8·0

0·4 ml distilled water
0·1 ml 50mM ADP
0·3 ml chloroplasts, 1 mg chlorophyll/ml.
Then: 0·1 ml 50mM K_2HPO_4, containing $25\mu Ci/ml$ ^{32}P.

Wrap one of the tubes in aluminium foil, and illuminate the other for 10 min.

Next, to each tube add 3 ml 10% trichloroacetic acid (TCA), stopper, mix and centrifuge at 4000 rev/min for 2 minutes. Into two test-tubes, labelled 'light' and 'dark', pipette 1·0 ml of supernatant from the appropriate centrifuge tube. To each test–tube add 1·0 ml of solvent-saturated water and 4·0 ml of isobutanol–benzene solvent mixture (1:1, saturated with water).

Stopper the tube with a polythene cap, mix well on a Whirlimixer and allow the phases to separate (this should take less than 2 min).

To each tube, very carefully add 0·5 ml of acid ammonium molybdate solution down the side of the tube to form a third layer, then mix the two lower aqueous layers with a swirling action and leave for 5 min. With the tube stoppered, shake vigorously on a Whirlimixer for 30 s, leave until the phases separate and suck off the organic layer using a filter pump.

Add 20 μl of non-radioactive 20 mM KH_2PO_4, mix, extract with a further 4 ml isobutanol/benzene mixture and once again remove the organic phase— *make sure this is complete.*

Preparation of samples for counting
Take 1 ml samples of the aqueous phase of the two tubes, as a measure of the phosphate esterified as ATP, and 0·1 ml sample of the TCA-treated incubation mixture, as a measure of the total phosphate added. In each case, make up to 10 ml with 0·1M KH_2PO_4 in scintillation vial.

Measure the counts/minute of the samples in a scintillation counter, and subtract the background count.

Before leaving the laboratory, check your bench and yourself with a Geiger counter and do not leave until you are satisfied that there is no radioactive contamination.

Recording results
Write up the procedure for extraction of esterified phosphate in the form of a flow-chart, explaining the purpose of each step. Calculate the percentage of the phosphate esterified for each reaction mixture.

Discussions
What is the significance of the counts measured in the dark tube? What experiments would you carry out to check the possible explanations?

Extensions and applications
The experiment can be used to examine various aspects of the photophosphorylation system; for example it is easy to demonstrate the effects of 1 μM dichlorophenyldimethyl urea (DCMU), a herbicide which inhibits photosynthetic electron transport, or 1 mM ammonium chloride, which uncouples electron transport from phosphorylation.

The method can be adapted to most systems where ATP is formed. A suitable system might be the oxidation of substrates such as succinate or β-hydroxybutyrate by isolated rat liver mitochondria. The measurement of phosphorylation can be combined with a manometric measurement of oxygen uptake to determine P : O ratios for these substrates.

Further reading
Using experiments of this type, Arnon, Allen and Whatley (Nature (1954) 174, 394) were able to demonstrate that ATP is generated by the photosynthetic electron transport system, and not as a side-effect of the reoxidation of NADPH.

For further reading on photosynthesis, see
Hall D.O. and Rao, K.K.
    1972 'Photosynthesis', Institute of Biology Studies in Biology No. 37, Edward Arnold, London.
Gregory, R.P.F.
    1971 'Biochemistry of Photosynthesis', John Wiley, London,

On the use of radioactive tracers in biology:
Thornburn, C.C.
    1972 'Isotopes and Radiation in Biology', Butterworths, London,

Hints for the teacher
The dispensing of concentrated ^{32}P solution and the disposal of radioactive waste should be done by a responsible person. If possible, the preparation of chloroplasts and setting up of incubation mixtures should be completed before the students are provided with ^{32}P.

R. CAMMACK
King's College
London

6 Investigation of the Calvin cycle using radioactive carbon dioxide

The experiment described demonstrates the method used by Calvin and his co-workers when they elucidated the carbon dioxide fixation cycle of photosynthesis. In using biochemical techniques and involving culture and manipulation of a unicellular alga the experiment helps to link the two fields of biochemistry and cell biology for the student.

Time required
The photosynthetic incorporation of $^{14}CO_2$ into the algae and preparation of the extract can be concluded in 3–4 hrs, but the chromatograms need 2–3 days for development. Preparing the chomatogram for autoradiography takes only about 30 min, but it requires 2–4 weeks for activation. Developing and fixing the film requires only 30 min; film drying about 1 hr.

The basic experiment therefore can be considered to be a single practical while the rest of the experiment must be carried out by the student either outside of normal class hours or in subsequent laboratory periods. If technical help is available then matters are greatly simplified. For example if the chromatograms are placed on racks it is a simple matter for a technician to change the solvent and to dry the papers when required.

Assumed knowledge
In order to carry out this experiment for himself or in a small group, the student needs to have a reasonable degree of manual dexterity and competence in the laboratory. Ideally he should be familiar with the techniques of paper chromatography and preferably have some knowledge of the principle if not the practice of autoradiography. Previous experience at developing photographs however is not really essential as this is very straightforward. Some background knowledge of the original experiments of Calvin's group is helpful in understanding the purposes of the various stages of the experiment.

Most HNC- and degree-level students are able to manage this experiment.

Theoretical background
For some time it has been known that photosynthesis consists basically of two stages. A light-dependent one, in which light is used to generate metabolically useful sources of energy (ATP) and of reducing power (NADPH); and a light-independent one, in which these components are used to convert CO_2 into a variety of photosynthetic products.

The incorporation of a simple inorganic molecule such as CO_2 into complex organic ones such as carbohydrates is obviously involved and its investigation proved difficult.

The discovery by Ruben and Kamen of the long-lived isotope ^{14}C provided a useful tool for the investigation of these pathways. Calvin and his co-workers realised that after exposing a photosynthetic organism (for convenience a unicellular alga) to $^{14}CO_2$ for various periods of time, the ^{14}C in each case is incorporated to different 'distances' along the metabolic routes concerned. When photosynthesis is stopped by immersion in alcohol, the pattern of labelling is frozen and then by determining *which* compounds are labelled, and the *quantities* (or specific activities) of labelling in each case, the sequence of intermediates is apparent. The earlier a compound appears in the metabolic pathway, the more intense its labelling should be at any given time when compared with compounds later in the sequence. Also as the period of exposure to $^{14}CO_2$ lengthens, additional labelled compounds appear as the ^{14}C moves down the metabolic pathways.

A major problem is to separate, identify and quantitatively determine the potentially large number of possible products of photosynthesis. The separation is done by two-dimensional chromatography and the compounds can be located and identified by using various spray reagents, by comparison of Rf values with published figures, and if necessary by elution followed by various chemical tests and co-chromatography with authentic compounds. Determining which of the chromatogram spots are radioactive cannot be done with accuracy by using a Geiger–Müller tube held over the paper, because of the weak nature of the β-emission of ^{14}C. The technique of autoradiography using a piece of X-ray film placed alongside the paper is used to locate the radioactive spots. The radioactive emission from these spots activates the film at the site adjacent to them and they appear as black areas after development. Non-radioactive spots do not produce such black areas.

The intensity of the radioactivity is determined by eluting the compound from the paper and measuring its activity in a scintillation counter.

Necessary equipment

For the photosynthesis
1 250 ml 3-necked Quickfit flask (Gallenkamp)
1 250 ml separating funnel
2 250 ml Dreschel bottles (Gallenkamp)
1 500 ml beaker
1 10 ml pipette
1 50 ml measuring cylinder
1   1 ml hypodermic syringe & 2 needles
1 Vaccine cap
1 Cone/tube bend with tap
1 vacuum pump
1 lamp
1 stirrer and follower
50 ml Absolute methanol
500 ml 10% Barium (or sodium) hydroxide
100 ml Buffer containing, 45 μm NaOH, 5 mg
   fumaric acid (BDH)
1 *Chlorella* culture grown in a simple medium such as
   Knops culture solution. 100μCi of NaH $^{14}CO_3$
   from the Radiochemical Centre, Amersham, Bucks
2 Paper tissues
100 ml Centrifuge tubes and centrifuge (MSE)
1 hot plate
1 evaporating basin
1 fume cupboard or large tray
2 lengths 30 cm pressure tubing
1 stop clock

For the chromatography
1 Sheet of Whatman No. 1 paper 20 cm square
1 developing tank suitable for ascending chromatography
10 μl micropipettes
Pencil
Sufficient of the following two solvents (freshly
   made)
(a) Butanol (4)/Acetic Acid (1)/Water (5)
(b) Phenol (225g)/Water (25 ml)/Ammonia (2·5 ml)

For the Autoradiography
1 Sheet of No-screen, ester base, one-sided X-ray
   film (e.g. Kodirex; Kodak)
Kodak 6B safelight
Large sheet foil
Sellotape and scissors
2 Glass-plates 20 cm square
1 storage box
1 spare sheet chromatography paper
2 developing troughs

Tongs, clock, running water, hanging clips
1 litre Kodak D19 developer
1 litre fixative (either proprietary or 170g sodium
   thiosulphate and 25g sodium metabisulphite per
   litre).

Type of group
The experiment can be approached in a variety of
ways. Ideally it should be carried out by pairs of
students. This requires a reasonable amount of space
and apparatus if a large class is involved, but this can
be minimised by setting the experiment in rotation
among a group of students as part of a 'circus' of
experiments.

Alternatively, the lecturer or selected students
could prepare a radioactive algal extract and individual
students or pairs of students run their own
chromatograms and autoradiograms using a portion
of this extract.

Instructions to the student
Assemble the apparatus provided as in Fig. 1 and
ensure that all the joints are securely made. If
Quickfit apparatus is not available, be sure to use
good-quality, well-fitting bungs and to seal all joints
with vaseline.

You will be provided with an actively growing culture of a unicellular alga such as *Chlorella* and will be
told its concentration of cells in terms of dry weight/
ml. Take sufficient volume to obtain 20 g of cells and
if dilution is necessary use the fumarate buffer provided. A final volume of 50 ml is most suitable.

Add the algal suspension to the round-bottomed
flask and ensure that it is stirred vigorously. Connect
up the rest of the apparatus, close taps A and B and
leave the vacuum pump switched off. Illuminate the
algae and allow to photosynthesise for 5 min. An
illumination intensity of approximately 1000 foot
candles is adequate, and this can be provided by a 60
W lamp at 9 in. While this is proceeding carry out the
following three operations.
1 Check the apparatus over and make certain it is
   not likely to come apart.
2 You will be provided with a solution of radioactive sodium bicarbonate (NaH $^{14}CO_3$). You
   will be told how much sample to take to
   obtain 100 μCi of ^{14}C. Take up this quantity in
   a hypodermic syringe, carefully remove the
   needle and connect the syringe to the needle
   inserted through the vaccine cap in the apparatus. Leave it there for the present. Ensure that
   the solution does not contact your hands and if
   any is spilt use a tissue and dispose of it

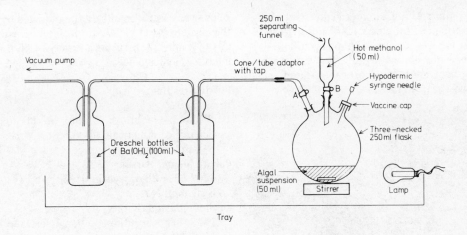

Figure 1 Apparatus for exposure of photosynthesising
algae to $NaH^{14}CO_3$

according to standard radiochemical procedures.
3 Have boiling on a hot plate a volume of methanol
equal to the volume of *Chlorella* cells to be used
(normally 50 ml). Inject the labelled bicarbonate
solution into the photosynthesising *Chlorella*
suspension and start a stop clock. Carefully place
the hot alcohol in the separating funnel and after
1–5 min *open* taps A and B, in that order.
Switch *on* the vacuum pump and switch *off* the
light. The alcohol will stop further photosynthesis by killing the algae and will extract the
cell contents. Draw air through the apparatus for
a minimum time of 15 min while maintaining
vigorous stirring of the culture. Residual $^{14}CO_2$
will be drawn off and safely trapped as barium
(or sodium) carbonate:

$$^{14}CO_2 + Ba(OH)_2 \rightarrow Ba^{14}CO_3 + H_2O.$$

Detach the flask from the rest of the apparatus.
Place the apparatus to one side, bearing in mind
that it will be contaminated. Centrifuge the
contents of the flask at 1000 g for 5 minutes and
remove the supernatant. Evaporate the supernatant on a hot plate using initially a 500 ml
beaker followed by an evaporating basin. Continue until a volume of 1–2 ml is obtained, *do
not* allow to dry out and be careful to avoid
spitting in the final stages.

Preparation of the autoradiograms
Take a piece of Whatman No. 1 chromatography

paper of suitable size (about 20 cm square) and mark
a pencil cross 2in in from one corner. Write on the
paper your name, date and the direction of development in the two solvents.

On the origin place 50 μl of the algal extract,
making the spot as small as possible. Develop the
paper in water-saturated phenol; dry it in a forced
draught for as long as practicable (1–18 hr) and develop the paper again in the second direction, using the
butanol, acetic acid and water solvent. In each case
develop the paper until the solvent is near the top
(about 10 hours).

Carry out the following operations in a dark room
using a suitable safe-light.

Take a piece of X-ray film provided, and using two
small pieces of sellotape, carefully fix the chromatogram to it by its edges. Handle the film by its edges
only and do not place on a rough surface. Place a
piece of chromatography paper over the film and
sandwich the whole between two glass plates to
ensure good contact between the film and paper.
Place in a suitable storage box (e.g., an old chromatography paper box) and ensure this is light tight by
wrapping in aluminium foil. Alternatively use a
light-tight cupboard. Leave for 1 month.

Development of the autoradiogram
In the darkroom, remove the X-ray film from the box
and separate the chromatogram from it. Develop the
film in the developer solution for 5 min using
constant agitation of the trough. Rinse for 10 s in

clean water, immerse in fixative and keep agitating the trough until the film clears (2–4 min). Wash in running water for 15–30 min and hang up to dry.

The origin of the chromatogram should be visible as an intense spot and several other separated spots should be apparent (see Plate 1)

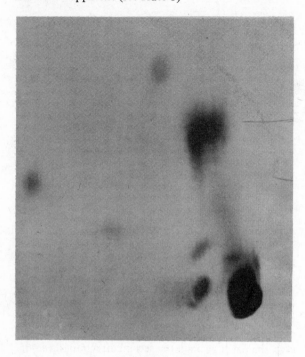

Plate 1

Discussion

In your experimental report discuss the theoretical basis and limitations of the techniques involved in the experiment and discuss also the results actually obtained by Calvin and his co-workers. Include both your chromatogram and autoradiogram.

Extensions and applications

Attempts can be made to identify the substances present on the chromatogram by various spray reagents. Details of these can be found in standard texts of paper-chromatography techniques. It has to be borne in mind, however, that autoradiography is a very sensitive technique and insufficient material may be present in the chromatogram to produce a visible colour reaction. Determination of the Rf values of the spots seen on the autoradiogram and comparison with published values can lead to tentative identification of the separated compounds.

If a large group of students or a prolonged project

is involved, useful results can be obtained by exposing the algae to $Na H^{14}CO_3$ for different lengths of time (5 s to 5 min), or by using a variety of photosynthetic inhibitors and algicides.

Further reading

Bassham, J.A.
    1962 'The Path of Carbon in Photosynthesis', Scientific American, 206, 88, W.H. Freeman. San Francisco.
Calvin, M. and Bassham, J.A.
    1962 'The Photosynthesis of Carbon Compounds' Benjamin; NY
Devlin, R.M. and Barker, A.V.
    1971 'Photosynthesis' Van Nostrand-Reinhold, London
Rabinovitch, E. and Govindjee,
    1969 'Photosynthesis' J. Wiley, London
Rogers, A.W.
    1967 'Techniques of Autoradiography' Elsevier, Amsterdam
Wang, C.H. and Willis, D.L.
    1965 'Radiotracer Methodology in Biological Science', Prentice Hall, Englewood Cliffs, NJ

Hints for the teacher

Several items will have to be prepared in advance. A suitable alga should be grown. *Chlorella pyrenoidosa* is ideal and since the culture does not have to be pure, a sample obtained from any biological supply house, e.g., Phillip Harris Ltd, will suffice. The culture can be grown in a simple medium, e.g. Knops solution, and on a windowsill if required. *Knop solution.* One of the simplest and most used solutions is the four-salt Knop solution. Make up five stock solutions as follows:

1   5 g KCl in 1 litre
2   5 g KH_2PO_4 in 1 litre
3   5 g $MgSO_4 . 7H_2O$ in 1 litre
4   20 g $Ca(NO_3)_2 \cdot 4H_2O$ in 1 litre
5   0·2 g $FeCl_3$ in 1 litre

50 ml of each of the stock solutions mixed together and made to 1 litre gives a Knop solution. (Loomis, W.E. and Schull, C.A. "Methods in Plant Physiology" McGraw Hill, New York, 1937). It is best if, before the class commences, the culture is centrifuged and the algae resuspended in the fumarate buffer. This removes salts present in the medium which may interfere with the chromatography at a later stage. It will be necessary to determine, by drying a sample of known volume, the quantity required to give approximately 20 mg of dried algae.

It is necessary to calculate the volume of radio-

ctive sodium bicarbonate solution to be taken in
rder to obtain 100 μCi of ^{14}C. For convenience of
andling by the student it is best if the solution is
iluted with water before use so that about 1 ml is
equired. Since the students will be working with an
sotope, albeit a relatively harmless one at the
oncentrations used in this experiment, it is advisable
o give them some elementary instruction in the
echniques of handling isotopes and the dangers
nherent in them. Suitable vessels for the disposal of
he contaminated solutions (e.g., clearly labelled old
Vinchester bottles) and for contaminated tissues,
eedles etc., should be provided. Washing the con-
aminated apparatus is perhaps best done by tech-
icians.

The entire experiment should ideally be carried
ut in a fume cupboard, but an enamel (or plastic)
ray on the laboratory bench is suitable.

Current regulations allow experiments to be car-
ied out in normal laboratories provided not more
han 100 μCi of activity is present in total. Thus a
ingle experiment could be set up as a demonstration
or students to make their own autoradiograms in a
on-radiochemical laboratory.

D.M. HAWCROFT
Lanchester
Coventry

7 A study of RNA and DNA synthesis in the salivary gland cells of dipteran larvae using autoradiography

Time required
The two experiments can be done in 2 days.
Autoradiographic exposures of 3 and 6 days are
envisaged. Together with the time required for
practising dissection, squashing etc., before doing the
actual experiment, and the time required for analysing and recording the results, the whole set of
experiments can be completed in about 4 weeks.

Assumed knowledge
Use of watchmaker's forceps for dissection. Making
squash preparations. Simple histochemical procedures
like staining and making permanent mounts for light
microscope examination. Some experience of photomicrography and dark-room procedures.

Theoretical background
Several larval tissues of Diptera possess giant chromosomes. These chromosomes are polytenic and have
resulted from repeated replication of chromosomes
without mitotic segregation. This is related to the fact
that in most organs of insect larvae, growth involves
increase in cell size rather than cell multiplication. On
account of their large size, the polytenic chromosomes, with over a thousand strands associated in
parallel, are favourable material for investigations on
chromosome structure and function, and on the
mechanism of gene action. Ordinary chromosomes
are metabolically active in the interphase stage so
familiar to cytologists, when the chromosomes are in
an extremely diffuse stage and hence most difficult
for microscopic visualisation. The polytenic chromosomes, on the other hand, clearly show their DNA-
rich band (chromomere) and DNA-poor interband
(interchromomere) organisation, and at the same time
are metabolically very active. The chromosomes also
possess nucleoli and can therefore be considered
equivalent to interphase chromosomes.

The arrangement of the bands and interbands is
unique for each chromosome of a complement and is

43

constant in all the cells of a species. An interesting aspect of the polytene chromosomes is the variation in the behaviour of different bands in (a) different cell types of an individual at a given period, and (b) a given cell type at different periods of development. The phenomenon is called puffing and the bands involved have a swollen appearance and stain very diffusely. In certain diptera (chironomids for example) puffing can be so extreme that the bands involved appear as huge structures which are known as bulbs or Balbiani rings.

There are several kinds of evidence which suggest that puffs represent the sites of active genes. In other words, puffing is regarded as a manifestation of gene activity. Current concepts envisage the gene as a stretch of DNA containing a specific sequence of bases that serve as template for synthesis of RNA— the primary gene product. It is now well established that puffs are very active sites of RNA synthesis. It is also known that an important amount of nuclear RNA is synthesised in the nucleolus. The formation of an RNA puff involves the following: (a) accumulation of acidic proteins, (b) *in situ* synthesis of RNA and storage of RNA. Puffing may be visualised structurally as uncoiling or unfolding of tightly packed DNA-histone fibrils in the band region. The interrelationship of the various processes mentioned above is very complex and still remains to be elucidated. Workers in several laboratories are carrying out experiments with various chemicals that either inhibit RNA or protein synthesis or alter the DNA-histone association, in order to gain further insight into the early events that occur during gene activation. There is no increase in the content of DNA and histone during RNA puff formation.

The bands of the polytene chromosomes are generally regarded as genetic units as well as units of transcription and replication. Most of the observations pertaining to these have been made by using autoradiography.

Autoradiography (radioautography) is a cytochemical method in which radioactivity incorporated into tissue components is visualised by photographic means. Like visible light, ionising radiation can produce an image in a photographic emulsion; the energy required to activate the silver bromide crystals in the emulsion comes in this case from the passage of charged particles instead of light photons. Most radioisotopes of interest to biologists emit beta particles and these form the main radiation for autoradiography (ARG). Some of the beta-emitting isotopes in common use are tritium (hydrogen isotope of mass 3), carbon-14, sulphur-35, iodine-13 and phosphorus-32. The physical properties of these radioactive nuclides are different, especially the half life and the energy spectra of the emitted radiation. The half-life of—P-32 is 14·3 days, but that of tritium and carbon are 12·3 yr and 5730 yr respectively. The consequence of these values is that there is a vast difference in the percentage of atoms that decay in each case over a given period of time, and this is an important factor for autoradiographic exposure. Tritium emits beta particles of unusually low energy and this property enables high autoradiographic resolution (of the order of 1 μm) in light microscope ARGs. The maximum energy of betas of P-32 is about 100-fold greater than that of tritium betas, and consequently the resolution obtained with P-32 is very poor (6-10 μm).

The specimen in an ARG may be a smear, squash or section of a tissue with incorporated radioactivity. In grain-density autoradiography, which is the most frequently used type, the specimen on a glass slide is covered on one side with a rather thin layer of a specially manufactured photographic emulsion and exposed in total darkness for a period of time during which the radioactive atoms decay. The silver bromide crystals that are hit by the emitted radiation undergo a change and they are then said to possess a latent image. This image is not directly visible, but when developed photographically, the latent image is converted into a true image (in the form of silver specks or grains) large enough to be seen with the ordinary microscope. Those bromide crystals that were not hit and hence not reduced to silver are dissolved out of the emulsion and lost during photographic fixation. In the final preparation, the pattern of silver grains represents the pattern of radiation that fell on the emulsion.

Unlike other methods of detecting radioactivity, the main advantage of ARG is that it enables study of the distribution of radioactivity within a specimen. It enables correlation of radioactivity with intracellular structures as small as nucleoli and isolated chromosomes, and in the case of polytenic chromosomes with individual bands or puffs.

The pattern of ARG experiments is basically the same as in all tracer experiments. Labelled precursors which are introduced into the cell get incorporated into macromolecular constituents synthesized in their presence. Free, unused precursors and low molecular weight intermediate compounds are usually washed out of the tissue during processing of the tissue in fixatives and various solvents. For investigating syntheses of RNA and DNA by autoradiography, the precursors used are uridine-5-^{3}H and thymidine (methyl-^{3}H) respectively. In these labelled precursors the position of the tracer isotope in the molecule is known with certainty and this is an important factor in their specificity.

Necessary equipment

Binocular microscope (for dissection and subsequent steps) with illuminator and variable magnification.
Watson Research Model with illuminator or Carl Zeiss stereomicroscope with epiilluminator on stand F.
A compound microscope (for examining autoradiographs) with attachments for photomicrography.
Zeiss research microscope or other similar make
Watchmaker's forceps Nos. 4 and 5—made in Switzerland—available at Arnold R. Horwell Ltd or Baird and Tatlock.
Refrigerator
Photographic darkroom with the usual facilities and a safelight fitted with Ilford filter 904F. Sliding weight beam balance (to weigh emulsion in the darkroom).
Water bath with thermostatic control (for melting emulsion in the darkroom). Grants Instruments Ltd., supply a small (13in x 6in) bath with lid which is quite handy and reliable. The indicator (red) lights on the bath should be disconnected to prevent possible fogging of the emulsion.
Hamilton syringe 25 or 50 μl capacity (to measure out tracer from stock ampoule)—V.A. Howe and Co. or Griffin and George. Alternatively, one can use fine calibrated microcapillary glass pipettes.
Microscope slides (76 x 25 mm; thickness 1·0–1·2 mm), from Chance Brothers Ltd.
Cover slips No.1,18 x 18 mm; No.1,22 x 22 mm, from from Chance Brothers Ltd.
Ordinary Slide trays—Griffin and George. Slide boxes for 20 slides from Griffin and George or Baird and Tatlock (These are for autoradiographic exposure).
Scotch vinyl plastic electrical tape No. 33. Made by 3M Company and obtainable at electrical shops. (This tape is used to seal the Clay Adams slide boxes to make them light and air proof.)
Racks for hanging slides. These are made in the lab (see Fig. 1). (used for drying slides after staining, emulsion coating etc.).

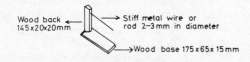

Figure 1

Hendon Photoclips (for hanging slides from the racks) obtainable at photographic dealers or from Hamilton Tait, Eastfield Drive, *Penicuik,* Scotland.
Coplin jar or a container for melted emulsion into which slides are dipped for coating. This is also quite easily made in any workshop with perspex, internal diameter approximately 6 mm; height 55 mm and width 30 mm.
Solid watch glasses or glass depression slides. (Baird and Tatlock, Griffin & George)
Soda glass tubing, 6–7 mm in diameter (Griffin and George) (used for drawing fine pipettes for transferring glands).
Rubber teats for above pipettes
Slide staining troughs to hold 10 slides (A dozen).
Coplin jars (6 of these). These and the troughs are obtainable from Griffin and George or Baird and Tatlock.
Miscellaneous items usually available in a laboratory such as graduated pipettes, beakers, measuring cylinders, thin glass rods, weighing bottles (good for keeping medium 199 during dissection), safety blades, spirit lamp, Whatman filter paper—Qualitative No. 1, disposable syringes (Gillette Surgical), Cotton wool, medical wipers, polythene sheeting, polythene bags etc.
Dry Ice.
Tissue culture medium 199·(1x) made by Bio-cult.
Decon 90 concentrate (for decontamination of radioactive glassware etc.) available at Medical Pharmaceutical Developments Ltd.
Silica gel (BDH) or Drierite (anhydrous calcium sulphate.) Repelcote (water repellant) (Hopkin & Williams Ltd.) A 2% solution of dimethyldichlorosilane in carbon tetrachloride. (This is used to siliconise the cover slips required for squashing. Dip the cover slips one by one in Repelcote, then dip in distilled water, shake off as much water as possible and leave to dry in a dust-free area).
L4 nuclear emulsion in gel form (Ilford).
Gelatin and chrome alum (to prepare 'sub' slides for good adhesion of squashed material). Make a 0·5% solution of gelatin. This is done by heating water to about 90°C and adding gelatin little at a time and stirring all the time. When dissolved, add chrome alum to make a 0·1% solution. Cool, filter and store at 4°C. Chemically cleaned slides are dipped in this solution and allowed to dry vertically on a hot plate. Store the gelatinised slides in a dust-free box. Number them with a diamond pencil. (A slide drying plate can be obtained from Griffin & George).
Glacial acetic acid, (BDH).

Trichloroacetic acid, (BDH).

Absolute ethyl alcohol.

Giemsa stain, 0·68% solution, in methanol/glycerol, (BDH). Chemicals for phosphate buffer (0·01M, pH 6·8—7·0) made as follows:

$$Na_2HPO_4 \quad 12H_2O \quad (0·18 \text{ g/50ml})$$
$$NaH_2PO_4 \quad 2H_2O \quad (0·08 \text{ g/50 ml})$$

Mix equal parts of above (1:1 solution) to 1 ml of the commercial Giemsa stain, add 40 ml of the above buffer. Stain slides in this for 4 minutes (duration of staining can be varied as required, but if the tissue is stained too heavily, it will be difficult to detect and photograph the overlying silver grains in ARGs). Wash in distilled water by leaving the staining trough or Coplin jar under a slow running tap. This mode of washing will prevent the floating metallic surface film in the stain from sticking to the slides. Hang the slides on rack to dry. Proceed to mount only when absolutely dry.

Methyl green—pyronin. Both these are available from Gurrs. To prepare the staining solution, make up the following: sodium acetate $3H_2O$ 3·42 g, concentrated HC1 1·01 ml, distilled water 125 ml. Check pH to 4·7. Make up to 500 ml with distilled water. Now add methyl green 1·1 g. When it is completely dissolved, pour the solution into a separating funnel. Add excess chloroform, shake vigorously, allow to settle and drain off the chloroform. Repeat such chloroform wash 5—6 times until the chloroform runs clear. Now add 1·25 g pyronin. When this is dissolved completely, filter the solution and store in a glass reagent bottle in the refrigerator (4°C). The staining solution should last 4—6 months. Stain ARGs for 1 hr in troughs. Save the stain by pouring it back into the stock bottle. Then pour 70% alcohol into the staining trough and drain off after 3—4 s. Do the same with absolute alchohol and then hang the slides to dry on racks with the photo clips. The stain can be used repeatedly but it is advisable to filter each time before use.

Mounting. The air-dried autoradiographs are mounted in XAM—a neutral mounting medium or other similar medium (Gurr). If the medium is too thick, it may be diluted with Xylol. Mount using cover slips No. 1, 22 x 22 mm.

Photographic developer (Kodak D-19b or D-19)

Photographic fixer. There are several in the market, but we have found Johnson Fix-Sol-hardening fixer concentrate most satisfactory. This is diluted 1:9 with distilled water just before use and shaken very vigorously to ensure thorough mixing.

Type of group

In my laboratory, I usually set a pair of students to do autoradiographic experiments. Groups of up to 4 may, however, do the work with each of them taking turns with the different steps. With a bit of planning, everyone should have the chance to do a part of every step in the experiment such as dissection, tracer incubation, squashing and emulsion coating.

Instructions to the student

Select larvae of *Drosophila, Chironomus* or *Smittia*. [We have cultures of *Drosophila* and *Smittia* (a chironomid) in our laboratory]. *Drosophila* have only three larval instars while chironomids have four. Younger larvae (early third instar of *Drosophila* or third instar of chironomids) are better for DNA replication studies and older ones (late third instar of *Drosophila* or fourth instar of chironomids) are better for RNA synthesis studies. Collect larvae well in excess of the actual number required for incubation—in solid watch glasses containing medium 199. Remove all adhering food particles etc. and transfer the cleaned larvae (3—4 per watch glass) into fresh medium. To dissect out the salivary glands, hold the larva by its head and part of first segment with one pair of forceps, and with the other hold the middle abdominal segments. Pull apart gently; with a little practise, the head with the pair of glands attached to it will come off from the rest of the larva. Discard the rest of the larva and also remove carefully as much as possible of the fat bodies attached to the gland border without injuring the glands. Injured areas will show up as milky white spots and if many of these develop discard the gland. If only one gland of a pair is so affected, it may be detached at the duct from the healthy one which may be taken up for further incubation. However, transfer the healthy looking gland to fresh medium to prevent the action on it of hydrolytic enzymes that may have escaped into the medium from the injured gland. The transfer of glands is best accomplished by means of a fine pipette, but if one is not careful there is the danger of losing glands by sticking to the walls of the pipette. At no time during dissection, cleaning, transfer and incubation should the glands be allowed to be exposed to air or to become dry. Each incubation is done preferably with 3 or 4 glands so that even if one or two are lost during the subsequent steps, there will be some left for squashing. During incubation the watch glasses or depressions in slides may be sealed with a lid or cover slip lightly smeared with a trace of petroleum jelly at the corners.

About 30–40 μl of tracer medium is sufficient to incubate 2 pairs of glands, therefore 1 ml of tracer medium should be enough for 30 or so incubations. A useful radioactive concentration for incubation is 60 μCi/ml of 199 medium, both for uridine-5-^{3}H, (sp. act.>20 Ci/mmole), and thymidine (methyl-^{3}H) (sp. act.>15 Ci/mmole), for the exposure times prescribed in the present exercise. Do about 20–25 incubations at about 24°C in each tracer; half of them for 15 min and the other half for 30 min duration. At the end of these periods, stop the incorporation by adding to the tracer medium a few drops of 10% (w/v) of cold (2–4°C) trichloroacetic acid. The glands will immediately go opaque. Pick up the glands (may use forceps here) and put them into cold 5% TCA in another watch glass for about 10–15 min. Then transfer them into cold 45% (v/v) of glacial acetic acid and distilled water for about 5 min and squash in a drop of this acetic acid on a 'sub' slide using a siliconised cover slip. There are numerous ways of squashing which simply involves application of uniform pressure on the tissue. The amount of pressure required to effect good squashing is determined by each worker in actual practice. Great care must be taken during squashing to avoid lateral movements of the cover-slip which will cause the tissue to roll up and not squash. Another very important requirement is that the slide must be free of dust particles, fine fluff etc. which will prevent proper squashing. Only a well-squashed tissue is likely to adhere firmly to the slide.

After squashing place the slide on the flat surface or a block of dry ice for 5–6 min or until the preparation is frozen. Then flip off the cover slip with the tip of a safety blade and put the slide into absolute ethyl alcohol. Remove the slide after a period of 10–15 min and dry in air and store until all the squashes have been brought to this stage. Alternatively, the slides may be taken through 90% alcohol and stored in 70% alcohol in the refrigerator. Check the slides at this stage. If the squashed material has fallen off there is no point in storing the slide or proceeding with it any further. When the required number of squashes have been accumulated (say 20–25 per experiment) place them in staining troughs or Coplin jars and add distilled water (or bring them to distilled water, through descending grades of alcohol if they had been stored in 70% alcohol). The squashes are now ready to be coated with photographic emulsion.

In the darkroom, weigh out under safelight (Ilford 904F) 10 g of Ilford L4 emulsion in a 150 ml beaker. Add 10 ml distilled water and hold the beaker in a thermostatically controlled water bath at 45°C for 10 min, stirring gently with a glass rod. The melted and diluted emulsion can now be poured into the container which is also kept in the water bath at 45°C. Coat the slides by dipping them, one by one, in the emulsion. Keep the slide in the emulsion for a second or two and withdraw gently and with uniform speed. Hang the coated slides on racks for about an hour and when dry, store them in light-proof slide boxes (Clay Adams slide box sealed with scotch tape) containing a sachet of desiccating agent. Leave the box in a refrigerator running at 4°C. The box may additionally be wrapped in aluminium foil before being taken out of the darkroom.

Expose half the number of preparations from 15 min incubations as well as 20 min incubations for 3 days, and leave the rest for 6 days. After the required exposure, develop the slides for 3 min at 18–20°C in Kodak D-19 or D-19b developer, rinse for 30 s in distilled water and fix in diluted Johnson Fix-sol for 8–10 min. All the photographic solutions and rinses must be at the same temperature. Wash the processed autoradiographs for about an hour under a slow running tap or in several changes of distilled water. Stain the ARGs with Giemsa or methyl green–pyronin. Finally, mount in XAM (Gurr) for other suitable medium e.g. Euparol (Hopkin and Williams) and examine the preparations next day in a light microscope.

Recording results

The number of grains present in an autoradiograph is taken as a measure of the amount of label incorporated. ARG is, at best, a semi-quantitative method and the amounts of radioactivity at different sites in a cell can be compared as the ratio of grain counts. If the labelled regions are of different sizes, the relative radioactivities are usually expressed as ratio of grain densities, that is the density of grains per unit area. Often, however, the degree of blackening of the emulsion over labelled structures is so markedly different that it is only necessary to assess the distribution of radioactivity by eye.

The analysis of the autoradiographs of salivary gland squashes can be carried out at different levels of sophistication, depending on the conditions of the experiment and the type of answers sought. If the students are familiar with the chromosome map of *Drosophila melanogaster* (or that of the other dipteran flies used in the experiment) the labelling patterns can be analysed with reference to particular bands or puffs in the chromosome in different tissues or at different developmental stages. The present exercises are largely intended to introduce the students to the material (polytenic chromosomes) and the technique of autoradiography as used for the study of macromolecular synthesis. The students should, in the first instance, note the sites of RNA and DNA syntheses as indicated by clusters of silver grains along the chromosomes. Between the 4 sets of autoradiographs with each tracer (15 and 30 min incubations and exposure times in each case of 3 and 6 days) a range of labelling sites and labelling intensity may be expected, giving ample scope for grain counting and other types of scoring of the results. Not all squashes or nuclei in a given squash will be suitable for analysis, and so only the useful ones must be chosen for study. See if any pattern of labelling emerges in any of the experiments. If so, describe it. Note the scatter of grains in heavily labelled bands and try to gain some idea of the autoradiographic resolution obtained. Most of the grains will be found confined to the limits of the bands. In not so heavily labelled chromosomes of well spread squashes, try to count the grains. A simple exercise would be to count grains in all the chromosomes of a complement and compare this with the count on nucleolus of the same nucleus. Choice of squashed nuclei is very important for this comparison. Are the nucleoli labelled in all nuclei and are they labelled to the same extent? What may account for the differences if any? Note the labelling of nucleolus organiser region, if it can be identified, and compare with that of the nucleolus.

Write a brief but comprehensive account of your observations from uridine labelled preparations. With glands incubated in thymidine, try to note the overall pattern of labelling. Check whether similar or different patterns occur in all the chromosomes of a given complement and in different nuclei of a given gland. At the present time, it is widely accepted that chromosomal DNA in eukaryotic organisms is made up of many segments, serially joined and each constituting an independent replication unit. It would also seem that adjacent replication units can have different replication times. Are your results consistent with these assumptions? Look for label within the nucleoli and on the nucleolus organiser. Write a brief account of the replicative organisation of DNA in the giant chromosomes.

The ARGs are usually examined in bright field of transmitted light. At low magnification, it may be possible to see in focus both the silver grains and the specimen (chromosomes, nucleoli). At higher magnification, however, the focal plane of the grains will be different from that of the specimen and hence the

two images can be studied only by changing the focus in the course of the examination. As far as possible, photomicrographs should be made at low, rather than high, magnification, consistent with proper clarity of silver grains. If, however, a compromise has to be made it may be better, as far as the present material is concerned, to focus for the grains and be satisfied with a slightly blurred image of the chromosome or nucleolus. But do photograph with the specimen in focus and compare the result. We need not concern here with any of the more complicated set up for photomicrography of autoradiographs.

Extensions and applications
The scope of ARG as a research tool in cell biology is very wide, and at present it is difficult to find a field of study in which it has not been applied. The giant chromosomes of diptera are also material of ever-increasing potential for investigations pertaining to gene action. The exercise that has been prescribed in this manual may be extended by using chemical inhibitors of nucleic acid and protein syntheses, and autoradiography may be combined with other procedures. It may also be useful to check the specificity of isotope incorporations by the use of appropriate nucleases. A few of the squashes may be treated just before emulsion coating with either RNase or DNase. Prepare a solution of 50 mg/100 ml crystalline pancreatic RNase in phosphate buffer (pH 7·4) and immerse the squashes for 1 hr at $37°C$. For DNasing, treat the preparations for 2 hr at $37°C$ in an aqueous solution of 50 mg/100 ml crystalline DNase in $2·5 \times 10^{-3}$ molar $MgSO_4$, pH adjusted to 7·0 with 0·1N NaOH. After each of the treatments, rinse the slides in distilled water and take them through a 15 min stay in 5% TCA solution (w/v) at $4°C$ to remove soluble nucleotides. Rinse in distilled water and coat.

As an alternative to the fixation procedure described earlier, the glands may be fixed for 4−5 min in 1:3: acetic acid: alcohol (called Carnoy's fixative), then stained for 15 min in acetic orcein or aceto carmine. The stained glands are then transferred to 45% acetic acid and squashed as described before and finally emulsion coated.

Further reading
1 For methods:
Baserga R. and Malamud
    1970 'Autoradiography', Harper and Row, New York
Rogers, A.W.
    1967 'Techniques of Autoradiography', Elsevier Amsterdam

2 For field of enquiry:
Articles on DNA and RNA synthesis and associated aspects are found in several journals, notably Chromosoma, Experimental Cell Research, Journal of Cell Biology and Journal of Cell Science.
See also: Plaut, W.
  1963 J. Mol. Biol., 7, 632
Plaut, W. et al.
  1966. J.Mol.Biol., 16, 85,
Huberman, J.A. and Riggs, A.D.
  1968 J.Mol.Biol., 32, 327
Berendes, H.D.
  1965 Chromosoma, 17, 35
Berendes, H.D.
  1968 Chromosoma, 24, 418

Hints for the teacher

It is important to see that the students familiarise themselves with dissection and squashing before proceeding to do the full experiment. Transfer of glands from watch glasses with a pipette seems deceptively simple, but in practice it is quite hazardous and may result in loss of considerable numbers due to the stickiness of the gland. Careful manipulation of the glands and liquid levels in the pipette is required to prevent this. This procedure also therefore must be practised beforehand. Some loss of glands at the various steps in the experiment is unavoidable and for this reason start with far more glands than the number of squashes required. It will also be helpful to dissect 10–15 pairs of glands before proceeding to set up the tracer incubations. The dissection should then be continued until the required numbers have been set up for incubation. Plan the experiment and brief the participants in advance so that each of them will be prepared to attend to the various steps that will overlap in time as the experiment progresses. (See the protocol shown in the table, which starts with the incubation in tracer).

Emulsion coating is simple and fast. With some care and elementary precautions, the background in the autoradiographs can be kept to a very low level.

Decontamination of glassware, etc: All glassware and instruments that came into contact with radioactive medium must be decontaminated before being returned to the general pool for use again. Soak them in a 2% solution of Decon 90 for 24 hr. This is then poured down a special sink if one is available for disposal of radioactive material. The glassware, etc., are then washed for several hours under a running tap. The radioactive solutions used for the experiment are absorbed on cotton wool, which is then sealed in polythene bags, wrapped in paper bags, and sent for incineration. The table or bench surface used for the experiment is covered, as a precautionary measure against accidental spillage of tracer, with polythene sheeting.

There are certain legal requirements to be met before carrying out experiments with radioactive substances. The supervisor of the experiments must therefore discuss the exercise with the local (University) Radiation Protection Officer and seek his advice on the precautions in the use of radioactive substances. The supervisor's attention is also drawn to the booklet *Code of Practice for the Protection of Persons Exposed to Ionising Radiations in Research and Teaching'* published by HMSO, London WC1.

J. JACOB
Edinburgh.

Tracer	Fixative	45% Acetic acid	Squash	Dry ice	Absolute Alc.	Dry in air
10·30	10·45	11·00	11·05	11·06	11·12	11·27
10·35	10·50	11·05	11·10	11·11	11·17	11·32
10·40	10·55	11·10	11·15	11·16	11·22	11·37
10·45	11·00	11·15	11·20	11·21	11·27	11·42
10·50	11·05	11·20	11·25	11·26	11·32	11·47
10·55	11·10	11·25	11·30	11·31	11·37	11·52

Section 3

Phase Microscopy and Measurement Techniques

1 Setting up the phase-contrast microscope: examination of oral epithelial cells and measurement of refractive index

Time required
About 2–3 hr.

Assumed knowledge
The student is assumed to be familiar with the interference of light waves and the effect of refractile bodies in slowing up or altering the *phase* or optical path of the light waves travelling through it[3,4].

Theoretical background
Figure 1 shows, and the text explains, the paths of the light rays through the phase-contrast microscope. It is not possible to give a complete explanation in the space available and the student is referred to the paper by Zernike.[5] Published in 1955 this is his address delivered on receiving his Nobel Prize in 1953 given for his discovery of phase-contrast in the 1930s.

Briefly biological objects, tissues, cells and organelles, are usually transparent in visible light but differ in refractive index from their surroundings. Such objects do not show up well in the ordinary light microscope and cells have to be fixed and stained before becoming visible, with the possibility of attendant changes. This is because the eye or photographic plate is not sensitive to the path differences or phase changes introduced into the light beam by transparent refractile objects. Before the invention of phase-contrast microscopy, some structure in cells could be made out by stopping down the iris, or aperture stop under the condenser, or by dark-field microscopy. In the phase-contrast microscope the phase changes, or path differences, are made visible by using interference of light waves[3,4]. The diffracted rays, dotted in Fig. 1, do not in the main pass through the $\frac{\lambda}{4}$ annulus in the phase plate and they

50

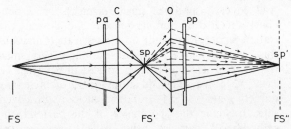

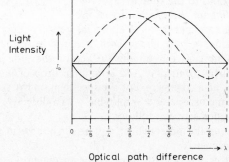

Figure 1 The condenser lens, C, forms an image, FS′, of the Fieldstop, FS (an illuminated lens if Kohler illumination is employed), on to the specimen, sp. The objective lens, O, forms an image of the specimen and FS′ at sp′, FS″. A phase annulus, (pa), is situated in front of the condenser at the so-called focal point. A phase plate, (pp), a plate with an annular depression $\frac{\lambda}{4}$ thick, is situated within the objective lens at the so-called back focal plane. The phase annulus produces a hollow cone of light, the direct beam, which passes through the phase plate. The dotted rays are those scattered or diffracted by the specimen and they fall over the entire surface of the phase plate.

Figure 2 The variations in light intensity (ordinate) as a function of optical path difference in wavelengths, (λ), which is introduced by the specimen (abscissa). These relationships are for a non-absorbing phase plate and are modified somewhat when the phase plate is absorbing. Phase plates are made to absorb the direct beam so as to increase the sensitivity of detection of thin specimens. I_b is the background intensity. Solid curve and dotted curves are for objects of optical path difference greater than and less than the surround, respectively. Thus thin refractile objects ($<\frac{\lambda}{4}$) refractive index look dark in this so-called positive phase-contrast.

interfere with the direct beam which does, to form an image in which the intensity varies with the phase change caused by the specimen according to fig. 2. In positive phase contrast, as the thickness increases from zero, the image first darkens; between an optical path difference of $\frac{\lambda}{8}$ and $\frac{\lambda}{4}$ the image lightens and appears with no contrast at $\frac{\lambda}{4}$; thereafter, with increase in path difference, the object appears brighter than the background. Optical path difference is defined as $(n_c\text{-}n_m)t$, when n_c and n_m are the refractive indices of the cell and surrounding medium respectively and t is the thickness of the cell.

The energy removed from the inner region of the image is distributed around it to form a phase-contrast *halo*. Energy can only be redistributed by interference phenomena. The refractive index n_c of an object can be measured by immersing it in a series of media of increasing refractive index. (When n_m is slightly less than n_c, the object is dark and when n_m is slightly greater than n_c the object looks bright. When n_m equals n_c the path difference is zero and the object disappears in the microscope. Hence the phase microscope can be used to detect the null point of zero path difference and the refractive index obtained by the so-called matching method.

The measurement of optical path difference, other than a zero value, cannot be made by phase-contrast

microscopy, but by so-called interference microscopes. The relationship between path difference, refractive index and dry mass and concentration of substances in cells is discussed elsewhere.[1,2]

Necessary equipment
1 One phase-contrast microscope and phase telescope
2 Slides and cover slips (No. 1½)
3 Blunt scraper e.g. spatula, teaspoon.
4 Refractive–index oils ranging from about 1·53 to 1·58 at 0·01 intervals. These may be obtained from Messrs R.P. Cargille Laboratories Inc., Cedar Grove, N.J. 07009 USA or MicroBio Laboratories Ltd. Alternatively they may be made by mixing the following liquids and the refractive index determined from an Abbé Refractometer (Gallenkamp). Range: 1·494–1·558, toluene and monobromobenzene (BDH); 1·5599–1·598, monobromobenzene and bromoform (BDH)
5 Petri dish, benzene and absorbent paper

Type of group
Groups of two or three per microscope

Instructions to the student
Setting up the phase microscope
Make a specimen of oral epithelial cells by scraping the inside of the cheek with a blunt instrument: spread the cells out on the surface of the slide and allow them to air dry. Add a minimal drop of water and put on a cover-slip. Put the specimen on the stage of the microscope and employing a x16 objective set up the microscope as follows:

1 Focus the objective on say an air bubble in the specimen.
2 Focus the field stop on to the specimen. Centre it by the screws on the condenser in microscopes with built-in illumination. Open up the field stop to just fill the field of view of the x16 objective.
3 Insert the phase telescope in the microscope tube, after removing the eyepiece. Or use the Bertrand lens (called an Optovar system in the Zeiss microscope), if provided, to convert the eyepiece into a telescope. Focus the telescope on to the phase plate which is situated in the back focal plane of the objective. (For bright-field observations the aperture stop or iris under the condenser would now be inserted, usually by rotating the mount carrying phase-annuli and iris, and would be approximately in focus. For bright field this iris would now be centred and closed down. The student should observe the cell in this so-called stopped-down illumination)

Next rotate the mount under the condenser so as to bring on to the optic axis the phase annulus which matches the phase plate in the x16 objective. Usually they are numbered. Centre the phase annulus with the second pair* of screws provided, so as to make it lie within the boundary of the phase plate. Now observe the specimen in phase contrast.
4 Repeat the above procedures for x40 and x100 objectives, each of which requires its own phase plate. It may be necessary to re-focus the condenser with change in objective: this is due to uncorrected spherical aberration in the lens.

Measurement of refractive index by the matching method
Make up six slides of epithelial cells which have oral bacteria associated with them and allow to air dry. Alternatively make air-dried smears of a large bacterium e.g., *B. megaterium*.

---

*This text has been written for microscopes provided with the centring adjustments described in it. It may need modifying for other microscopes.

Add to each specimen small fragments of carbon-black obtained by burning benzene and allowing the carbon smoke to deposit on a glass plate. Carry out this operation in a Petri dish burning a *very* small fragment of absorbent paper soaked in benzene. To each successive slide add drops of oil of increasing refractive index, 1·54–1·58, and enclose with cover-slips. Examine the appearance of the bacteria, the carbon-black particles being added to make the bacteria easy to find, since they will lie in nearly the same plane. When the oil is of the same refractive index as the bacteria, they appear with approximately zero visibility and they appear either dark and light with oils of lower or higher refractive index. The refractive index is usually about 1·54–1·56. This experiment is intended to familiarise the student with the use of the microscope rather than to determine a biologically exciting parameter, the refractive index of air-dried protoplasm. If the refractive index of a water-containing living cell can be measured,[1],[2] then this is simply related to the concentration of total dry material present.

Discussion
As you focus through the oral epithelial cells at x100, note and discuss the nature of the cell surfaces. Can you tell if the bacteria are inside or outside the cell? Why are the concentrations of material in the cell and cell mass interesting parameters? What advantages does phase contrast have over dark-field?

Extensions and applications
Phase-contrast microscopes are used for studying the distribution of organelles, nuclear structure etc., in living unstained cells and can be used to observe changes during growth and division. This is often done by coupling a cine-camera to the microscope.

Further reading
1 Barer, R.
1956 'Phase Contrast and Interference Microscopy in Cytology', in 'Physical Techniques in Biological Research' 3, 30, ed. Oster, G. and Pollister, A.W., Academic Press, N.Y.
2 Davies, H.G.
1958 'The Determination of Mass and Concentration by Microscope Interferometry', in 'General Cytochemical Methods', pp. 55–161, ed. Danielli, J.F. Acad. Press, NY.
3 Jenkins, F.A. and White, H.E.
1957 'Fundamentals of Optics', 3rd ed. McGraw-Hill, New York

4 McKenzie, A.E.E.
 1962 'A Second Course of Light',
 Cambridge University Press.
5 Zernike, F.
 1955 'How I Discovered Phase Contrast',
 Science, 121, 345—9.

Hints for the teacher
Make sure that not too much benzene is used to make the carbon black—say less than $\frac{1}{50}$ ml: otherwise it is a fire hazard.

H. G. DAVIES
King's College
London.

2 Variation in cell size and cell enumeration

Time required
3 hr.

Assumed knowledge
Elementary microscopy

Theoretical background
Human red blood cells are biconcave discs which carry oxygen around the body. Some diseases result in red blood cells of irregular size and shape. There may also be a variation in the normal concentration of cells. The size, shape and number of cells per unit volume and their haemoglobin concentration possess diagnostic value. Haemocytometry is the determination of the number of cells in a known volume of liquid using a haemocytometer of known depth and with engraved lines of known dimensions upon its surface.

Necessary equipment
1 Uniform bore capillary tubing and teat
 Microburner (Gallenkamp)
 Ruler calibrated in mm (or smaller divisions)
 Griffin Christ Simplex II centrifuge with capillary tubing head *or* Swing-out bench centrifuge plus adaptors
 Uncoagulated human blood
2 Microscope with mechanical stage and micrometer eyepiece, preferably binocular (Objectives: low, high and oil-immersion)
 Micrometer slide
 Toisson's red cell diluting fluid (Gurr)
 50 ml volumetric flask
 0·25 pipette
 Pasteur pipette and teat
 Improved Neubauer haemocytometer (Section 13.7) and cover slips (Gallenkamp)
 Soft cloth
3 Immersion oil
 Stained blood smears A and B (Leishman; BDH)
 The 'fresh' blood samples may be obtained as expired blood from the Blood Transfusion Service or Pathology Department. The latter can usually provide the stained slides, given sufficient notice.

Type of group
The experiments are intended to be carried out individually.

Instructions to the student

Packed cell volumes
Mix the sample of human blood thoroughly and suck (by means of the teat and adapter) the blood sample up the capillary pipette to within about 2 cm of the top. Seal the top end in a microburner and allow to cool. Invert and centrifuge at about 3000 rev/min. Measure the height of the liquid column and that of the packed cells at intervals until a constant volume of packed cells has been obtained.

Haemocytometry
Take a sample of well-mixed uncoagulated human blood and dilute this 1 in 200 with red cell diluting fluid. Mix well and rapidly flood the chamber of the haemocytometer so that the well is fully filled but no liquid has overflowed into the side wells. The special cover-slip should have been previously slid onto the haemocytometer using two hands so that coloured Newton's rings are visible at each side. Allow the cells to settle and view under the microscope. Note whether the distribution of cells is relatively uniform. If it is not, clean and dry the haemocytometer and start again. When a relatively uniform distribution has been achieved, count the number of cells in a known area of the chamber (about 500 cells). Count only those cells not touching the ruled lines within the squares and also include any cells touching any two of the four ruled sides of each square you are counting. It is as well to note the numbers of cells in each large square as this will give an indication of irregular distribution. Calculate the number of red blood cells per mm^3, knowing the depth, and area of squares, of the haemocytometer (values inscribed on latter), and the dilution of the blood (see Section 13.7). Repeat the entire procedure and compare results.

Estimation of mean cell diameter
Calibrate the micrometer eyepiece of your microscope using the calibrated microscope slide and the oil-immersion objective. Determine the number of eyepiece divisions for the maximum possible length of calibrated scale on the slide. This will ensure the greatest accuracy. Insert the blood slide A, which has been stained with Leishman's stain, and measure the diameter of the longest axis of 50 cells. Repeat the process using stained slide B. If time permits examine the types of white blood cells present. These are the cells possessing nuclei. It is also possible to carry out this experiment with greater accuracy using a projection microscope.

Recording results
1 Using your packed cell volume value and the red blood count, calculate a value for the volume per average cell.
2 Compare your two cell count results.
3 Prepare a histogram of frequency of a particular size (or size range) against cell diameter (x-axis) for both blood samples (both on the same paper in different colours). Comment upon the results and calculate the mean cell diameter (which is really cell length in asymmetric cells) and the standard deviation in each case.

Discussion
1 Discuss the validity of the value which you have obtained for the average cell volume.
2 Comment upon possible sources of error in the method.
3 Comment upon the appearance, size and shape of the red blood cells and indicate whether either sample was obtained from a normal person and if so, identify it. Why was the largest cell axis measured rather than any other measurement? Is it valid to use stained preparations?

Extension and applications
Haemocytometry may be used to enumerate any particle population of appropriate size range (e.g. bacteria, yeast, algae., etc.), provided that the cells do not adhere and can be dispersed in uniform suspension. If the cells are motile it will be necessary to add an appropriate motility inhibitor. The fixative formaldehyde is often used for this purpose. The method has been widely used for the enumeration of blood cells.

Examination of stained blood smears is used extensively in the diagnosis of diseases affecting the cells of the blood.

Further reading
Britton, C.J.C.
   1969 'Disorders of the Blood', 10th edn, J. & A. Churchill, London
Dacie J.C. and Lewis, S.M.
   1968 'Practical Haematology', 4th ed. J. and A. Churchill, London

Hints for the teacher
1 The packed cell volume sedimentations have in

the past been determined using a centrifuge specially designed for the purpose, but an ordinary swing-out bench centrifuge may be used if adapters are made either from a block of plastic with holes drilled in it or from appropriate sized pieces of rubber pressure tubing.

2 The cost and fragility of the items should be explained to the students. It should be emphasised that adequate initial mixing and then rapid flooding of the haemocytometer are most important. One should be aware of the possibility of spreading disease when using untreated blood samples and act accordingly. The blood dilution may be carried out in haemocytometer dilution pipettes (Gallenkamp) if these are available, but dilution in volumetric flasks is equally effective. Neubauer haemocytometers are the best for red-cell enumeration but any other type may be used, provided that it has small marked areas and a similar depth. If these areas are larger, it may be possible to obtain reasonable results using a greater dilution. Beginners tend to get rather poor results initially and it may be necessary to repeat the determinations several times to give consistent results.

3 Try to obtain a series of stained slides from the same blood sample in each case. Use blood from normal persons and from several different abnormal cases, as different as possible to the normals. If the projection microscope is used, fading of the stained slide in the area irradiated by the xenon arc may occur, but it is easy to move to another area.

T.R. RICKETTS
Nottingham.

3 The measurement of nuclear and cytoplasmic volume fractions by a point-counting technique

Time required
This experiment requires approximately 3 hr. If the analysis is to be undertaken on photographs, and the transparent grids used are made by the class, then a further $1\frac{1}{2}$ hr would be required.

Assumed knowledge
It is assumed that the students can use the microscope correctly, including use of the oil immersion objective and are capable of recognising the composition and cellular components of the tissue chosen for analysis.

Theoretical background
Stereological procedures aim at obtaining information about the three dimensional structure of complex objects from the two-dimensional flat images presented by the microscope from thin slices of finite thickness. By measurement and counting, much quantitative information may be obtained about the original object. For the simple determination of volume fractions we may use the fundamental premise of stereological methods—the De Lesse principle which states that the volume fraction V_{Vi} of a component in the tissue (Fig. 1) is equal to the measured area fraction A_{Ai} occupied by that component.

$$V_{Vi} = A_{Ai}$$

(A full theoretical proof of this relationship will be found in Reference 5.)

For stereological principles to apply, sections must be random, their thickness must be negligible in relation to the size of the objects under study and they must be prepared to standardise methods in order to avoid artifacts and to ensure reproducibility of tissue structure and dimensions. Stereological studies only give statistical estimates of the parameters chosen and hence the accuracy may be improved by increasing the sample size and increasing the number of sample points counted.

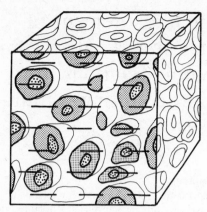

Figure 1 Diagram of a multipurpose grid superimposed on one plane of a cube of tissue, showing the test probes falling on various components being counted.

Key

Cut section of cytoplasm in the plane being counted.

Cut section of nuclei in the plane being counted.

Outline of cells in the cube not in the plane being counted.

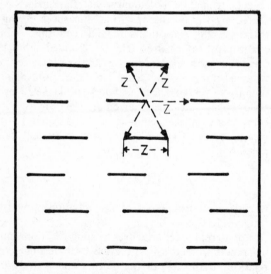

Figure 2 Weibel's multipurpose grid showing 21 lines of equal length (say Z) of equidistant spacing of Z units in all directions. This diagram should be used for making transparent grids with Chartpak Rotex if this method is being adopted (see 'Photography').

In practice the estimation of A_{Ai} is achieved by the superimposition of a series of lines or points (the test 'probes') on the image of the structures. Various probe systems have been developed for different purposes but for this exercise Weibel's multipurpose system of short linear probes (Fig. 2) has been chosen. The total set of probes contains 21 lines of equal length (say Z) which are in 7 equidistant and parallel rows with the interline spacing of Z units in all directions. The image of the probes is superimposed on the material and a hit is scored when the end of one of the lines falls on the component being counted, e.g., a nucleus. In this case, the required area fraction is then equal to the number of hits on the component (say P_{NUC}) divided by the number of hits on the nucleus together with the number of hits on the cytoplasm ($P_{NUC} + P_{CYT}$); then

$$A_{NUC} = \frac{P_{NUC}}{P_{NUC} + P_{CYT}}$$

Extrapolating from the De Lesse principle we may write

$$V_{NUC} = \frac{P_{NUC}}{P_{NUC} + P_{CYT}} \qquad \text{Equation 1}$$

or, as a percentage,

$$V_{NUC} = \frac{P_{NUC} \times 100}{P_{NUC} + P_{CYT}} \qquad \text{Equation 2}$$

As already stated, the sections may have a finite thickness which is considerable, with respect to the objects of interest. Projected images from such thick sections would show areas much greater than would be expected from a knowledge of their true volumetric composition. This effect, (the Holme's effect, Fig. 3) can be minimised by firstly working with sections cut as thin as practicable and secondly by counting only those points which lie on nuclei clearly visible at one plane of focus, and by ignoring any points that lie on out of focus images of nuclei which are lying higher or lower in the section.

Necessary equipment

The image of the grid, with its test points, must be superimposed upon an image of the section. This may be done in several ways.

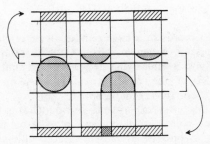

Projection of area
from thin section

Projection of area
from thick section

Figure 3 Diagram of Holme's effect showing that the projected area from a thin section of tissue is considerably less (and therefore more accurate) than projection from the whole of a thick section.

1 Direct microscopy

By the introduction of a graticule bearing the test points into the focal plane of the microscope eyepiece. A Weibel graticule Code No. 105·844 may be used with a standard eyepiece but better results are achieved with the Weibel eyepiece (Sold by Wild, Code No. 255·507) which incorporates the grid together with a screw-mechanism for focusing it sharply.

Suitable agents for the supply of the Weibel eyepiece are Micro Instruments (Oxford) Ltd., 7 Little Clarendon St, Oxford, OX1 2HP; Wild UK, Revenge Road, Lordswood, Chatham, Kent, ME5 8TE, or (overseas) Wild Heerbrugg Ltd., CH-9435 Heerbrugg, Switzerland.

The Weibel graticule above costs approximately £7 whilst the graticule mounted in a special focusing eyepiece is approximately £23·00. This method allows direct counting from the microscope, as the image of the section appears in focus at the same time as that of the grid.

2 Microprojection

The test probes may be drawn out on a sheet of paper and the microscope image projected onto the paper by means of a prism or front-surfaced mirror mounted at 45° over the exit pupil of the microscope eyepiece.

If a projection head is available for the microscope, then the grid may be drawn out on transparent acetate sheet and fixed with adhesive tape to the projection screen face.

3 Photography

The microscope fields may be photographed or drawn with a camera lucida and the test probes, drawn out on a sheet of transparent acetate may be directly superimposed on the photograph.

The transparent test probe grids may be made very easily from plastic sheet and Chartpak Rotex $\frac{1}{16}$ in black adhesive strip (both obtainable from large stationers or art supply shops). The transparent plastic sheet may be placed over the diagram (Fig. 2) and held in position with sellotape. Strips of the Chartpak are then cut with fine dissecting scissors and rubbed down over the printed lines to outline the test grid. Two hand tally counters are a great help in counting the 'hits' on components, eliminating much arithmetic!

Suitable material for determination of nuclear and cytoplasmic volume ratios is furnished by

(a) Pancreatic exocrine cells, compared with the endocrine cells of the Islets of Langerhans which are always present on the same slide.

(b) Cells of the zona reticularis of the adrenal cortex, compared with the cells of the adrenal medulla.

Counting sheets (see Results) should be available before the point counting on the sections or photographs actually begins.

Type of group

This experiment is best done with the class working in pairs; it is possible to use groups of four, but this is not desirable as much time is wasted. It is possible, with sufficient equipment and/or very small classes to allow students to work individually.

Instructions to the student

1 Superimpose the image of the test grid upon that of the section; both must be in sharp focus at the same time.

N.B. The focus of the microscope system must be set before counting of a given area begins and must not be altered during the counting of that area.

2 Scan along the lines of the test points and register, on the two tally counters, each time the end of a line falls upon nucleus or upon cytoplasm respectively. Any end points which fall upon intercellular spaces or blood vessels should be ignored.

3 At the conclusion of each scan of the test grid, enter the two totals on the counting sheets described under Results and then reset the counters to zero.

4 Repeat the process with another randomly chosen field. As many fields as time allows should be scored for each tissue to be investigated, in order to lessen the sampling errors. It is preferable to rotate the grid between successive counts by some arbitrary amount (say 45°) in order to avoid possible errors due to orientation of the cells or nuclei.
5 Repeat the experiment for the comparison tissue.
6 Calculate the results as shown in the next section.

Results

It is suggested that the results should be tabulated as shown.

	Tissue A		Tissue B	
	'Hits' on Nucleus	'Hits' on Cytoplasm	'Hits' on Nucleus	'Hits' on Cytoplasm
Trial No. 1				
2				
3				
4				
etc.				
Totals	P_{NUC_A}	P_{CYT_A}	P_{NUC_B}	P_{CYT_B}

At the end of the counting, total each column. It is these figures which are used in the calculations. The total number of test points is given by summing P_{NUC_A} and P_{CYT_A}. The volume fractions of the nuclei of tissues A and B are then given by the equations (1 or 2). The volume fraction of cytoplasm of tissue A and tissue B can be calculated by deduction.

Discussion

It should be remembered that the procedure is mainly a statistical one and that therefore the result is only an approximation to the true value. In order to obtain an estimate of the reliability of the result, the standard deviation σ is used; this may be calculated from the formula for tissue A:

$$\sigma NUC_A = \sqrt{\frac{V_{NUC_A} \cdot (1 - V_{NUC_A})}{(P_{NUC_A} + P_{CYT^A})}}$$

If, for example, $V_{NUC_A} = 0.3$ and

$$(P_{NUC_A} + P_{CYT_A}) = 450$$

then $\sigma NUC_A = 0.021$, so we may write

$$V_{NUC_A} = 0.3 \pm 0.021.$$

In other words in a large number of determinations on the same specimen, 68% of the determinations would give a value lying between 0.321 and 0.297 and 95% of the values would lie between $\pm 2 \times \sigma NUC_A$, i.e., 0.342 and 0.258.

In this class experiment the actual number of sampling points is far too few. In an actual research project in progress at present, a grid with 168 end points is in use and 144 separate trials are made, so that the total number of points sampled is 168 x 144 = 24 192.

Other sources of error can occur from any bias introduced by, e.g., choice of fields to sample and from any spatial organisation or 'anisotropy' of the nuclei, and in your discussion you should indicate possible precautions which might minimise such errors. Consult a standard text book of statistics (see Further Reading), and apply a suitable test to determine whether the differences in nuclear and cytoplasmic ratios which you find between tissue A and tissue B are significant, and at what level of probability.

Extensions and applications

The same multipurpose grid can be used to determine surface area/volume ratios of cells or parts of a cell. This parameter can be a very useful physiological pointer, since it gives the proportion of the cell which is in contact with its surroundings. To determine surface area/volume ratios the number of end point hits on the component or cell are scored as above but in addition, the number of intersections of the actual probe lines with the surface contour of interest is also recorded. The surface area/volume ratio is then given by the formula

$$\frac{S}{V} = \frac{4 \times N_T}{Z \times P_T}$$

where N_T = number of line intersections with the surface, P_T = the number of 'hits' of the end points on the component or cell, and Z is the length of each test line on the graticule for the magnification used in the experiment. By using a different type of test grid, the point-counting method may also be used to measure the percentage volume of scarce components (such as lysosomes) which are present in the cytoplasm of cells.

58

These methods together with other automated techniques are currently being used by the authors in large-scale study of the differences between normal and carcinomatous cells. Both optical and electron micrographs of biopsies are being evaluated for such features as percentage volume of nucleus and of intercellular space, percentage volume of cytoplasm occupied by tonofibrils, number of ribosomes per cubic micrometre of cytoplasm, surface area/volume ratios of the cells and the percentage of the cell membrane occupied by desmosomal specialisations. Some very significant differences have already been observed between squamous cell carcinomata of the uterine cervix and the corresponding normal tissues; the carcinomata show much less intercellular space, have larger, more variable nuclei and have far fewer desmosomes than the corresponding normal tissues.

Further reading

1 Aherne, W.
   1967 'Methods of Counting Discrete Tissue Components in Microscopical Sections', Journal of the Royal Microscopical Society, 87, 493–508
2 Freere, R.H., and Weibel, E.R.
   1967 'Stereologic Techniques in Microscopy', Journal of the Royal Microscopical Society, 87, 25–34
3 Philp, J.R. and Buchanan, T.J.
   1971 'Quantitative Measurement on Finite Tissue Sections', Journal of Anatomy, 108, 89-97
4 Siegel, S.
   1956 'Non Parametric Statistics for the Behavioural Sciences', McGraw-Hill, New York
5 Weibel, E.R.
   1963 'Morphometry of the Human Lung', Heidelberg, Springer-Verlag; New York, Academic Press Inc
6 Weibel, E.R., Kistler, G.S. and Scherle, W.F.
   1966 'Practical Stereological Methods for Morphometric Cytology' Journal of Cell Biology, 30, 23–38

Hints for the teacher

The only requirement for point counting is a suitable means of superimposing the image of the test probes upon that of the object to be analysed. Several methods have been suggested in the section on equipment; the one chosen depends very much on the equipment available in any particular institution.

Use of the special eyepiece is convenient but becomes expensive with large classes, whilst if prepared grids on acetate are used, then much time will be required to prepare these in advance of the first class, together with the photographs or camera lucida drawings on which they are to be used. Once prepared, however, the material is available for use in subsequent classes.

If the Weibel eyepieces are to be purchased, then they should be ordered from Wild distributors several months in advance of the date on which they will be required.

S. BRADBURY
G. WIERNIK
MARY PLANT
Oxford

Section 4

Cytochemistry

1 Alkaline phosphatase in tissue sections

Time required
About 3½ hrs.

Assumed knowledge
Some tissue morphology of kidney, liver and intes tine.

Theoretical background
If alkaline phosphatase is present in cells of a tissu section, incubation with a substrate containing phos phate will liberate phosphate ions. If calcium ions ar also present in the substrate, the phosphate ions wi combine with calcium ions to form insoluble calciur phosphate which is precipitated in the sections. Th precipitate, being white, must be visualised. This i done by replacing the calcium by cobalt and precipi tating the cobalt as the dark sulphide. Alkalin phosphatase is specific for phosphomonoesters and i most active at pH 9.

Necessary equipment
1 Four slides—kidney (prepared by freeze substitution or alcohol fixation)
2 Four slides—intestine (prepared by freeze substitution or alcohol fixation)
3 Two slides—liver (prepared by freeze-substitutio or alcohol fixation)
4 One slide of superimposed kidney
5 One slide of superimposed intestine
6 Sets of Coplin staining jars (Gallenkamp) eacl containing 100 ml; 18 in all as follows:
    2 x xylol
    2 x absolute alcohol
    1 x 90% alcohol
    1 x 70% alcohol
    1 x 50% alcohol
    1 x 30% alcohol
    1 x H_2O (distilled)
    H_2O (distilled) at 37°C
    1 incubation medium at 37°C
    1 Control medium at 37°C
    2 H_2O (distilled)
    1 x 2% Ca $(NO_3)_2$ To each add 2 drops
    1 x 2% Co $(NO_3)_2$ 2% barbitone (BDH)
    1 $(NH_4)_2$ S (2 drops $(NH_4)_2$ S made up to

100 ml with H_2O)

1 H_2O (distilled)

7 Incubation medium

20 ml 2% barbitone (BDH)

20 ml 2% Na glycerophosphate (BDH)

10 ml 2% calcium nitrate

5 ml 1% magnesium chloride made up to

100 ml with glass distilled water.

8 Control medium: as above, but leave out Na glycerophosphate

9 Glassware and others:

(a) 15 cover slips

(b) 1 diamond pencil

(c) water bath at 37°C

(d) Canada balsam (BDH, Gurr) or mounting medium.

Type of group

Students work singly.

Instructions to the student

You are provided with sections of rat kidney, liver and intestine.

1 Dewax the slides with xylol, bring down through the alcohols to water; i.e., pass from 100% alcohol, to 90% and so on to water. (Keep one jar of xylol and one of absolute alcohol for later use.)

2 Place in distilled H_2O at 37°C for 2 min.

3 Incubate in buffered sodium glycerophosphate at 37°C for the following times per tissue:

(a) rat kidney and intestine for 1, 8 and 15 min.

(b) rat liver for 2 hr.

At the same time place control slides in buffer lacking the glycerophosphate substrate at 37°C. Ideally you should run a control slide for each incubation time, but select 8 min for kidney and intestine, 2 hrs for liver.

4 Place in calcium nitrate, 2–5 min.

5 Place in cobalt nitrate, 2–5 min.

6 Place in two washes of distilled H_2O, 1 min each.

7 Place in ammonium sulphide, 2 min.

8 Wash in water, 2 min.

9 Dehydrate, clear and mount, using clean absolute alcohol and xylol.

Because the reaction releases phosphate ions which are able to diffuse until 'trapped' by combination with calcium ions, distribution artefacts may occur. Some check on the amount of diffusion in the tissue can be obtained by 'superimposing' active material on inactive material. You will be supplied with sections previously inactivated in hot water with additional active material superimposed, so that about half the inactive material is still visible. Dewax these sections and run through the schedule using 15 min for the incubation period. Examine your preparation for diffusion of active material into the inactive section.

Results

Make detailed and annotated drawings to show distribution of the enzymes in tissues. Be sure to look at your controls. Note the staining intensity carefully.

Discussion

The variation in the times of incubation show the patterns of distribution and penetration of the components of the assay medium. Careful note should be taken of these changing patterns. Be sure of your identification of the structures in the tissue sections; check with morphology and physiology text books. Why do certain tissues and parts of the tissue remain unstained? Explain in some detail the results from the superimposed active and inactive sections. What function does alkaline phosphatase have in the cell and how is it usually distributed within the cell?

Extensions and applications

Alkaline phosphatase is a good diagnostic criterion for certain types of cells, e.g., intestinal, and also for intercellular structures. It is a rapid and reproducible histochemical test and is widely used for both research and clinical purposes.

Further reading

Pearse A.G.E.

1968 and 1972 'Histochemistry: Theoretical and Applied Vol I and II', Churchill, London

Giese, A.C.

1970 'Cell Physiology' W.B. Saunders, Philadelphia

Davson, H.

1970 'General Physiology', Churchill, London

Hints for the teacher

Some time is needed for the preparation of the tissue sections but the procedures are standard and not difficult. The experiment usually works very well.

Superimposed slides are prepared by taking slides of kidney and intestine, dewaxing and rehydrating, followed by immersion in hot (80°C) water for a short period. The slides are dried and then fresh tissue sections placed on the slide so as to 'overlay' part of the inactivated tissue.

Adapted from a practical of Dr. J.F. Danielli.

D.O. HALL
King's College
London

2 Reconstruction of cytochemical patterns from a number of methods

Time required
3 hr, excluding the written discussion of results, has proved rushed. Probably best in an all-day class. Can be broken off and resumed, but this may make it more difficult for the student.

Assumed knowledge
1 How to use the light microscope at all magnifications with an histological section.
2 An understanding, in principle at least, of the way in which a section is made from a piece of tissue.
3 Chemistry to A-level, or reading to bridge the gap.
4 In general this is an exercise for cell biology teaching, in the first term, say.

Theoretical background
Cytochemical methods are used to demonstrate where in a cell a particular compound or class of compounds is found. A reaction of the compound under examination is used to produce a colour at the same site. The specificity of the reaction may be narrowed by the nature of the material; thus the periodic acid–Schiff reaction (see below) is not specific for carbohydrate when one is examining all the range of compounds which a chemist may have in a test-tube, but it is so specific in an histological section of biological material. Cytochemistry merges imperceptibly into non-specific staining, as can be seen by considering a series made up of methods used in preparing slides for this exercise.
1 The reaction is known chemically, is specific for one class of compounds found in cells, and leaves coloured material proportional in amount (and therefore in measurable light absorption) to the compounds being demonstrated; the Feulgen reaction is of this sort.
2 Although the reaction is chemically defined and specific, it can be used only qualitatively, because the extent to which the reaction proceeds depends on many circumstances, not all controllable; the periodic acid-Schiff reaction is of this sort.
3 The reaction is not for a compound as such, but for sites of an enzyme activity, such as alkaline phosphatase.
4 A staining procedure is known to be specific although the chemistry is not precisely known or although there is adsorption of dye rather than a chemical reaction; this applies to the methyl green-pyronin method.
5 The location of a dye such as the aniline blue of azan, is known empirically to indicate a class of substances, with a specificity which depends upon the observer's knowledge of the range of material which might take up the dye.
6 Although the attachment of any dye must depend upon chemical characteristics, a stain may in fact be cytochemically almost valueless; such is haematoxylin and eosin, as used for this exercise.

Necessary equipment
1 Microscope, including mechanical stage, oil immersion lens and eyepiece graticule
2 Micrometer slide (may be shared)
3 Slides bearing comparable sections, but stained differently. It is suggested that vertebrate gut epithelium should be used, and the small intestine of the newt or the Chinese hamster are suitable since both species have large cells. The numbered slides, of which one set will be put out for each student, will have sections stained as follows: (Stains from Gurr, techniques from Pearse, Histochemistry 1968.)
 (a) Haematoxylin and eosin. Several suitable fixatives, but avoid over-fixation with most
 (b) Heidenhain's azan, or other blue trichrome. A mercuric fixative is best, but Bouin may prove satisfactory. (If not azan, modify Section b under 'Instructions to the Student' below as necessary.)
 (c) Periodic acid–Schiff. Use fast green or light green counterstain.
 (d) Stained for mitochondria by Bharadwaj and Love's method (Stain Technol., 34, 331–4 1959). Omit potash digestion of nucleic acids; this appears not to affect the staining appreciably, and loosens sections. Inspection suggests that the structures stained darkly are those rich in lipoprotein, but there is no established chemical specificity.
 (e) Feulgen and fast green (or light green).

62

(f) Methyl green and pyronine. Use 1% of each in buffer at pH 4·5, stain for about 1 hr (but increase if necessary), and dehydrate with tertiary butanol or acetone, to avoid loss of stain. Fixation in solvent-grade methanol is suitable, since penetration is not a consideration.

(g) Alkaline phosphatases by Gömöri's method (see Section 4.1)

4 Labelled photographs or drawings should be provided, either to individual students or in a way that provides access for all to them, to help students find the features mentioned in the exercise; it is assumed that the students are familiar neither with the parts of a cell nor the landmarks of a section such as these.

5 Squared arithmetic paper with ¼in (5 mm) squares. (At least 2 sheets for each student.)

ype of group

tudents work individually. The grouping of individuals does not depend on the exercise.

nstructions to the student

ou are given a set of seven slides each as listed bove. . . . thick, from the same piece of the . . . . of . . . . .(You will be given the information to fill in hese blanks.) Find the columnar epithelium in the ections, using all magnifications of your microscope, nd making sure that you are clear in your mind bout the correct procedure for using it. Seek help vhenever you do not understand the instructions.

Section a

ind the columnar epithelium and examine it. At the reatest resolving power of the microscope, with the il-immersion lens, note the strands and granules of n epithelial cell nucleus.

This section has been stained with haematoxylin dark blue) and eosin (shades of red and pink). The aematoxylin, as used in this preparation, is a stain or cell nuclei, although certain other material may lso be coloured by it. Haematoxylin and eosin is a ommon combination of stains, convenient in use.

Section b

his section is stained by Heidenhain's azan, a triple tain in which azocarmine (red) stains nuclei, orange ; (orange or light red) stains cytoplasm and aniline lue (blue) stains collagen. Collagen is a fibrous rotein laid down by cells and found outside them. It s one of the commonest extracellular materials in nimals; it not only forms tendons, the fibres of

bone, the white of the eye etc., but also binds cells together in, e.g., muscle.

The blue colour may also be seen in the free border of the epithelium. (There is, both in collagen and at the cell surface, carbohydrate, discussed below in relation to Slide c; this visibly affects the staining most readily at a brush border, which consists of microvilli, parallel projections of the cytoplasm, presenting much cell surface per unit area of section.) In examining the cell nucleus you may find that the nucleolus (ribonucleoprotein) is orange, although the chromatin granules and strands (DNA) are red. This difference corresponds to that shown, with greater chemical reliability, by Sections e and f.

Section c

The purple-red colour shows the presence of mucopolysaccharides, while the green is a general protein stain to show up other structures. The purple dye is basic fuchsin which, with the use of sulphurous acid to prevent attachment elsewhere, combines with the aldehyde groups of the section, and shows their position. The sections have been submitted to oxidation by periodic acid. This yields an aldehyde group from two $-OH$ groups or an $-OH$ and $-NH_2$ group attached in the cis-configuration to adjacent carbon atoms. In animal tissues, the only insoluble material with these configurations is carbohydrate. The glycogen has been removed by diastase (BDH) and the purple colour therefore shows the site of mucopolysaccharides, i.e. polysaccharides in which many (at least 4%) of the component monosaccharides are amino-sugars. (A control experiment showed no reaction in unoxidised material; therefore all the aldehyde groups taking up basic fuchsin are due to oxidation.) The stain solution, which is used here is called Schiff's reagent. Thus the reaction giving the colour is called the periodic acid-Schiff (PAS) reaction.

Some of the sites of mucopolysaccharides (or mucoproteins, in which they are combined with proteins) are:

1 Mucus, probably absent here.

2 Collagen. The proportion of carbohydrate to protein is highest in the fine collagen called reticulum. (All collagen begins as reticulum, and some remains as such throughout life.) Thus reticulum is PAS-positive, but more mature collagen is not.

3 The extracellular ground substance in which both cells and fibres such as collagen are embedded. The concentration of carbohydrate may be insufficient to give a visible reaction in an histological

section, but in some situations e.g. the matrix of cartilage, the PAS reaction is strongly positive.

4 At cell surfaces. The layer of PAS-positive material is very thin, but may sometimes be detected by the light microscope at the free surfaces of epithelia and in intracellular cement. At the brush border, the configuration of the surface makes the purple-red colour very evident indeed. (cf. note on Section b)

When inspecting this section with the oil-immersion objective, find out whether the microvilli of a brush border can be distinguished with the light microscope.

Section d

The darkly stained structures are those rich in lipoproteins. Among these structures are mitochondria, seen as rods inside the cells. (Other material darkly stained includes the nucleoli, as the ribonucleoprotein of these has lipoprotein as a constituent.)

Make sure that you can distinguish the mitochondria of the epithelium, using the oil-immersion lens.

Section e

The purple-red colour corresponds to sites of DNA the section having been stained by the Feulgen reaction. This is similar to the PAS reaction in that Schiff's solution of basic fuchsin with sulphurous acid is used to colour sites of aldehyde groups. In the Feulgen reaction, however, the aldehyde groups are produced by hydrolysis in warm acid. The only aldehyde groups formed, and still in the section, arise from the desoxyribose of DNA where its linkage to the base has been broken by the hydrolysis. The counter-stain is the dye light green (or fast green), as in Slide c.

Examine an epithelial nucleus. The Feulgen-positive material corresponds to the chromosomes of mitosis, which are also Feulgen positive, and the two are interconvertible. The threads (chromonemata) and granules (chromocentres) are parts of chromosomes. Note that the nucleolus is Feulgen negative. If a cell such as one of these is examined alive, and then fixed and subjected to the Feulgen reaction, the pattern of chromocentres and chromonemata remains (with suitable fixation) unchanged. That is, these structures are present in life. After digestion by an enzyme specific for depolymerising DNA and making it soluble, a cell is entirely Feulgen negative.

Section f

The colours in this section are red due to pyronin, which colours RNA, and green due to methyl green which colours DNA. Given the conditions under which staining was carried out, the colours are reliably specific for the two nucleic acids. However the attachment of the pyronin is loose, since it is easily washed off by water or ethanol, and its combination with RNA is not stoichiometric.

Examine epithelial cell nuclei, and note the respective positions of RNA and DNA. The nucleoli contain RNA and the chromocentres and chromonemata contain DNA. Has the nucleolus any special relation to DNA, and does this section demonstrate any nuclear RNA that is not in the nucleolus?

Section g

This section has been stained by a reaction for alkaline phosphatase. This is a collective name for enzymes splitting phosphate ester linkages and having an alkaline optimum pH. Since such enzymes may split high-energy phosphate bonds, they are important in the liberation of energy so stored. But, it should be remembered that phosphatases which are not rendered insoluble and held in position by the histological fixative, or which are inactivated by the preparative techniques, will not be seen in the section.

The black, brown or grey deposits which indicate phosphatase activity consist of cobalt sulphide. The section has been incubated with beta-glycero phosphate in the presence of calcium ions. The liberated phosphate has been precipitated as the calcium salt in the presence of a considerable excess of calcium ions. This salt is converted by cobalt chloride into cobalt phosphate, also insoluble, and this in turn converted to cobalt sulphide by ammonium sulphide.

Examine the localisation of alkaline phosphatase activity in the epithelial cells of your section.

Recording Results

1 Write such notes, comments and answers as may arise from going through the seven sections with the preceding notes.

2 Use either Section (a) or Section (b) to make a map of the main features of an epithelial cell as seen with the oil-immersion lens. Make the drawing on squared paper. Use your eyepiece graticule to note distances. Choose a scale conveniently relating graticule divisions to squares on the paper so that an epithelial cell is at least half the page in height. With the graticule you should now be able to draw on the squared paper the main features of the cell at the correct

distance apart. Parts of adjacent structures should be included; you are unlikely, in any case, to distinguish entire intercellular boundaries. Use the micrometer slide to calibrate the eyepiece graticule, and so add to your drawing a scale of micrometers. Title the drawing, label its features, and make a note of the fixative, stain and lenses used.

3 Repeat (2) using Section (d).

4 In order to relate your inspection of the slides to the features shown in (2) and (3), prepare a table with the following 12 rows:

> fixative (information from lecturer);
> nucleus (nucleolus);
> nucleus (chromocentres);
> nucleus (chromonemata);
> nucleus (elsewhere than in these three situations);
> nuclear membrane;
> mitochondria;
> brush border;
> lateral cell boundaries;
> basal cell boundaries;
> elsewhere in cytoplasm, diffusely;
> elsewhere in cytoplasm, localised (specify position as nearly as possible.)

Give the table the following 7 columns:

> DNA (Feulgen);
> DNA (methyl green);
> RNA (pyronine);
> PAS reaction;
> alkaline phosphatase;
> lipoprotein (as roughly indicated in Section d); aniline blue staining with azan.

In the table, use a plus sign to indicate presence and a minus sign for absence. For alkaline phosphatase and RNA indicate relative intensities by one, two or three plus signs. Note that a negative result may be true only under the conditions of this experiment; you may have to record the absence of some substances or activity in an organelle where other tests might show them to be present. Therefore the positive results you enter on the table are probably valid (although fallacies due to diffusion etc. would have to be excluded in a fuller investigation) but the negative findings are less certain. (e.g. although it was reported previously that all DNA is nuclear, cytoplasmic DNA has been demonstrated microscopically, as well as by other means.) Nevertheless, in practice, your observations should be a reliable guide to the principal sites of the reacting materials.

5 Write short notes describing the nucleus, the cytoplasm, and boundaries of the cell, as they have been shown in this exercise.

Discussion.
In the discussion of results, it may be helpful to consider the following points.

1 Possible fallacies

2 Correlations between this exercise, which is concerned with spatial arrangement, and any relevant biochemical findings, known to you, from disrupted cells

3 Any other functional interpretations of the spatial relations you have found

4 Any conclusions you can draw, after suitable reading, if necessary, of text-books, on how your findings might be constant or variable among different tissues, different vertebrates, different animals or different eukaryotes.

Extensions and applications
The general usefulness of cytochemical methods may appear from the discussion which you have written. However, the exercise points more to the physiology of the cells selected than to comparison of these with others, and it can be seen that cytochemical procedures can also elucidate differences between cells, enabling one to obtain information about special organelles (e.g., nuclei of ciliates) or abnormal material (e.g., amyloid substance). Further, cytochemistry of cells at different stages of growth or of some other process (e.g., viral invasion) can elucidate the chemistry of those processes. Cytochemistry is therefore a general tool of very wide application.

Further reading
On cytochemistry in general, and the possibility of dependence of results on fixation, see Figs. 1–14 and 1–15 in the 9th edition of Bloom W. and Fawcett D.W., 'Textbook of Histology', W.B. Saunders, Philadelphia 1962 with the accompanying text, or read similar passages in the earlier parts of any of several cell biology and histology books. In such chapters note the principles of autoradiography, electron-microscope cytochemistry, quantitative histochemistry, and quantitiative studies, especially by interferometry, on living cells. Specialised books are:

Pearse, A.G.E.,
1968 'Histochemistry', Churchill, London
Ross, K.F.A.,
1967 'Phase Contrast and Interference Microscopy for Cell Biologists' Edward Arnold, London

Wied, G.L.,
    1966 'Introduction to Quantitative Cytochemistry',
    Academic Press

Hints for the teacher
This exercise takes much preparation for the first run. Given histological assistance, a clear academic year may reasonably be allowed, since any display material to help the students will need to be based on the preparations which will in fact be used. After the first run, the same preparations can be used repeatedly, but it is probable that modifications will be made year by year, in the light of experience. Because the exercise is designed to introduce a group of methods and to lead the student to correlate findings (as well as to give practice in microscopy), rather than to lead to a single result, its effectiveness is probably enhanced by tutorial support and correlation with lectures. [Dr. Boss uses photographs and diagrams for his teaching; advice can be obtained from him (Eds.)] .

J. BOSS
Bristol.

3 Observation of some changes taking place during cellular differentiation in rat kidney

Time required
Half-day class: three hours. The lecturer can suggest more recording of intermediate stages of development (see below) if more than three hours can be used.

Assumed knowledge
1 How to use the light microscope at all magnifications with an histological section.
2 An understanding, in principle at least, of the way in which a section is made from a piece of tissue.
3 Although a knowledge of histology of the kidney may make the exercise more interesting. no knowledge of histology is necessary.
4 Experience of measurement and drawing with the microscope is desirable but not essential.

Theoretical background
During the development of a multicellular organism, cells of many kinds are descended from the zygote, so that cells which are sisters or cousins to each other come to differ in chemical abilities, physiological functions and microscopical appearance. This process of differentiation commonly presents itself to the observer as a set of simultaneous changes. For example, the development of contractile fibrils in a skeletal muscle cell causes the displacement of the nuclei to one side, and the eccentric position of the nuclei is itself a distinctive characteristic of vertebrate skeletal muscle. Or again, if a protozoan animal can be changed experimentally from a ciliate to a non-ciliate form, not only will the cilia be lost, but the shape of the cell will also change. These examples may *seem* simple to the point of triviality, but in this exercise you will probably observe changes contemporary with each other, but with causal links which are not at all clear.

Necessary equipment
1 Microscope, including mechanical stage, 2 mm oil-immersion lens and eyepiece graticule

2 Micrometer slide (may be shared).
3 Histological sections, 5 μm thick, of kidneys from rats at birth and at 6, 11, 15 and 32 days of post-natal age, and stained
 (a) by azan
 (b) by PAS
 (c) by Feulgen
 (d) for mitochondria
 (e) by an alkaline phosphatase reaction
These are five of the methods used for preparations in Section 4.2. If the stains are numbered 1–5 in this order, and the age in days added, each of the 25 slides has a unique number, thus: 1/0, 1/6, 1/11, 1/15, 1/32, 2/0, 2/6 etc. There should be one set of slides for each student.
4 A card, accessible to each student, with a diagram of a mammalian nephron and its relation to the zones of the kidney, and with a series of sketches to show how the nephron, and Henle's loop and Bowman's capsule in particular, develop
5 A card accessible to each student, with labelled micrographs to assist in identifying parts of the nephron in the azan-stained section of 32-day kidney, and in distinguishing the most immature renal corpuscles in the new-born.

Type of group
Students work individually. The grouping of individuals does not depend on the exercise.

Instructions to the student
You have 25 histological sections, 5 μm thick, of rat kidney. The first figure on each slide indicates the stain: 1–azan, 2–PAS, 3–Feulgen, 4–for mitochondria, and 5–for alkaline phosphatase. The methods of preparation are discussed in Section 4.2 and you should refer to these if you have not done that exercise. The second figure on each slide refers to the post-natal age of the rat in days. New-born is indicated by O, and days 6 and 11 fall in the period of obligatory unsupplemented suckling; between 15 and 32 days the young are gradually weaned, and forced weaning is possible at any time in this period. The kidney at birth is very immature, and becomes roughly similar in development to a new-born human kidney only at about the time when weaning begins. The names of parts of the nephron (kidney unit) used in this exercise can be found on the cards provided, from which you will see how they originate and how to recognise them. You should note that the primordium of the proximal and distal convoluted tubules at first provides a direct connection from Bowman's capsule to the collecting duct system, but the two become distinct, and Henle's loop descends from their junction with each other. In the exercise you will trace how cells differentiate from each other to become characteristic of the proximal and distal convoluted tubules, and how each reaches its fully developed appearance. (In both tubules there is reabsorption of the fluid which is formed by the filtration of plasma into Bowman's capsule, in the process of urine formation. However, the two tubules differ in the amount and kind of substances reabsorbed or passed into the lumen; for further information, see textbooks of vertebrate physiology.)

1 Find in 1/32 the cortex and medulla, and a glomerulus in the cortex. Near the glomerulus note the sections of tubule. Use the indicator cards to assist you in finding a glomerulus, if necessary, and with the aid of the labelled photomicrograph distinguish proximal from distal convoluted tubules. The proximal, being longer, are the more abundantly represented in your sections.
2 Use all the 32-day sections to note differences between the two kinds of tubule cell.
3 In Section 1/0 find in the outer cortex an immature renal corpuscle distinguishable by the cuboidal epithelium, as shown on the card with the photomicrographs. With this as a landmark, and bearing in mind that the renal corpuscle and tubules of the same nephron are near each other, examine the tubules nearest to the immature renal corpuscle. Use any or all of the slides of the new-born to examine these immature tubules.
4 Use sections from rats of ages between birth and 32 days to find intermediate stages in the development of proximal and distal tubule cells from an ancestor in which the differences between the two types are not apparent. It may help you to note that the most mature nephrons are those with their renal corpuscles nearest to the medulla.

Recording results
Clearly there are many ways in which your observations might be recorded, and there is scope for initiative. However, the following form might be helpful to those who wish to use it. Whether you make drawings may depend on your experience (cf. 'Assumed knowledge' above), and how much attention you pay to the intermediate stages of development will depend on the time you have. The suggestions which follow assume both the experience

and the time; you should leave out or curtail items as appropriate.

1 Make a table with five columns corresponding to the five methods of preparation and ten rows, two for each of
   (a) nuclear size, shape and content,
   (b) density and size of mitochondria
   (c) arrangement and position of mitochondria
   (d) form and chemical character of the free surface, and
   (e) any other observable characters.

   Each of these topics has two rows because it is subdivided for proximal and distal tubule cells. You should then enter in the table notes on the two kinds of cell, as seen at 32 days, so as to compare the various features as seen by the preparative methods used.

2 Make a similar table, but with a single row for each of the five topics, to summarise the features of the immature tubule cells in the new-born.

3 Make a table with the same five rows, but with three columns, for immature tubular cells, mature proximal tubule cells and mature distal tubule cells. This should show how the various features change as the two mature types develop; reference to methods of preparation can be made in the body of the notes incorporated in the table which is made through a consideration of the first two tables.

4 Comment, in brief notes, on the timing of change in the cells, noting whether the changes are simultaneous, and whether any of them are gradual, showing intermediate stages. If there are any intermediate stages, describe them.

5 Illustrate any points with scale drawings, made accurately from observation, wherever you think these would be useful, and refer, as appropriate, to your own drawings when you are compiling tables or making other notes.

Discussion

In your discussion of the results you might consider what is happening in differentiation from the cellular point of view and in relation to the multicellular organism. Concerning the former, what are the problems which arise if one tries to interpret the differentiation you have been examining in terms of changes in the parts of the genome being transcribed? Within the organism as a whole, can you, after looking up the kidney in histology and physiology books, say how the cellular features arising in the differentiation of these tubule cells might be related to the development of function?

Extensions and applications

Histological methods provide only one set of approaches to differentiation. Problems of differentiation provide the principal meeting of molecular biology with cell biology—or do they?

Further reading

1 Whittaker, J.R. (ed.),
   'Cellular Differentiation', Dickenson, Belmond, California, is a collection of original papers by leading workers in the field.

2 Seek in the larger histology and human (or other embryology books descriptions of histogenesis in selected tissues and organs. Good examples might be the kidney (since it has been the subject of this exercise), skeletal muscle and bone.

Hints for the teacher

The remarks on Section 4.2 apply here. However, further development of the exercise after the first run may require no new preparations since you may wish to pursue further the possibilities of existing sections. For example, this material presents opportunities for exercises on the differentiation of Bowman's capsule or the juxtaglomerular complex and can be used for problems on the relation of mitosis to differentiation; mitosis is abundant especially at the younger stages.

[Dr Boss uses photographs and diagrams for his teaching; advice can be obtained from him (Eds.)].

J. BOSS
Bristol

4 Autoradiographic study of the mitotic cycle

Time required

This exercise can be performed in 3 hr, including write-up time. If the practical is to be expanded for the students to make their own autoradiographs, then 3 periods of 3 hr, 1 hr and 3 hr are needed.

Theoretical background

The students should be familiar with the concept of autoradiography, the cell cycle, ^{3}H-thymidine as a marker of DNA turnover, and mitosis.

Necessary equipment

Equipment involves a microscope with x40 and x100 oil-immersion bright-field objectives.

Plants of *Vicia faba* (Broad bean) have their roots fed with ^{3}H-thymidine (Radiochemical Centre, Amersham), a specific precursor for DNA, at a concentration of 5 μCi/ml for 1 hr. The plants are then transferred to a non-radioactive solution and lateral roots are removed at 3 hr, 6 hr and 11 hr after feeding. The roots are fixed for 3 hr in acetic acid—absolute ethanol (1:3) and are rehydrated via alcohol. Squash preparations of the root apices are made after hydrolysis for 8 minutes in 1N HCl at 60°C and autoradiographs prepared. The autoradiographs are exposed for 7 days, developed and stained.

Type of group

The class can be run with any size of group, although 2–3 students prove easiest to handle.

Instructions to the student

By counting the numbers of labelled mitoses at each feeding time, it is possible to determine:
1 The length of the mitotic cycle
2 The length of the cell cycle
3 The lengths of the period of the DNA synthesis
4 The timing of DNA synthesis during interphase.
From your preparations, count 50 mitoses for each feeding time.

Recording results

Plot a graph of the percentage of labelled mitoses versus time. Determine the length of mitosis and the length of DNA synthesis in the following way.

Normally the mitotic cell can be divided temporally, as shown in Fig. 1,

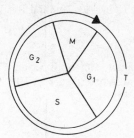

Figure 1

Where M = mitosis, S = duration of DNA synthesis at interphase, G_1 and G_2 are two periods of interphase where DNA is not synthesised, and T = the total duration of the cell cycle.

In an ideal system, the wave of labelled mitosis corresponds to successive divisions of that part of the cell population which was in S during the availability of the labelled precursor. Therefore, the fraction of the labelled mitoses is 0 until the end of G_2 when it rises rapidly to 100% in the period equal to M.

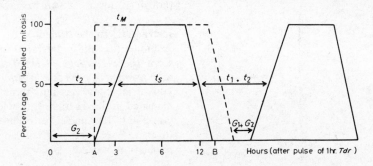

Figure 2

The labelled fraction remains at 100% for a time equal to the duration of S less the time taken for M, and eventually falls to 0. If all of the mitoses are scored regardless of the stage of division, then the time from the mid-point of the labelling-pulse to that at which 50% of the mitoses are labelled is equal to $G_2 + \frac{1}{2}M$, and hence M can be calculated.

$$t_2 = G_2 + \frac{1}{2}M$$

$$t_1 = G_1 + \frac{1}{2}M$$

$$G_1 = (t_1 + t_2) - (M + G_2)$$

$$S = T - (G_1 + G_2 + M)$$

For these calculations, it can be considered that T for *Vicia faba* is 15 hr, A = 2 hr and B = 15 hr.

Further reading

Cleaver J.E,
1967 'Thymidine Metabolism and Cell Kinetics' North Holland, Amsterdam, (pp. 104-7 and 112-15)

P.B. GAHAN
Queen Elizabeth College
London.

5 Autoradiographic study of cell proliferation and differentiation

Time required
This exercise as given can be performed in 3 hr including write-up time.

Assumed knowledge
The students should be familiar with the concepts of autoradiography, the use of ^{3}H-thymidine as a marker of DNA turnover, the concept of cell movement, and the histology of the small intestine.

Necessary equipment
Equipment required includes a bright-field microscope with x40 and x100 oil-immersion objectives and the prepared autoradiographs given in the schedule. Male mice, e.g. C57 strain, are injected intraperitonally with 20 μCi in 0·1 ml of an aqueous solution of ^{3}H-thymidine (^{3}H-TdR), Radiochemical Centre, Amersham. The mice are killed 40 min, 4 hrs, 24 hr and 37 hr after the injection. Pieces of small intestine are removed immediately and fixed in glacial acetic acid: absolute ethanol (1:3) for 1 hr followed by 18 hrs in 10% formalin. The tissues are embedded in paraffin wax and transverse sections are prepared. Autoradiographs were made and were stained with haematoxylin and eosin (Gurr) (see Pearse, 1968).

Type of group
The class can be run with any size group of students though groups of 2–3 are easy to handle.

Instructions to the student
Examine the autoradiographs of the sections and *score the percentage of labelled epithelial nuclei in each segment at each feeding time.* It is suggested that you divide the villus into three arbitrary regions a indicated in the diagram. It is suggested that for speed you count 100 nuclei for each observation, in order to arrive at a percentage of labelled nuclei within a population of cells.

Recording results
Plot the results you obtain in the form of a graph of percentage labelled nuclei per segment versus time.

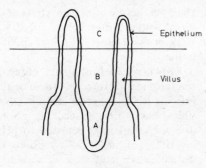

Figure 1

scussion

at can you conclude from these data relating to
ll proliferation and differentiation in the intestinal
thelium? Can you observe any differences in the
ensity of labelling (blackening) of the nuclei of the
pt region (A) between sections from animals killed
40 min and 37 hr? How do you explain your
servations? Compare the percentage of labelled
ls in the region A of the epithelium with the
jacent muscle cells. How do you explain your
servations?

rther reading
eson, C.R. & Leeson, T.S.
1970 'Histology' pp. 307-8, 155 et seq, 2nd edn.
om W. and Fawcett, D.W.
1962 'Textbook of Histology', 9th edn., W.B.
  Saunders & Co., Philadelphia
ghes, W. et al.
1958 Proc. Natl. Acad. Sci. USA, 44, 476.

P.B. GAHAN
Queen Elizabeth College
London.

6 Fluorescence microscopy using amoebae and other protozoa

Time required
A minimum of 2 hr.

Assumed knowledge
Some facility with doing laboratory work and using a
good microscope—this practical can be used to teach
students how to use a fluorescence microscope.

Theoretical background
Fluorescence microscopy was introduced by Carl
Reichert in 1911, who designed the first instrument.
The method makes use of the properties of fluorescent dyes. The specimen with bound fluorochrome is
irradiated with a band of selected wavelength of near
ultra-violet light which is absorbed, and then a major
part of the absorbed energy is emitted as radiant
energy of rather lower frequency (i.e., longer wavelength) than the absorbed energy. This is the visible
fluorescent radiation, which is specific for the dye-
cell material complex.

The wavelengths used are determined by exciter
filters placed between the ultra-violet source
(generally a super-pressure mercury burner) and the
specimen. Since some ultra-violet light passes through
the specimen and will cause *instant* and *severe retinal
damage,* one must *never* look at material without *first*
checking that the protective barrier filters are in place
between the eyepiece and objective.

Acridine orange (a diamino acridine) is strongly
metachromatic and a cationic dye. In weak solutions
it fluoresces green, but in concentrated solutions the
fluorescence is orange-red, due to the formation of
dimers and polymers of dye occurring in higher
concentrations. The dye is strongly bound to nucleic
acids and the colour seen depends on the quantity of
dye bound. Thus DNA, being highly polymerised,
offers fewer binding sites and fluoresces green, while
RNA, being less highly polymerised, fluoresces red
because it offers more sites for binding acridine
orange and for the formation of dimers and polymers.

Necessary equipment
Each student will need the following.

A culture of *Amoeba* (or *Tetrahymena, Stentor, Spirostomum*)

Slides and cover slips

Teats

Pasteur pipettes drawn a second time to make a finer pipette

45% acetic acid—5 ml; together with wide-bore Pasteur pipette

Dry ice (small piece)

Diamond pencil

6 staining troughs with lids, to contain
Acetic acid-ethanol 1:3
Absolute ethanol

Acridine Orange (0·03 mg/ml in 0·1 M acetate buffer, pH 4·5) (Gurr)

2 x Acetate buffer
(0·1 N acetic acid, 102 ml) and (0·1 M sodium acetate 98 ml) approx. pH 4·5

Distilled water

Prepared slides of cells (2 per student) to be placed in ribonuclease for 4 hr at 38°C (ribonuclease conc. 0·3 mg/ml in distilled H_2O adjusted to pH 7 with $Na_2 HPO_4$)

A fluorescence microscope

Immersion oil

Type of group

Individual students, any size class, but problems arise with only one fluorescence microscope if class is too large.

Instructions to the student

Preparation of material

(a) Using clean slides, place some amoebae in a drop to one end of the slide and allow them to become attached. Mark slide with diamond pencil (to indicate side and cells when staining). (If motile ciliates such as *Tetrahymena* are used, place a concentrated suspension on slide; *Stentor* will attach to the slide.)

(b) Withdraw some liquid from the drop using a fine pipette.

(c) Take a cover slip with a drop of 45% acetic acid and fix—flatten the cells.

(d) Place on dry ice (5—10 min) so that cover slip can be 'flicked off'.

(e) Place in acetic acid—ethanol (1 : 3) for 5 min.

(f) Place in absolute ethanol for 5 min, drain and air dry. Check that you have a reasonable number of cells present.

(g) Place in acridine orange (0·03 mg/ml) 0·1 M acetate buffer, pH 4·5 for 30 min.

(h) Wash in two changes, 5 min each, of acet buffer.

(i) Mount in clean buffer and observe und fluorescence microscope. These cells will full of RNA; in order to see DNA so slides have been placed in ribonucle: (0·3 mg/ml in H_2O adjusted to pH 7 w $Na_2 HPO_4$; 4 hr at 38°C). Rinse these distilled water and place in acridine oran (continue from (g) above onwards).

Fluorescence microscopy

(a) Check that the barrier filters are present microscope.

(b) Switch on uv source at least 15 min bef required, keeping all exciter filters in pla to reduce stray irradiation. Do not swit off until all students have finished, beca the lamp must be allowed to cool for 2 before further use.

(c) Leave exciter filter BG12 (e.g. Zeiss) in pl (acridine absorbs at 380—400 nm) and move others to one side.

(d) Place immersion oil between slide and c denser (this enhances fluorescence—why this?) and set up the microscope using f bright-field and then dark-ground illumi tion (providing specimen is fluoresc enough).

(e) Look at orange-red fluorescence due to R and notice that in general the nucleus obscured. Compare with the ribonucle treated slides. In amoebae notice nucl and smaller nuclei from undigested fc organisms; in ciliates look for macro- a micro-nuclei, in *Stentor* and *Spirostomu* long chain of macronuclei, in *Tetrahym* one macronucleus.

Extensions and applications

Fluorescence microscopy is widely used in biology and immunology. By coupling fluorochron to antibodies, the specific location of antigens in cell can be determined. Recent use of this technic in cell biology have been in studies of membr fluidity following cell fusion in mouse-human hete karyons[2] and the accumulation of human nucl specific antigens in the chick nucleolus following formation of HeLa-chick erythrocyte hete karyons.[3]

Further reading

Bertalanffy, L. von

1963 'Acridine Orange Fluorescence in Cell
Physiology, Cytochemistry and Medicine'.
Protoplasma, 57, 51

Frye, L.D. and Edidin, M.

1970 'The Rapid Intermixing of Cell Surface
Antigens after Formation of Mouse—human
Heterokaryons'. J. Cell Sci., 7, 319

Ege, T., Carlsson, S-A and Ringertz, N.R.

1972 'Immune Microfluorimetric Analysis of
the Distribution of Species Specific Nuclear
Antigens HeLa-Chick Erythrocyte Heterokaryons'. Exp. Cell Res., 69, 472

Hints for the teacher

1 Emphasize the danger of eye damage from
ultraviolet radiation and ensure that a demonstrator remains with the microscope. Tape barrier
filters in position if there is any likelihood of
their being moved. If possible have the microscope in a dark room or dark corner of a room.

2 All protozoa can be obtained from Cambridge
Culture Collection, 36 Storey's Way, Cambridge
and grown up to give sufficient numbers. *Stentor*
will feed on many ciliates while *Spirostomum*
prefers mouldy conditions—a few boiled wheat
grains in Chalkley's medium (See Section 8.2)
which have been set up for a while will produce
adequate bacteria. *Amoeba* can be grown in
Chalkley's medium and fed on ciliates (see
practical on Pinocytosis Section 7.4).

3 Slides for ribonuclease treatment can be prepared
several days previously and stored in the refrigerator in the presence of silica gel.

SHIRLEY E. HAWKINS
King's College
London.

Section 5

Chromosome Cytology

1 Chromosomes of cultured animal cells

Time required
Not less than 3 hr

Assumed knowledge
Students do not need any specialised knowledge ⸱
experience to succeed with this experiment. The⸱
require only to be familiar with the use of Paste⸱
pipettes, centrifuges and microscopes. Some unde⸱
standing of the theory of cell division and cell cycl⸱
is obviously an advantage.

Theoretical background
The demonstration of the chromosomes of anima⸱
has become a relatively straightforward procedu⸱
now that efficient tissue-culture methods are avai⸱
able. Preparations can be made from the blood ⸱
humans or animals, or from tissue cultures set u⸱
from skin, or other, biopsies. Preparations can also b⸱
made directly from established cell cultures an⸱
today's experiment is an example of this. The stage⸱
involved are:

1 Accumulation of mitotic cells, using colchicin⸱
   to block cells in metaphase from further progre⸱
   through the cell cycle. This may not be necessar⸱
   with rapidly dividing cultures.
2 Separation of these mitotic cells, usually b⸱
   shaking.
3 Hypotonic treatment to swell the cells.
4 Fixation.
5 Spreading; the swollen, fixed cells are 'burst⸱
   onto a glass slide so that the chromosomes sprea⸱
   out.
6 Staining.

Necessary equipment
1 Generally available:
   37°C water baths.
   Centrifuge (bench top) (calibrate 1000 rev/min⸱
   Colcemid 0·004% (CIBA—Geigy)
   Sodium citrate 1%
   Fixative methanol—acetic acid 3:1
   Giemsa stain (BDH)
   Buffered distilled water to pH 6·8 with phos⸱
      phate
   Mountant, such as DePex (Gurr)

74

2 Each student (or pair, see below) will require for each preparation:
1 bottle of healthy dividing cells
1 1 oz. screw cap bottle
1 15 ml centrifuge tube
1 1 ml pipette
Several Pasteur pipettes and rubber teats
Several clean glass slides (3in x 1in) and cover slips
1 staining jar or tray
1 Bunsen burner.

Type of group
Students may work as individuals or in pairs. Limiting factor is probably ability to produce large numbers of bottles of suitable tissue culture cells.

Instructions to the student
Shake the bottle of cells (as demonstrated). This tends to dislodge the mitotic cells from the glass surface. Pour the suspension into a 1 oz screw cap bottle. Add 0·3 ml of 0·004% colcemid, shake gently. Incubate in the 37°C water bath for about 1 hr.

Pour the cell suspension into a 15 ml centrifuge tube and spin at 1000 rev/min for 5 min.

Pour off the supernatant and resuspend the cells in about 5 ml of 1% (hypotonic) sodium citrate.

Incubate for a further 12 min at room temperature.

Centrifuge at 1000 rev/min for 5 min.

Tip out the supernantant. Tap the tube vigourously to suspend the cells in the small amount of citrate that drains back to the bottom of the tube. Add 5 ml of fixative (1 part glacial acetic acid, 3 parts methanol) down the side of the inclined tube, slowly at first, then more rapidly. Pipette very slowly.

Leave for 5 min.

Centrifuge at 1000 rev/min for 5 min

Carefully remove the fixative. Add a few drops of fresh fixative and suspend the cells by tapping the tube. Add more fixative until the cell suspension is slightly opalescent.

Leave for 5 min and then make chromosome spreads as follows.

Drop 3 or 4 drops of cell suspension from a height of about 1in onto a *clean* 3in x 1in glass slide. The fluid should quickly spread over the whole surface of the slide. Pass the slide through the bunsen flame so that drying is rapid.

Stain the slides in 5% Giemsa at pH 6·8 for about 6 min. Wash in buffered distilled water and dry.

Count the number of chromosomes in each of 20 cells and make a frequency-distribution table of your results.

Recording results
This depends to some extent on the facilities available. If arrangements can be made to photograph good-quality chromosome spreads, the students can present these and can prepare karyotypes by cutting out the chromosomes and arranging them in a logical sequence. Otherwise counts can be presented as a frequency-distribution table. Counts can be sub-divided into the different chromosome groups (metacentrics, sub-metacentrics, acrocentrics and telocentrics).

Discussion
How much variation do you find in your counts? Is the cell line diploid or aneuploid? How does the modal count vary from the diploid number for the species? Has the cell line any abnormal characteristic (marker) chromosomes?

Extensions and applications
The most obvious application is the use for detecting chromosomal abnormalities in humans. This can now be achieved on foetal cells which have been isolated from aspirated amniotic fluid and then cultured, as well as on adult blood cells or skin fibroblasts. The method is very widely used in research to characterize cells and follow variation and selection acting on those cells.

Further reading
1 Ford, C.E.
1968 'General Pathology', 4th edn., by H.W. Florey, Lloyd–Luke, London,
2 Yunis J.J.
1965 'Human Chromosome Methodology' 4th edn. Academic Press New York
3 Swanson C.P., Merz, T. and Young, W.J.
1967 'Cytogenetics' Prentice-Hall New Jersey.

Hints for the teacher
The method can be applied successfully to most established cell lines. We have used Chinese hamster fibroblasts, A9 cells (subline of L cells), Sminthopsis fibroblasts, HeLa cells etc. (Biocult, Flow). It can be easily adapted for cells such as Ehrlich ascites tumour cells.

We find that a 4 oz medical flat bottle at an appropriate stage of growth (½–¾ confluent, actively dividing) usually provides sufficient cells on vigorous shaking, although trypsinisation may be required, e.g., with Sminthopsis cells.

It is a good idea to have other things to do during incubation periods. We demonstrate many different karyotypes, normal and abnormal human, animal, hybrid cell etc. We get students to prepare a karyotype from a photograph of a human spread provided (XX, XY, XO, XXX, XXY, etc.). We demonstrate fluorescent banding patterns with Quinacrine, and also Y bodies, and stain Barr bodies.

E. SIDEBOTTOM
Oxford.

2 Demonstration of banding pattern of mammalian chromosomes

Time required
2–3 hr, excluding microscopy and drawing.

Assumed knowledge
The majority of Polytechnics and some Universities that offer sandwich courses are keen that their students spend at least part of their course receiving industrial training. There are numerous advantages to be gained from such a venture, in addition to the fact that these students give a better account of themselves on returning to their courses.

Students afforded this facility are usually in the second year of their course and, therefore, a knowledge of biology to at least this standard is assumed. A considerable amount of the students' time is normally spent on this project, which usually commences with the student familiarising himself with the various publications on the subject.

Theoretical background
In recent years much interest has been focused on the DNA composition of chromosomes, including their heterochromatin content. The specific binding of heterochromatin with fluorochromes has been widely investigated and more recently, experiments on DNA hybridisation have stimulated many publications on the banding patterns of chromosomes. The original 'banding' experiments were done by Jones and Corneo (1971) and Pardue and Gall (1970), and both involved alkaline denaturation of the chromosomal DNA followed by renaturation in a suitable buffer at the correct temperature and pH.

The technique of Drets (1971) involves denaturation followed by incubation in saline—citrate solution and staining in Giemsa. Other methods involve variations of the pH of the stain or the type of denaturation—incubation processes in order to evaluate substantially the various procedures and to assess the best method for consistent reproduction of results.

The principles involved in the banding techniques are:

1 Denaturation of part or all of the DNA/histone content, by
   (a) Alkaline Treatment
   (b) Enzyme Treatment
   (c) Heat Treatment
2 Renaturation of part of the DNA if denaturation was complete by incubation in a suitable buffer
3 Staining—Giemsa or Leishman is the usual stain. The staining technique can be varied by altering:
   (a) pH of the stain
   (b) Temperature at which slide was stained
   (c) Time of staining
   (d) Concentration of the stain.

The heat-denaturation method seemed to be the most promising and experiments are being continued involving this process. The effect of heat on chromosomes is termed by some authors as being one of differential denaturation (Summer *et al.*, 1971). Chromosome preparations are incubated in a suitable buffer for a specific time and temperature before being stained. This is done because it is thought that heat, in addition to incubation, causes partial denaturation, which is effective at the areas of less repetitive DNA base sequences in intact and stained material. As yet, this explanation is an hypothesis and the explanation is only tentative. Further investigations are required to elucidate the problem further.

Necessary equipment

Solutions

0·2M NaH$_2$PO$_4$	)	Sörensen buffer
0·2M Na$_2$HPO$_4$	)	made up at pH 6.8

7% Giemsa stain—made up with 7 ml stock Giemsa (Gurr)
20 ml Sörensen buffer
73 ml distilled water.

Other equipment

3 bottles with ground glass stoppers for making up the buffer
1 pH meter and accessory equipment (distilled water, beaker)
A volumetric flask
A magnetic mixer (optional). (Gallenkamp)
A conical flask (size according to volume or buffer made up)
Measuring cylinder and pipette
Water bath set up at 80°C
Coplin jars or other glass staining dishes (Gallenkamp)
Forceps for removing slides from jars after incubation
Wash bottle containing distilled H$_2$O

DePex mounting medium in suitable dispensing bottle (Gurr)
Clean cover slips
Bunsen burner (only when alternative method is used)
Acetic orcein stain (2% lacto—acetic orcein)

2 g orcein (Gurr))	
50 ml acetic acid)	stir and warm
50 ml lactic acid)	

Microscope with x100 oil-immersion objective
Culture of diploid animal or human cells (Biocult or Wellcome)

Type of group
Individual students

Instructions to the student
1 Chromosome preparations are obtained from either human or mammalian peripheral blood or monolayer cultures which are harvested by standard procedures, *without* the addition of colchicine
2 The chromosomes of the cells are fixed in suspension in ethanol—acetic (3:1) mixture, and slides made by air drying.
3 They are then placed in histological jars containing Sörensen buffer (pH 6·8), and incubated at 80°C for 1½ hr.
4 The slides are then stained for 6 min in a solution containing 7 ml Giemsa stock and 20 ml Sörensen buffer made up to 100 ml with distilled water. (Temporary preparations can be made using acetic—orcein stain.)
5 After a brief rinse in distilled water the slides are dried and mounted.

Discussion
The results will show that all the *prometaphase* spreads are consistently banded (persistent 'knobs' of heterochromatic regions along the length of the chromosomes). Because of this consistent observation we omit the addition of colchicine before harvesting, since this causes certain adverse condensation effects resulting in less discernible banding patterns. Similar bands can be obtained from our routine cell-suspension cultures (harvested without colchicine) by just dropping onto a hot slide and fiercely but carefully heating to dryness. The slides are then stained as above for 6 min. This procedure is a simple adaptation of our routine chromosome staining and has been found to be very important in the study of banding patterns of chromosomes.

By using the technique described above, we

obtained reproducible results and further found that the heating process is important in bringing them about. It has been noted that heat is the essential factor causing the banding patterns; a process termed as differential denaturation. We also consider the ommission of colchicine to be important in producing banding because of the less condensed nature of the chromosomes.

The techniques for identifying mammalian chromosomes by means of their fluorescent bands have improved so much in the last few years that recommendations have now been made that, in the future, the standard for identifying human and other mammalian chromosomes should be based on the fluorescent karyotype. Similarly, several Giemsa staining procedures which preferentially stain centromeric heterochromatin have been described and as a result examination of the resultant banding patterns in human chromosomes appear identical to those given by the fluorescent techniques.

The results that are now accumulating by these techniques not only give some insight into the organisation of mammalian DNA and chromosomes, but also provide a simple technique whereby chromosomes could be identified with more certainty. The techniques make use of that property of chromosomes which have regions of highly repetitive DNA sequences, to show up on subsequent staining.

Extensions and applications

One of the greatest problems confronting the cyto-geneticist today, is identification of chromosomes, especially the occurrence of abnormal markers which occur in various congenital and neoplastic conditions. Marker chromosomes occur in a variety of conditions and if it can be shown through fluorescence and the banding techniques that these markers are basically the same, then a very important breakthrough would have been made regarding the relationship of marker chromosomes to the malignant state. This would be especially so where marker chromosomes are evident in pre-malignant states.

Further reading

Drets, M.E. and Shaw, M.W.
    1971 'Technique for Demonstrating Banding Patterns of Chromosomes', Proc. Nat. Acad. Sci. US, 68, 2073–5
Jones, K.W. and Corneo, G.
    1971 'Location of Satellite and Homogenous DNA Sequences on Human Chromosomes', Nature (London) New Biology, 233, 268–71
Mammalian Chromosomes Newsletter,
    1972 Volume 13, No. 1 or consult authors.
Pardue, M.D., and Gall, J.G.
    1970 'Chromosome Localisation of Mouse Satellite DNA', Science, 168, 1356–8
Pearson, P.
    1972 'The Use of New Staining Techniques for Human Chromosome Identification', Journal of Medical Genetics, 9, 264
Sumner, A.T., Evans, H.J., and Buckland R.
    1971 'New Technique for Distinguishing between Human Chromosomes', Nature (London) New Biology, 232, 31–2

N.P. BISHUN
D.C. WILLIAMS
P. DOYLE
Marie Curie Memorial Foundation.

The cytology of *Cepaea nemoralis* with emphasis on recording chiasma frequency

Time required
Two 3 hr practicals (+ discussion period)

Assumed knowledge
1 The mechanics and significance of meiosis and mitosis
2 Elementary chromosome methodology: squashes, staining, and the use of the light microscope
3 Familiarity with the internal anatomy of pulmonate snails is useful for the dissection
4 Elementary statistics: calculation of $\bar{x}$, s^2, comparison of means.

Theoretical background
The material
Cepaea nemoralis is a hermaphrodite colony-forming terrestrial snail, the panmictic unit (breeding population) usually being taken to be 30 metres linear. The animal is ecogenetically interesting because of the well developed colour and banding polymorphism of the shell. There are at least 9 loci involved in the polymorphism; 6 are more or less strongly linked. The most important of these loci are the C locus (colour) of which there are probably 9 alleles and the B locus (banding) of which there are 2 alleles. There is considerable intercolony variation in the proportions of the different phenotypes and in the degree of linkage disequilibria; many populations show an excess of a particular phenotype indicative of strong linkage. Monomorphic colonies are rare.

Many polymorphisms entail linkage between the relevant loci between which intra-chromosomal recombination by crossing over is reduced. It is therefore reasonable to examine the chromosomes of *Cepaea* to determine whether there is any obvious cytological basis for strong linkage. The haploid chromosome number in *Cepaea nemoralis* is 22; there is one large chromosome (the A chromosome) and 21 very much smaller chromosomes. It is not known upon which chromosome lie the loci for the shell polymorphism linkage group. The most suitable cells for the study of the chromosomes are dividing spermatogonia (for mitosis) and spermatocytes (for meiosis). Metaphase is the best stage in mitosis at which to observe chromosome morphology, and diplotene and early diakinesis are the best meiotic stages at which to observe the frequency (and distribution) of chiasmata.

The method
1 Two alternative schedules are given; which is used will depend on the equipment available and the time of year.
2 Pretreatment of dividing cells. Hypotonic solutions such as 0·42% KCl cause cytoplasmic swelling so that the chromosomes or bivalents become separated from each other. KCl has the additional property of preventing coagulation of sperm so that these can be separated from dividing cells by centrifugation (Schedule A), or by repeated washing (Schedule B).
3 Fixation of the cells is carried out with freshly prepared 3:1 absolute ethanol–glacial acetic acid.

Necessary equipment
For the dissection
Cepaea (1 per student) plus a few spare
Kleenex tissues
Dissecting scissors (1 per student)
Fine scissors (1 per student)
Scalpel with no.11 blade (1 per student)
Pins (10 per student)
Waxed dish—not less than 8 x 10 cm, with 1–2 cm. depth of wax (1 per student)
Mollusc saline (200 ml per student)
Low power binocular microscope with bench lamp (1 per student)
Solid watch glass (1 per student)

For both Schedules
Pasteur pipettes 5¾in (2 per student)
0·42% KCl (50ml per student)
100 ml measuring cylinder (1 per pair of students)
250 ml beaker (2 per student)
Absolute ethanol (100 ml per pair of students)
Glacial acetic acid (100 ml per pair of students)
Acid 70% alcohol for cleaning slides (1 litre per class)
   3 ml HCL per 100 ml 70% alcohol
Diamond pencil (1 per 6 students)
76mm x 38mm slides (4 per student)
70mm x 35mm cover slips (4 per student)
Michrome essence (enough to fill one large-sized staining jar per 4 students)

Michrome Mountant and glass dropper (1 per student)

Staining rack large enough to hold 16 slides (1 per 4 students)

Light microscope with mechanical stage and immersion objective. (one per student)

For Schedule A

Centrifuge (1 per 2 students is ideal, but 1 shared between 4 is adequate) Both the Gallenkamp semi-micro centrifuge CF–200 and the Griffin–Christ S17–730 Simplex II fitted with S17–718/01S adaptors are suitable.

2 ml centrifuge tubes (Gallenkamp CJ–005) (2 per student)

Mounted needles (2 per student)

Small piece of polystyrene or plasticine in which to stand centrifuge tubes (1 per student)

5% lactic-propionic orcein

45% acetic acid

45% acetic acid–cellosolve 1:1

Cellosolve (2 lots) (BDH)

Cellosolve-michrome essence (Gurr) 1:1

Set of large staining jars to take slide rack large enough to hold 16 slides (1 set per 4 students)

Lacto-proprionic orcein–Orcein 2 g, Lactic acid 50 ml, Proprionic acid 50 ml forms stock solution which is diluted as stock 45 ml: distilled water, 55 ml for working solution (Dyer, A.F. Stain Technology 38, 85, 1963)

For Schedule B

Hot plate (1 per pair of students)

Watchmaker's forceps (2 pairs per student) (One pair of fine forceps and a mounted needle can be used instead but are not as effective.)

60% acetic acid (50 ml per student)

Gurr's R66 Giemsa (Make up a 10% solution at pH 6·8 from stock bottle) (enough to fill one large staining jar per 4 students)

Gurr's pH 6·8 Buffer tablets

Solid watch glasses (6 per student)

Covers for solid watch glass (1 per student)

1 ml centrifuge tube (Gallenkamp CJ–005) (1 each)

Rubber bung for centrifuge tube (1 per student)

Type of group

Individual students in a large class.

Instructions to the student

Removal of the shell and ovotestis (15–20 min)

1 Select an adult animal and dry the shell with

Kleenex tissues to make it easier to hold.

2 Using scissors, cut through the upper limit of the lip where it joins the body of the shell and continue the cut round the whorls, breaking the shell away from the underlying soft parts by hand. Remove the entire shell taking care not to tear the soft parts during removal of the columella and the last part of the shell. (It does not matter if you cannot remove all the columella but it makes the next stage easier if you do so.

3 With the visceral hump uppermost, secure the animal in a waxed dish with one pin through the head and another placed so that the visceral hump is secure. Avoid putting pins through the tip of the visceral hump or you will damage the gonad.

4 Cover the animal with mollusc saline. Place the dish under a low-power binocular microscope, arrange a good source of light, and focus. With a 3rd pin uncoil as much of the visceral hump as you can so that the columellar muscle (on the inner surface of the visceral hump) is exposed.

5 Cut along the junction between the upper edge of the columellar muscle and the mantle with a scalpel so that the digestive gland is exposed. Use more pins to secure the hump as you do this. Attempt to keep the scalpel blade parallel with the surface of the animal so that you do not cut into the digestive gland.

6 The ovotestis can be identified by its pale colour and translucent appearance (the digestive gland is brown). The gonad opens into a tightly coiled hermaphrodite duct which is nearly always full of sperm. Take the hermaphrodite duct with fine forceps and dissect away the gonad with a scalpel. Transfer the gonad to a solid watch glass containing 2·5 ml mollusc saline. With fine forceps cut through the hermaphrodite duct where it joins the gonad.

Pretreatment, fixation and preparation of the slides

Schedule A (2–2½ hr)

This scheme involves teasing the gonad in saline, changing the fluids by centrifugation, and making air-dried films on cold slides.

1 With mounted needles tease the gonad into as many small pieces as possible so that the cells go into suspension. It takes at least 20 minutes to release sufficient cells to make several slides. It is preferable to tease only a part of the gonad really well than to break up the whole gonad only roughly.

2 With a pipette, transfer the cells to two 2 ml centrifuge tubes, ensuring that the amount of fluid in each tube is approximately the same. Resist the temptation to transfer unteased pieces of tissue; take only cells in suspension into the centrifuge tubes.

3 Spin the tubes for approx. 30 s at the lowest speed on the centrifuge.

4 The cells will have been thrown down into a small pellet; remove and discard the supernatant fluid with a pipette. (You can check that this contains mainly sperm by mounting a few drops on a slide.)

5 Add 0·42% KCl so that each tube is half full. Resuspend the cells by careful flushing with the pipette. Stand for 10-15 min. Make up the fixative during this time; pour about 50 ml into a small beaker.

6 Centrifuge for 30 s and remove most of the KCl. Leave just enough so that the cells can be resuspended by very careful flushing.

7 Using a clean pipette add fixative drop by drop to the cells in suspension and then more quickly until the tube is nearly full. Flush the cells quickly to prevent clumping and repeat on the other tube. Centrifuge for 20 s immediately and resuspend the cells in fresh fixative.

8 Change the fixative again. Allow the cells to stand for 30 min in suspension. Mark four slides with a diamond pencil so that you can identify them later. Clean the slides with acid 70% alcohol and put them in a row on the bench. Revise your knowledge of meiosis and mitosis.

9 To make air-dried films take up some of the cells in suspension into a pipette and allow 2 drops to fall cleanly on to each slide from a height of about 5 cm so that the drops do not overlap. The fixative will evaporate leaving the cells stuck to the slide. Repeat until all the fluid has been used up. Do not rush this stage by adding more fluid before the previous drops have evaporated.

10 Stain the slides in 5% lacto-proprionic orcein for 30 min

11 Rinse in 45% acetic acid for 20 s.

12 Rinse in 45% acetic acid–cellosolve 1:1 for 20 s.

13 Rinse in 2 changes of pure cellosolve for 10 sec each.

14 Rinse in cellosolve–michrome essence 1:1 for 10 s.

15 Transfer to michrome essence.

For stages 10–15 collaborate with several other students and use a staining rack with the reagents in large staining jars.

16 Check which side of the slide the film is on and mount with michrome under a clean cover slip.

Schedule B (2–2½ hrs)

This scheme involves tearing the gonad into pieces in hypotonic KCl, releasing the cells in 60% acetic acid, and making air-dried films on warm slides.

1 Transfer the gonad to 2·5 ml 0·42% KCl in a solid watch glass and pull it into about a dozen pieces using 2 pairs of watchmakers' forceps.

2 Using the forceps, transfer the pieces to 0·42% KCl in a second watch glass and tear into still smaller pieces. Do not attempt to *tease* the gonad.

3 Transfer the pieces to a third watch glass containing 0·42% KCl, where they are left until the total time in the KCl reaches 15 min.

By using the 3 changes of KCl tubule fragments and most of the sperm are left behind.

4 Make up fixative and pour some into two watch glasses.

5 Take each piece of gonad in turn and using the fine forceps dip each piece into fixative and then transfer it immediately to the second watch glass. Repeat with the other pieces of gonad.

6 When all the pieces of gonad have been transferred, break them into yet smaller pieces, the aim being at this stage not to separate individual cells but to ensure quick penetration of the fixative.

7 Cover the watch glass with a square of glass and leave for 45 min. Mark 4 slides with a diamond pencil so that you can identify them later and put them on the hot plate. Check that the hot plate is at 60°C, and also that it is absolutely level. Revise your knowledge of meiosis and mitosis.

8 Transfer 3–4 pieces of tissue to an empty 1 ml centrifuge tube. Add 0·75 ml 60% acetic acid and stopper the tube with a rubber bung.

9 Shake the tube vigorously and/or tap it on the side of the bench for 20 s. The pieces of tissue will become transparent and most of them break up so that the cells fall into suspension. Remove and discard any remaining lumps of tissue.

10 Take up some of the suspension and put 3–4 drops on one of the slides. Immediately suck the drops back into the pipette and put them on another part of the slide or on a different slide. The logic of this treatment is that the cells in suspension migrate to the edges of the drop so that as the fluid is withdrawn the cells are air dried on to the slide. Repeat the process

several times and discard the drop.

It is important that stages 9 and 10 are completed within 5 min. After this time the acetic acid may cause disintegration of the cells.

11 Repeat the procedure using more gonad taken from fixative. Care should be taken not to put too much material on the slide; if the cells overlap, detail will be obscured. It takes about 15 min to prepare 4 slides.

12 Stain the slides for 20 mins in Giemsa R66 which has been diluted to a 10% solution with distilled water buffered at pH 6·8.

13 Rinse the slides thoroughly in running tap water for about 10 s.

14 Put the slides on filter paper (film uppermost) and carefully blot them dry with Kleenex tissue. Let the slides air dry.

15 Put the slides in michrome essence for 5−10 min and mount them in michrome under clean cover slips.

Recording results
Identification of stages in cell division
Confirmation of the chromosome number can be obtained from mitotic metaphases (PlateI,1) and from diplotene or diakinetic cells (Plate I, 3−6). Cells in leptotene and zygotene are usually abundant and easy to identify (Plate I,2). Oil immersion is required to see the chromomeres and the pairing of the chromosomes. At these stages it is not possible to distinguish between the different bivalents. Diplotene and diakinetic cells can be identified even with low-power objectives. In both types of cell the frequency and distribution of chiasmata can be worked out using oil-immersion objectives.

Variation in chiasma frequency
Use only diplotene or early diakinetic cells for this as the contraction of the chromosomes in late diakinesis makes scoring unreliable. The 21 small bivalents nearly always show only 1 chiasma. The large A group bivalent shows variation between cells in one animal, between different animals in the same population, and between different populations. Up to 6 chiasmata have been recorded in this bivalent.

1 Examine the slide systematically starting at one corner using the low-power objective. Transfer to oil-immersion objective when a suitable cell is seen.

2 Score the number of chiasmata in all the bivalents in 30 different cells, ensuring that you do not score any overlaps in the A-group bivalent (Plate I, 5) as chiasmata (Xta). Record the data for the A-group bivalent separately from the others, as shown in Table 1.

Table 1

Cell No.	Total Nos. of Xta in the 21 small bivalents	Nos. of Xta in the A-group bivalent
1	21	1
2	22	3
3	21	2
etc.	etc.	etc.

3 Collate the data for the entire class as shown in Table 2.

4 If 2 populations have been studied it is possible to compare inter-colony variation in chiasma frequency in the A-group bivalent. Calculate the mean and variance for each population for the

Table 2 Population I

Animal No.	Nos. of cells examined	Total Xta in all bivalents	Mean Xta per cell	Analysis of A-group bivalent							
				Total Xta	Mean Xta per cell	Nos. of cells with the given Nos. of Xta					
						1	2	3	4	5	6
1	30	730	24·34	102	3·40	0	1	18	9	2	0
2	30	683	22·77	53	1·77	8	21	1	0	0	0
17	510	11908	396·93	1140	38·00	105	225	141	33	6	0
Means	30	700·47	23·35	67·06	2·23	6·18	13·2	8·29	1·94	0·35	0

82

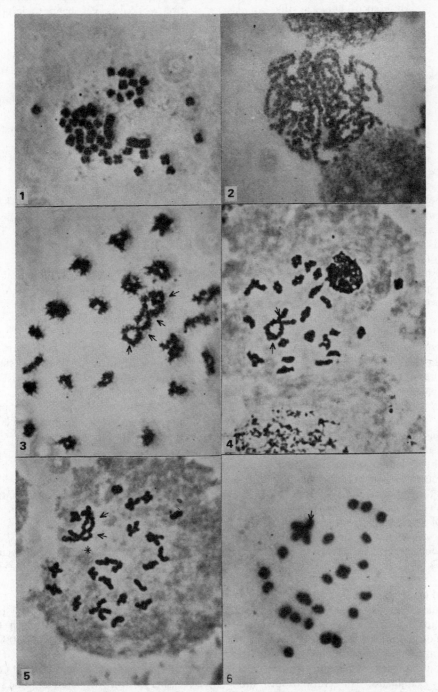

Plate 1 The chromosomes of *Cepaea memoralis*. **The positions** of chiasmata are indicated **by arrows**. The asterisk indicates an overlap. 1 Mitotic metaphase. **The large A** chromosomes are conspicuous. 2 Zygotene. Chromomeres and the pairing of the chromosomes are visible. 3 Early diplotene. The A-group bivalent has 4 chiasmata. 4 Diplotene/early diakinesis. The A-group bivalent has 2 chiasmata. 5 Diplotene/early diakinesis. The A-group bivalent has 2 chiasmata and one overlap. 6 Late diakinesis. The A-group bivalent has one chiasma.

results from 30 cells per animal. Use a *t*-test to compare the differences between the populations. The results from 3 different populations are shown in Table 3.

Table 3.

Population	Nos. of animals examined	Mean Nos. of Xta in the A-group bivalent in 30 cells per animal	Variance
I	17	67·06	273·0
II	20	50·55	75·0
III	16	38·19	91·5

Discussion

1 The purpose of the study was to examine the chromosomes of *Cepaea* to determine whether there was any obvious cytological basis for strong linkage. Do you see any? Do you find any evidence that chiasmata are restricted in number and/or position in any way?

2 If you have compared chiasma frequency between different populations, do you find significant inter-colony variation in the A-group bivalent? If so, does this relate to
   (a) the size of the population?
   (b) the degree of polymorphism and linkage disequilibria shown by the population?

The 3 populations recorded above are ecogenetically very different. Population III is very dense (2 adults per m^2), is highly polymorphic and shows strong linkage disequilibrium in that yellow shell colour is strongly linked to unbanding. Population is equally dense but is monomorphic for colour. Population I is very much less dense (1 adult per 4m^2) and is monomorphic for colour.

3 What do you know of the heritability and stability of Xta frequency?

4 What do you know of any experimental procedure which result in alteration of chiasma frequency?

Extensions and applications

1 The techniques are of wide application in cell biology since they can be adapted to the study of the chromosomes of a wide variety of organisms. They permit the treatment of chromosomes after the films have been made and allow a more thorough examination of chromosome morphology to be made than can be carried out using simple squash techniques.

2 It is usually held that chiasma formation results in genetic recombination. Although natural selection may act on chiasma frequency just as on any other phenotypic variable, insufficient comparative work has yet been undertaken on inter-population variation for any firm conclusions to be reached. Shaw (1972) considers that in some cases variation in chiasma frequency may be of neutral adaptive significance. What evidence would *you* accept that a character was adaptively neutral? The data for the 3 populations presented here suggest that small populations of *Cepaea* show increased recombination compared with large polymorphic ones.

Further reading

On the material

Cain, A.J., Sheppard, P.M., and King, J.M.B.
1968 'The Genetics of some Morphs and Varieties of *Cepaea nemoralis* (L)', Phil. Trans. R. Soc Lond. B 253, 383—96

Williamson, M.H.
1972 'The Analysis of Biological Populations', Special Topics in Biology Series, Arnold, London

On the techniques

Luciani, J.M., Devictor-Vuillet, M., and Stahl, A.
1971 'Hypotonic KC1: an Improved Method of Processing Human Testicular Tissue for Meiotic Chromosomes', Clinical Genetics II, 32—6

Meredith, R.
1969 'A Simple Method for Preparing Meiotic Chromosomes from Mammalian Testes', Chromosoma, 26, 254—58

On chiasmata

Bantock, C.R.
1972 'Localisation of Chiasmata in *Cepaea nemoralis* (L)', Heredity, 29, 213—221

Henderson, S.A.
1969 'Chromosome Pairing, Chiasmata and Crossing Over', In 'Handbook of Molecular Cytology' (ed. A. Lima-de-Faria), 15, 326—57, North Holland, Amsterdam

Hewitt, G.M.
1965 'Population Cytology of British Grasshoppers II, Annual Variation in Chiasma Frequency', Chromosoma, 16, 579—600

Shaw, D.D.
1972 'Genetic and Environmental Components of Chiasma Control. II, The Response to Selection in *Shistocerca*', Chromosoma, 37, 297—308

Zarchi, Y. et al.
1972 'Chiasmata and the Breeding System in Wild Populations of Diploid Wheats', Chromosoma, 38, 77—94

Hints to the teacher

Collecting of Cepaea

If it is intended merely to observe the chromosomes and score chiasma frequency in one population, then the origin of the snails is not particularly important. If the class is large enough it is more interesting to use individuals originating from different colonies.

Maintenance of Cepaea

30–40 snails can be kept indefinitely in a large plastic box 16 x 25 x 10 cm with a few holes drilled in the lid. Moist (but not wet) filter paper should be put into the box and the snails fed on carrot. Natural chalk should always be available. Boxes need to be cleaned out about once a week.

Activity of the gonads and the alternative schedules

The cycle of events in the wild appears to be as follows. Mitotic divisions occur throughout the year but at a higher frequency in the summer months. The bulk of the meiotic divisions occur between May and August; after this time the gonad is full of sperm and there are fewer divisions. The sperm passes into the reproductive tract during the winter. Animals collected in the winter can be brought on by keeping them at room temperature so that meiotic cells are available from February onwards. Even during the summer the yield of dividing cells is higher in material that has been maintained in culture for a few weeks; sometimes more than 150 diplotene cells can be found on slides prepared from one animal.

Schedule B can be used at any time, and is the only procedure which can be used during the late summer and autumn when the gonad is full of sperm.

Dissection

Discard the animal if the gonad is dark brown in colour; such animals rarely have many divisions.

Reagents

1 Molluscan saline:

5·7 g NaCl		Store at 4°C and
0·15g KCl	per litre.	allow to come to room
1·11g CaCl$_2$		temperature before use.

2 Giemsa

This is best diluted just before the class starts; it seems to deteriorate after a few days. It is slightly less permanent than orcein.

The schedules

1 It is possible to interrupt both schemes if this is necessary. Fixed tissues or cells in suspension can be stored indefinitely in stoppered centrifuge tubes. Store at 4°C. Slides can be stored indefinitely before staining; keep them free of dust.

2 Impress on students the undesirability of attempting to make slides from the whole gonad.

3 With Schedule A the centrifuge speeds are not all critical provided the cells are thrown down.

4 It is possible to tease the gonad in 0·04% colchicine in saline with Schedule A to see if there is improved chromatid separation, though there does not appear to be any consistent effect of colchicine.

Alternative emphasis

1 Scores of the distribution of chiasmata as well as their frequency are interesting but unreliable in a class of students since it is difficult to standardise a scoring method.

2 4N, 6N and 8N cells are not uncommon on most slides. Since the A chromosomes are so large it is easy to assess the degree of ploidy from different metaphases.

3 Chromosome morphology is difficult to study in Cepaea since most of the chromosomes are very small and not many slides will have really suitable metaphases. Discourage students from drawing chromosomes as the results are invariably valueless. Ideally good preparations can be photographed and idiograms prepared by cutting the chromosomes out and attempting to arrange them in pairs. This could be quite a useful exercise if one half of a class worked with C. nemoralis and the other half with the very similar C. hortensis. There is very little difference in chromosome morphology, and none in chromosome number.

Other organisms

Other molluscs could be used, e.g. Nucella and possibly Littorina. For marine snails use sea water to dissect rather than mollusc saline; ½ strength seawater rather than hypotonic KCl; and use schedule B, since in Nucella sperm are very large.

C.R. BANTOCK
D.J. PRICE
Polytechnic of North London.

Section 6

Viruses

1 Separation of viruses by sucrose gradient centrifugation

Time required
First day, 1 hr (demonstration, fractionation, assay).
Second day, 1 hr (collection of results, preparation and examination of electron microscope grids).

Assumed knowledge
Accurate pipetting ability.

Theoretical background
Since most viruses have a regular shape and size they have characteristic sedimentation rates. This property can be exploited in purifying or identifying them. The virus suspension is layered in a narrow band on top of a centrifuge tube containing a sucrose gradient, i.e. a steadily increasing concentration of sucrose from top to bottom of the tube. The purpose of the gradient is simply to stabilise the column of liquid and reduce the risk of disturbance by convection or jolting. The gradient is then centrifuged in an ultracentrifuge and the virus particles move in a band down the tube. Impurities will, with luck, move at different speeds so that by fractionating the contents of the tube a band of relatively pure virus can be obtained.

Calculation of approximate S-value
The sedimentation coefficient, S is defined as the speed of sedimentation in water at 20° under the influence of unit centrifugal field.

The centrifugal field is $r\omega^2$ cm/s² if r = distance in cm from the centre of rotation and ω is the angular velocity in rad/s. With this particular centrifuge rotor (Beckman SW 50·1) r = 8·3 cm at the centre of the tube and at 40 000 rev/min ω = 2π x 40 000/60 rad/s, so the centrifugal field = $1·45$ x 10^8 cm/s².

The speed of sedimentation is inversely proportional to the viscosity of the solution, which for 13% sucrose is 1·45 times that of water. Therefore

$$S = \frac{\text{distance moved x } 1·45}{\text{time} \quad \text{x} \quad 1·45 \text{ x } 10^8}$$

It is usually multiplied by 10^{13} to give a more manageable answer.

With this fractionator and recorder 1 cm on the chart = 0·53 cm down the centrifuge tube, and the time of centrifugation was 75 x 60 s, so

$$S = \frac{\text{distance on chart x } 0·53}{75 \times 60} \times 10^{8} \times 10^{13}$$

$$= \text{distance on chart x 12.}$$

Necessary equipment

1. Ultracentrifuge and swinging-bucket rotor, e.g. Beckman SW 50.1.
2. Sucrose gradient-making device, e.g. Britten and Roberts, Science 131 32 (1960) or Fenwick, Biochem. J. 107, 851 (1968) (MSE manufacture such a device)
3. Phages, e.g. RNA phage R17 and filamentous DNA phage ZJ2.
4. Gradient-fractionating apparatus, e.g. ISCO, Model 180 (Shandon) with automatic monitoring of optical density
5. Electron microscope
6. For each student: 1st day
   4 bottles containing 9·9 ml of sterile broth (Difco, Oxoid)
   4 sterile pipettes capable of measuring 0·1 ml
   2 sterile Pasteur pipettes delivering drops of approx. 0·025 ml
   2 Petri dishes of dried nutrient agar (Difco, Oxoid)
   2 ml of *E. coli* Hfr for assay
7. For each student: 2nd day
   1 E.M. grid with carbon coated formvar film (TAAB).
   1 Petri dish with piten paper disc to receive grid.
   Filter paper to drain grid.
   Access to fine forceps, 1% uranyl acetate solution (BDH).

Type of group

Individual students in large classes (results combined).

Instructions to the student

Gradient centrifugation

This will be used to separate two bacteriophages of different size and shape. The preparation of gradients will be demonstrated and 0·2 ml of a mixture of phages R17 and ZJ2 will be layered on a 5 ml gradient of 30 to 8% sucrose containing 0·01 M tris buffer, pH 7·2.

After centrifugation at 20° for 75 min at 40 000 rev/min (centrifugal force 160 000 g at the centre of the tube) the gradient will be fractionated with an automatic fractionator. The contents of the centrifuge tube are passed slowly through a spectophometer cell with quartz windows while the optical density is continuously monitored by a photocell and ultraviolet lamp and recorded on a chart. The positions of the bands of virus particles will be revealed since nucleic acids have a high absorbance at 260 nm. The monitored effluent is then collected in a series of tubes.

Each person will be given one fraction of the gradient, diluted with 1 ml of water, for titration of virus by plaque assay on *E. coli* to confirm the optical density recording.

Plaque assay

Label four bottles of broth 10^{-2}, 10^{-4}, 10^{-6}, 10^{-8}. Pipette 0·1 ml of gradient fraction into the first bottle containing 9·9 ml sterile broth with a sterile pipette and mix well. With a fresh pipette transfer 0·1 ml of this to the next bottle and so on until a dilution of 10^{-8} has been achieved.

Prepare two Petri dishes of nutrient agar by seeding them with lawns of bacteria as follows. With a Pasteur pipette apply about 2 ml of a suspension of *E. coli* to one dish and rock it to spread the bacteria evenly over the surface of the agar. Then tilt the dish, remove excess liquid with the pipette and apply it to the second plate, again draining off the excess. Leave the plates for at least 5 minutes level on the bench with the lids slightly open to dry the surface. Write the number of your fraction on the underside of each dish, divide it into three sectors and label them -4, -6, -8.

With a sterile Pasteur pipette which has been cut to form drops of approx. 0·025 ml, take a sample of the phage in the last dilution bottle (10^{-8}). Holding the pipette vertically, allow one drop to fall on the appropriate sector of each of the duplicate plates. The plates may be placed on a levelling table to avoid too much spreading of the drops. Empty the pipette and use it to place drops of the 10^{-6} and 10^{-4} dilutions in turn on each plate. Always work up the concentration range, not down. Leave the plates until the liquid has soaked into the agar (5–10 min) and then place them in an oven at 37°C for overnight incubation.

On the next day count the plaques and calculate the titre of the original gradient fraction in p.f.u. (plaque forming units)/ml. The fractions containing the highest concentrations of phage will be examined in the electron microscope.

Negative staining and electron microscopy
Hold an E.M. grid in a pair of fine forceps and with a Pasteur pipette place upon it a small drop from one of the selected sucrose gradient fractions. Tilting the grid over a beaker or sink, superimpose four successive drops of 1% uranyl acetate. Drain off the excess liquid with filter paper and allow the grid to dry. Examine it in the E.M. Those virus particles that remain stuck to the grid will be surrounded by the uranium salt, which is opaque to the electron beam. The electron-transparent particles stand out against a dark background–'negative stain'. Examine also someone else's sample from the other peak.

Recording results
Take a copy of the chart recording of the optical density in the sucrose gradient and calculate the approximate sedimentation coefficients of the two peaks.

Enter the results of your plaque assay on a communal table and obtain from it other people's results with the other sucrose gradient fractions. Plot them on the optical density recording, using a linear scale of p.f.u./ml

Take copies of the electron micrographs of the two phages, record the magnifications and determine the approximate dimensions of the particles.

Discussion
The particles of R17 contain about 30% RNA of molecular weight 1×10^6. Those of ZJ2 have 12% of DNA of molecular weight 2×10^6. Although the particle weight of ZJ2 is some five times that of R17, the electron microscope reveals that the faster sedimenting of the two is R17. Why is this?

Could you separate these two phages by conventional centrifugation of a mixed suspension?

Extensions and applications
Sucrose gradient centrifugation is widely used as a preparative or analytical technique for purifying and separating macromolecules or particles of regular size and shape, i.e. of characteristic homogeneous sedimentation rate.

Further reading
Martin and Ames
    1961 J. Biol. Chem., 236, 1372
Noll H.
    1967 Nature, 215, 360

Hints for the teacher
High-titre stocks of the phages are prepared by infecting growing cultures of *E. coli*, Hfr, about 10^8 cells/ml, with about 1 p.f.u./cell and incubating for 4 hr or until prior lysis of the culture with R17, or overnight with ZJ2 (filamentous phages such as ZJ2 do not lyse the cells, and with R17 the extent of lysis is variable).

Bacteria and debris are removed by centrifuging for 10 min at 10 000 rev/min and virus sedimented in 1 hr (R17) or 2 hr (ZJ2) at 50 000 rev/min and resuspended in 1/10 of the original volume. Titres of about 10^{12} p.f.u./ml (R17) and 10^{13} p.f.u./ml (ZJ2) have yielded good optical density peaks using 0·1 ml samples of each on a 5 ml sucrose gradient. The detailed procedure will depend on the technique of measuring optical density. Many variations on the described theme can be devised. *E. coli* may be obtained from National Collection of Industrial Bacteria, Aberdeen.

The bacteria (Hfr) for the plaque assay are grown to about 10^8 cells/ml. Freshly poured plates are unsuitable for this assay. Nutrient agar plates are stored at room temperature for 3-4 weeks. In this extra dry condition they absorb small drops of phage suspension within a few minutes. Nevertheless, a very level surface is desirable to avoid undue spreading. A hole drilled in a metal plate with a number 55 drill forms a convenient gauge for cutting pipettes.

If an automatic gradient fractionator is not available, fractions can be collected manually and the optical density of each measured. Sedimentation coefficients can be estimated from the fraction numbers (see Martin and Ames).

M.L. FENWICK
D.KAY
Oxford.

2 Growth of bacteriophage on bacterial lawns

Time required
3 hr on day of experiment, followed by 1 hr the following morning for counting of plaques.

Assumed knowledge
Sterile technique.

Theoretical background
Phage, or bacteriophage, are viruses which attack bacteria, and they consist essentially of a nucleic acid core within a protein coat. A particular phage is highly specific for the strains of bacteria which it will attack, and particular strains of bacteria will grow up a range of phage types. Within the gut of an omnivorous animal like the pig (and presumably this also applies to man) the coliform bacillus *E. coli* is present in great abundance, as are a range of phages which attack it. A continual evolutionary struggle between evolving bacteria and their phages must occur in the lower gut of such mammals.

If faecal material from the pig is cleansed of its bacteria (chloroform is used for this purpose in this experiment) then we can assay its phage population for various strains of *E. coli*. By growing the bacteria in a lawn, phage will betray their presence by the production of a clear plaque, each plaque resulting from the lysis of millions of bacteria following the multiplication of many phages from one original phage.

This method of phage assay has been utilised widely in studies of phage and bacterial genetics and in assays of water pollution.

Necessary equipment
Per student
8 sterile Petri dishes
1 x 500 ml conical flask, with 500 ml beaker
1 x 250 ml medical flat bottles (Gallenkamp)
10 g Oxoid nutrient agar
Pig faeces extracted with distilled water (see 'Hints for the Teacher')
8 sterile McArtney bottles (Gallenkamp)
2 x 10 ml centrifuge tubes
Bunsen burner
1 x 100 ml beaker
Test tube rack

2 x teats
Plastic covered rack for McArtney bottles
6 agar tablets

In the laboratory
Balance to two decimal places
1 can sterile Pasteur pipettes
Bench centrifuge, x 5
Water bath at 46°C, x 5
Sterile graduated pipettes, 1 ml x 50
Magic markers
Tin foil
Distilled water
Measuring cylinders, 250 ml x 10
100 ml cultures of *E. coli* B and *E. coli* K_{12}, grown in nutrient peptone to maximum concentration.

Type of group
Individual student or pairs. No obvious limitation on class size.

Instructions to the student
1 Preparation of Petri dishes: 6·73 g Oxoid nutrient agar is weighed out, added to 240 ml dist H_2O and left for 15 min to dissolve, then placed in a 500 ml conical flask (capped with foil), and autoclaved at 15 lb for 20 min. After autoclaving while still hot, pour approx. 30 ml into each sterile Petri dish and leave to set. Keep the Petri dishes sterile. Invert the dishes when the agar is hard (after 30 min). Dry the dishes with lids slightly off in 40°C oven for 1 hr.

2 Preparation of soft agar: add 6 agar tablets to a 250 ml medical flat bottle, make up to 100 ml, leave to dissolve for 15 min, autoclave as for nutrient agar, and on removal from autoclave, place in water bath at 46°C to keep agar liquid.

3 Isolation of phage: place approx 2 ml of pig faecal extract in a centrifuge tube, add 1 drop of chloroform, agitate by sucking in and out of Pasteur pipette for 2 min, then centrifuge for 5 min at 500 g.
The object of this exercise is to kill the bacteria in the faeces by chloroform, to retain the phage in a viable state, and to eliminate the chloroform before contact with *E. coli* bacteria.

4 Preparation of inoculum: you are provided with six empty sterile McArtney bottles. Place them in the water bath to warm them up, and to three add 0·2 ml of *E. coli B* and to the other three add 0·2 ml of *E. coli K12*. Mark the strain on the bottle and do the transfer with sterile technique. Add from the bottle of soft agar approx. 10 ml

to each bottle and keep in water bath. Number the bottles and add to one of each strain of bacterium 0·02, 0·05 and 0·1 ml of chloroformed faecal extract by sterile graduated pipette.

5 Pouring the plates: decant 1 lot of 10 ml soft agar + bacteria and phage on to each Petri dish and leave to set. Then incubate at 30°C overnight. The plaques must be examined and scored the following morning. The bacteria grow in the soft agar as a lawn and single phage multiply by cell lysis to produce clear plaques in the bacterial lawn. Score the plates and determine the number of B and K12 strain plaque forming units in the faecal extract.

Recording results
Tabulate the class results for different dilution on the two bacterial strains.

Discussion
Does dilution produce proportionate reduction in plaque number? Why does plaque size vary? Why does the number of plaque forming units vary for the two strains of bacteria?

Extensions and applications
Determination of sewage contamination in the sea has been carried out in this way.

Further reading
Adams, M.H.
1969 'Bacteriophages', Interscience, New York

Hints for the teacher
It is desirable to run through the practical in one's own laboratory a week beforehand. A few students should run control plates (1) without any bacteria or phage inoculum to test the sterility of the agar, (2) with bacteria alone to assay for purity of the bacterial culture, and (3) with chloroformed phage extract alone to test the efficiency of the chloroform treatment.

Bacteria: *E. coliB* and K12 may be obtained from the National Collection of Industrial Bacteria.

Pig faeces are much the best source for the phage and a friendly farmer the best contact. Other faecal material we have tried with little success includes rabbit and dog.

N. MACLEAN
Southampton.

3 Purification of turnip yellow mosaic virus

Time required
This and the following three experiments are intended to take two 3 hr practical classes, separated by a break of at least 1 hr or preferably 1 or more days. Although the four experiments are described separately, they are intended to run concurrently.

Assumed knowledge
It is assumed that the student is reasonably adept in biochemical laboratory procedures. Experience with centrifuges, and if possible an ultraviolet spectrophotometer, would be an advantage.

Theoretical background
The requirements for the successful purification of a plant virus are as follows.
1 An adequate concentration of virus in the initial infected plant material,
2 A difference between the virus particles and the host plant material so that these may be separated
3 That the virus particles are stable during whatever separation procedure is adopted.
The first stage of purification is to produce sap from infected plants by homogenizing the tissue. Commonly plants are homogenized with buffers containing chelating agents such as EDTA or reducing agents such as 2-mercaptoethanol. Sap is then clarified to remove the bulk of the host components, usually by freezing and/or treatment with organic solvents (ethanol, chloroform, butanol, etc.). The clarified extract is then treated to separate the virus particles from the residual contamination. Usually one or more of three methods are used:
1 salting out the virus with ammonium sulphate,
2 precipitating the virus or the host components by altering the pH, and
3 alternate centrifuging at high and low speeds.
The degree of purity of a preparation may be assessed by testing for non-viral components by centrifugation or electrophoresis, or by comparing the absorption spectrum with that of pure virus.

Necessary equipment

Non-consumables

Atomix-type blendor (MSE);
Bench type centrifuge, preferably refrigerated (MSE);
Rough balance
Ice bath
Ultraviolet spectrophotometer, preferably recording, with 4 x 1 cm pathlength cells;
Centrifuge tubes 8 x 50 to 100 ml and 2 x 10 ml;
Filter funnel and stand
Measuring cylinders 3 x 250 ml, 2 x 100 ml, 1 x 50 ml and 1 x 25 ml;
Stoppered flasks; 4 x 250 ml
8 test tubes
Calibrated pipettes; 4 x 1 ml
Calibrated pipettes; 2 x 5 ml

Consumables

2 pieces of muslin, 1 m^2
100 ml 0·01 M sodium phosphate buffer, pH 7
100 ml 90% ethanol (ice cold)
100 ml saturated ammonium sulphate solution
Approximately equal amounts (20 to 40 g) of healthy and TYMV infected chinese cabbage leaf (*Brassica pekinensis*, seed from Thompson and Morgan Ltd., Ipswich.) Infected leaf should be used when most of the leaf area is white, perhaps one month after inoculation
1 ml purified turnip yellow mosaic virus in phosphate buffer at E_{260} = about 0·6 to 0·7

Type of group

Students working in pairs is probably most efficient.

Instructions to the student

Weigh the healthy leaf tissue, place in the blendor and add enough 0·01 M-phosphate buffer to cover the blades (measure the volume used). Reduce the leaf tissue to a thick slurry with several short bursts, and filter the slurry through 4 thicknesses of muslin, squeezing the retained leaf material to extract all the sap. Cool the filtrate in an ice bath for 15 min and add *exactly* 0·3 times the volume of the filtrate of cold 90% ethanol. Mix and keep ice cold. Repeat the procedure with the infected leaf tissue using the same volume of buffer in the blendor and the same apparatus as above (briefly rinsed between extractions). Centrifuge the ethanol-treated extracts in the cold for 10 to 15 min at 3 000 to 5 000 rev/min. Measure the volumes of the supernatant fluids and add to each in a stoppered flask, 0·5 volumes of saturated ammonium sulphate solution. Mix and leave as long as possible at 4 to 6°C. Centrifuge the

resulting suspensions for 15 min at 3000 to 5000 rev/min, decant the supernatant fluids and resuspend the pellets in 4 ml phosphate buffer. Centrifuge these suspensions and retain the supernatant fluid. That from the infected leaf tissue is a preparation of partially purified virus, presumably contaminated with whatever material is in the preparation from the healthy leaf tissue. Prepare 4 ml of a 1/10 dilution in buffer of both preparations and measure the absorbance at 260 nm using the buffer as the reference. Dilute the virus if necessary to E_{260} = 0·3 to 0·9, and determine the absorbance ratios E_{260}/E_{280}, E_{260}/E_{242} and E_{260}/E_{320}. If a recording spectrophotometer is available, record the absorption spectra between 230 nm and 320 nm and calculate the ratios from the traces. Measure the purified TYMV preparation supplied in a similar manner.

Recording results

Compare the absorbance ratios of the purified virus with those of the partially purified virus, and those of the preparation from the healthy tissue.

Discussion

Assess the purity of the virus preparation obtained and discuss possible ways of increasing its purity. Consider the application of this technique to purifying other viruses, in particular why is it sometimes inapplicable? Discuss the limitations of spectrophotometry for assessing the purity of a virus preparation, and possible alternatives.

Extensions and applications

This and subsequent experiments cover some of the main approaches to studying plant viruses. Although plant virus purification is easier than that of animal viruses, many plant viruses have not yet been purified. Thus developing purification methods and biological assay systems for a virus are fundamental to many virus problems. Further, electron microscopy has proved essential in describing virus particle morphology and can often precede purification. The detergent-phenol method is probably the method of choice for isolating RNA from most viruses, and with recent refinements in separation techniques, RNA is being isolated from an increasing number of plant viruses.

Further reading

General textbooks covering these experiments up to an advanced level are as follows.

Bawden, F.C.

1964 'Plant Viruses and Virus Diseases', Ronald Press Co., New York.

Matthews, R.E.F.

1970 'Plant Virology', Academic Press, New York and London.

Specific references are as follows.

Matthews, R.E.F. and Ralph, R.K.

1966 'Turnip Yellow Mosaic Virus', Advances in Virus Research, 12, 273–328.

Ralph, R.K. and Bergquist, P.L.

1967 'Separation of Viruses into Components', in 'Methods in Virology' vol. 2, Eds. Maramorosch K. and Koprowski H. Academic Press, New York and London, pp. 463–545.

Horne, R.W.

1967 'Electron Microscopy of Isolated Virus Particles and their Components', in 'Methods in Virology', vol. 3 Academic Press, London pp. 521–74.

Hints for the teacher

Sources and purification of viruses: Sufficient tobacco rattle virus for preparing negatively stained samples can be obtained from the author. Turnip yellow mosaic virus and tobacco mosaic virus can be obtained as inocula of infected leaf tissue from Dr M.W. Johnson, John Innes Institute, Colney Lane, Norwich, and Prof. E.C. Cocking, Department of Botany, Nottingham University, respectively. TYMV can be kept as an isolate in chinese cabbage or turnip plants. Batches of chinese cabbage can then be inoculated when required, using infected leaf tissue ground in water (Section 6.4). Purified virus can be prepared by several cycles of ammonium sulphate precipitation (this experiment). Matthews and Ralph (1966) (see 'Further reading') discuss several alternative methods. TMV is best kept as frozen sap from infected tobacco (*Nicotiana tabacum*) plants and purified virus can be prepared from such sap by alternate cycles of high and low speed ultracentrifugation. Glasshouse facilities are required for propagating plants, although it is inadvisable to put TMV infected and healthy plants into a glasshouse together as the virus spreads very readily.

M.A. MAYO
Scottish Horticultural Research Institute
Dundee.

4 Infectivity assay of tobacco mosaic virus (TMV)

Time required
See Section 6.3

Assumed knowledge
See Section 6.3

Theoretical background
One essential requirement for any study of th behaviour of a virus is a quantitative assay techniqu The most sensitive and useful techniques common used are serological and infectivity assays. Both ca be highly specific tests for virus in the presence o much contamination, but infectivity assay is the on test of a virus as a biologically competent entity Infectivity is assayed either by determining th highest dilution which will infect a particular plan or by assessing the number of infections induced i an inoculated leaf. Infections can be detected a brown (necrotic) or pale green (chlorotic) spots o the inoculated leaves. In such assays the fundament concept is that the number of infections induced proportional to the concentration of infectious viru in the inoculum. Individual plants and also the leave on any one plant can vary quite markedly in the susceptibility to infection, and, with some viruses, th infectivity may not dilute in proportion to th dilution of the inoculum, that is a multi-hit syste may be operating or inhibitory substances may b present in the undiluted sample. Thus it is necessar to design assay experiments so that statistical analys can be used to minimise errors.

Necessary equipment
Non-consumables
8 test tubes
Calibrated pipettes; 8 x 1 ml., 1 x 2 ml
12 watch glasses

Consumables
12 squares of muslin, 5 cm
600 mesh carborundum (Hopkin and Williams Ltd in a dispenser (a convenient dispenser can be mad by placing carborundum about 2 cm deep in a ja

with several thicknesses of muslin across the top)
Soap for washing hands
3 ml purified tobacco mosaic virus at a known E_{260} of about 0·05
50 ml 0·01 M-sodium phosphate buffer, pH 7
12 *Nicotiana glutinosa* L. plants. These are ready for use when at least two leaves are fully expanded, about 2 to 3 months after sowing. Seed may be obtained from the University of London Botanical Supply Unit, Elm Lodge, Englefield Green, Surrey, or the University Botanic Garden, Cambridge. Further seed can be obtained by allowing plants to fruit.

Type of group
The quantities described are for an individual student.

Instructions to the student
Dust the leaves of the *N. glutinosa* plants with carborundum to obtain a *light* cover (avoid inhaling the dust). Prepare 2 ml each of a 1/10 and a 1/100 dilution of the tobacco mosaic virus supplied using phosphate buffer, pH 7. Transfer the remaining virus and the dilutions to watch glasses and, starting with the most dilute solution, wet a muslin pad with virus and rub the largest two leaves of 4 plants with the pad. Avoid damaging the leaves by excessive pressure or by repeated rubbing. After inoculation wash the leaves with tap water, and then wash your hands to inactivate any virus retained on the skin, and repeat the inoculation with the next solution on four new plants. After 4 days look for brown spots (lesions) of about 2 mm diameter on the inoculated leaves. Virus induced lesions are distinguished from damage due to the rubbing by their regularity.

Recording results
Record the lesion counts from the upper and lower leaves of each plant separately, and plot the results as lesion counts per leaf against the logarithm of the dilution factor. Calculate the mean specific infectivity of the virus preparation used as the number of lesions per leaf induced by a solution containing 1 μg/ml ($E_{260} = 0·003$).

Discussion
Estimate how many virus particles are required to initiate a single infection in *N. glutinosa* leaves assuming a virus particle (molecular weight of 4×10^7 daltons Avogadro's number = 6×10^{23}). Assess the error in this assay technique and consider how different treatments could be distributed among the plants to give a more accurate comparison of their infectivities. Compare the sensitivity of infectivity assay with that of other possible assay methods. Consider how a plot of infectivity against dilution might be used to determine how many types of particle are needed to initiate a particular virus infection.

Extensions and applications
See Section 6.3

Further reading
See Section 6.3

Hints for the teacher
See Section 6.3

M.A. MAYO
Scottish Horticultural Research Institute
Dundee.

5 Preparation of tobacco mosaic virus (TMV) RNA

Weighing bottle
Bottle for waste ether
Pipette fillers to fit all the pipettes used (Gallenkamp)

N.B. Glassware should be very clean to avoid possible nuclease contamination. Autoclaving is recommended but not essential.

Time required
See previous experiment

Assumed knowledge
See previous experiment

Theoretical background
The great majority of the plant viruses which have been studied comprise protein and RNA, in many cases one molecule of RNA and a number of identical molecules of protein. These are arranged around the RNA either in an isometric shell or as a helically coiled rod. A number of treatments break the non-covalent bonds between the protein and the RNA, the commonest being to shake a virus suspension with water-saturated phenol. Others include incubation with detergents, treatment with high salt concentrations, high or low pH, and heat. RNA isolated by such procedures, and containing virtually no protein, retains the ability to infect of the original virus, producing nucleoprotein virus particles in infected plants. Protein would therefore appear to function mainly, if not solely, as a coat protecting the infectious nucleic acid from degradation. Unlike tobacco mosaic virus particles isolated RNA is susceptible to ribonuclease such as that found in plant sap, skin secretions or saliva. Prevention of ribonuclease contamination is thus vital in this experiment.

Necessary equipment
Non-consumables
Bench-type centrifuge, preferably refrigerated
Ultraviolet spectrophotometer, preferably recording, with 4 x 1 cm pathlength silica cells
Ice bath
Ground-glass stoppered tubes; 3 x 50 ml.
Centrifuge tubes; 6 x 50 ml
Separating funnel; 1 x 250 ml
Measuring cylinder; 1 x 25 ml
Flask; 1 x 50 ml
Calibrated pipettes; 6 x 1 ml, 3 x 2 ml, 1 x 5 ml, 5 x 10 ml
6 test tubes
4 watch glasses
Small pestle and mortar

Consumables
12 *Nicotiana glutinosa* plants (see previous experiment)
600 mesh carborundum in a dispenser (see previous experiment)
6 disposable plastic gloves
Twelve 5 cm square muslin pads
About 2 g *healthy* tobacco leaf tissue (*N. tabacum*)
10 ml 0·01 M sodium phosphate buffer, pH 7, containing 0·1% sodium dodecyl sulphate (SDS from BDH specially purified grade)
25 ml water-saturated phenol (A.R.) containing 0·1% 8-hydroxyquinoline (BDH)
50 ml 0·05 M sodium phosphate buffer, pH 8
250 ml A.R. diethyl ether
50 ml A.R. absolute ethanol
Solid sodium chloride
A suspension of tobacco mosaic virus at a *measured* concentration of about 1%—the students require 10 mg per extraction ($E_{260}^{1\%} = 30$)
Soap for washing any phenol burns.

Type of group
Students in pairs is probably best, although the experiment can be done by an individual.

Instructions to the student
N.B. DO NOT mouth pipette any of the solutions used in this experiment. Take 10 mg of tobacco mosaic virus by pipetting an appropriate volume of the measured virus suspension provided into a ground-glass stoppered tube, and make the volume up to 10 ml with 0·1% sodium dodecyl sulphate in 0·01 M-phosphate buffer, pH 7. Add 10 ml water-saturated phenol containing 0·1% 8-hydroxyquinoline and shake the tube vigorously for 5 min (wrap a cloth around the stopper to prevent any possible escape of phenol). Centrifuge the resulting emulsion, in the cold if possible, for 10 min at 3000 to 5000 rev/min and pipette off the phenol phase (lower yellow layer) and discard. All waste phenol should be flushed away with large volumes of cold water. Add 0·3 g sodium chloride and 10 ml fresh phenol solution to the remaining aqueous layer, shake vigorously in a stoppered tube and centrifuge again. Pipette off the

aqueous (upper) phase into a separating funnel. Do not contaminate this solution with either phenol or any of the material collected at the interface; if this happens the mixture should be centrifuged again. Extract the clean aqueous phase 3 times with 5 volumes of cold ether. Collect the lower aqueous phase each time in a flask kept cold in an ice bath. Put the extracted sample in a stoppered tube and precipitate the RNA by thoroughly mixing with 2·5 volumes of absolute ethanol. After at least 1 hr at 4° or at −15° to −20°, centrifuge the suspension for 15 min at 5000 to 10 000 rev/min, preferably at 5°. Decant the supernatant ethanol (with particular care if the centrifugation was at less than 10 000 rev/min), drain the tube and dry any ethanol present on the inside lip of the tube with a tissue. Dissolve the pellet, which should be white and small, in 1·5 ml 0·05 M-phosphate buffer, pH 8, and prepare 2 ml each of a 1/10 and a 1/100 dilution in pH 8 buffer. Grind the tobacco leaf tissue provided in the pestle and mortar in about 4 ml tap water and use the resulting sap to make a second 1/10 dilution of the RNA preparation. Transfer each dilution to a watch glass and inoculate 2 leaves on each of 4 plants of *N. glutinosa* with each solution. Lightly dust the leaves with carborundum before inoculation and wear the plastic gloves. Inoculate by gently rubbing the leaves with a muslin pad wet with RNA solution, and in the order; 1/100, 1/10 and last the 1/10 in sap. Do not put dirty pads in the inoculum and use a new pad for each solution. Wash the leaves after inoculation with tap water. Prepare 4 ml of a 1/5 dilution of the RNA preparation in pH 8 buffer and measure the absorbance at 260 nm. Adjust the dilution so that $E_{260} = 0\cdot3$ to $0\cdot9$ and measure the absorbance ratios E_{260}/E_{280} and E_{260}/E_{235}, or if a recording spectrophotometer is available, record the absorption spectrum from 210 nm to 320 nm and calculate the ratios from the trace obtained.

Phenol burns should be washed *immediately* with soap and water.

Recording results
Calculate the yield of RNA, assuming that TMV comprises 5% RNA by weight, and $E_{260}^{1\%}$ for pure RNA = 250. Calculate the specific infectivity of the RNA preparation as the number of lesions induced per leaf at a concentration of 1 μg/ml and compare this with that calculated for virus inocula in the previous experiment. Assess the purity of the RNA preparation by comparison with the absorbance ratios for pure RNA of about 2 for E_{260}/E_{280} and E_{260}/E_{235}.

Discussion
Discuss the difference between the specific infectivities of virus and its constituent RNA, suggest reasons for any discrepancies found. Consider the significance of the finding that protein-free RNA is infective.

Extensions and applications
See Section 6.3

Further reading
See Section 6.3

Hints for the teacher
See Section 6.3

M.A. MAYO
Scottish Horticultural Research Institute
Dundee.

6 Measurement of particle lengths of tobacco mosaic virus (TMV) and tobacco rattle virus

Time required
See previous experiment

Assumed knowledge
See previous experiment

Theoretical background
Virus particles can be observed in an electron microscope by negative staining or sometimes metal shadowing. Negative staining with reagents such as potassium phosphotungstate or uranyl acetate is obtained when a solution is dried down to form an electron-optically dense film around the virus particles, which are then outlined as white areas on the final micrograph print. Two main types of particle structure can be distinguished; rod-shaped particles, either straight rods or flexuous filaments, and isometric particles appearing almost spherical. Rod-shaped viruses are assembled as a helix of RNA with protein sub-units attached to this helix, close packed to form a tube-like structure. Frequently some serration can be detected on the outer surface of the tube, and the centre is often hollow, penetrated by the stain. The length of the virus rods appears to be determined by the length of the nucleic acid helix, and thus rod-shaped viruses which have their genome divided into more than one piece of nucleic acid have more than one characteristic particle length. The simplest isometric viruses are assembled as icosahedrally symmetrical shells of protein molecules and there are therefore a multiple of 60 protein molecules in each shell. Nucleic acid is packed within this shell, although with some isometric viruses nucleic acid-free empty shells are found in purified virus preparations. Isometric viruses containing different amounts of nucleic acid are not clearly distinguishable by electron microscopy, but have differing densities and/or sedimentation coefficients.

Necessary equipment
Prints of electron micrographs of tobacco mosaic virus and tobacco rattle viruses negatively stained with potassium phosphotungstate, pH 7. Prints should include a total of about 500 of each virus particle. Micrographs should be at a plate magnification of about 40 000x and the print magnification should be 4x, with the total magnification stated. An accurately known microscope magnification should be used. Horne (1967) (see 'Further Reading') reviews negative staining techniques. Optional demonstrations could include metal shadowed tobacco rattle virus and negatively stained turnip yellow mosaic virus.

Type of group
Recording particle lengths is much easier in pairs.

Instructions to the student
Measure the prints of electron micrographs of tobacco mosaic and tobacco rattle viruses using a mm scaled ruler. Record the lengths of at least 200 virus particles to the nearest mm. Measure all the particles on each print used. Examine the prints carefully for any evidence of fine structure on the virus particles. Examine any demonstrations available.

Recording results
Plot the results for each virus as a histogram of the frequencies of each particle length. Calculate the actual size of the modal particle lengths using the magnification value provided.

Discussion
Discuss the significance of the difference between the two virus-particle length distributions. Discuss possible ways of demonstrating that a multi-modal distribution reflects a divided genome.

Extensions and applications
See Section 6.3

Further reading
See Section 6.3

Hints for the teacher
See Section 6.3

M.A. MAYO
Scottish Horticultural Research Institute
Dundee.

Section 7

Membranes
and Surfaces

1 The Gorter-Grendel experiment: lipid content of erythrocytes

Time required
One session of 3 hr.

Assumed knowledge
Use of microscope and haemocytometers. Use of separating funnels. A little about membrane properties: the barrier is mainly oil-like, monolayers of lipids spread with the fatty acid chains vertical and head-groups down; lipids are soluble in $CHCl_3$, proteins are not.

Theoretical background
In 1925, Gorter and Grendel knew of the Collander experiment in which it was shown that the molecules which get inside cells have to penetrate a hydophobic phase. They also knew that membranes contain lipids, which are primarily hydrophobic molecules. From Langmuir's work it was clear the lipids form layers one molecule thick on surfaces. Assuming therefore that a membrane has layers of lipid in it, they asked how many layers there might be. However the analytical techniques for identifying and measuring the amounts of the lipids did not then exist.

To get round this problem, Gorter and Grendel extracted the lipid from a known number of cells and spread aliquots on a Langmuir trough. They noted the area at which pressure began to develop and hence were able to calculate tha area of a monolayer formed from the total lipid from one cell. This they compared with the actual surface area of an erythrocyte as measured optically. As we know, they found a 2:1 ratio.

They were very lucky, for the following reasons
1 The extraction was only some 60% efficient.
2 They underestimated the area of the erythrocyte.
3 The surface pressure taken was very arbitrary.

Today we will repeat the Gorter-Grendel experiment, with the following improvements.
1 The extraction will be far more complete.
2 The monolayer area will be measured at a defined and appropriate pressure.
3 The erythrocyte area can be checked by the students, but we will take an area obtained by

97

more sophisticated techniques. (This is a crucial step and difficult to do accurately.)

Necessary equipment
Requirements for each group are as follows.
1 *Human blood*—about 100 ml whole blood. (Blood packs of outdated blood from a blood bank.)
2 *Centrifuge* (MSE Super Minor) *and large tubes* 50 ml at least but larger if possible. Tubes must be suitable for spinning with $CHCl_3$ (i.e., not polycarbonate) glass or polythene will do.
3 10 ml, 100 ml, 250 ml *measuring cylinders*
4 10 ml, 50 ml *volumetric flasks*
5 1 ml, 25 ml *pipettes.*
6 250 ml. flat-bottomed stoppered *flasks.*
7 Beakers, 2 x 50 ml 2 x 250 ml, 1 x 500 ml
8 Filter-pump, 500 ml *Buchner flask* and *Filter-funnels* about 8 cm. Appropriate filter papers (Gallenkamp)
9 *Spatula.* Glass *stopper* with flat end, or large cork for pressing down on filter pad
10 Microscope: eyepiece graticule and stage micrometer.
11 Haemocytometer. (I prefer this to the centrifuge method since it is more direct.)
12 Tally-counter
13 Microlitre syringe: 50 *or* 100 μl (S.G.E., Hamilton)
14 *0·9% NaCl*. Also *distilled water*
15 *$CHCl_3$ and CH_3OH* (A.R.) (BDH)
16 Petri dish ground flat at the edge and coated with paraffin-wax
17 2 Teflon barriers (Klinger)
18 Indicator oil (Edwards Vacuum pump oil)
A note about preparing troughs.
10cm Petri dishes are used. The edges are ground flat and then the dishes are either covered in paraffin-wax (a tedious business) or coated with teflon with an adhesive backing sheet. This can be obtained from Richard Klinger (Klingerit Works, Sidcup, Kent). Care must be taken to keep the edges smooth. A well is made in the centre by stretching the sheet until it sticks to the glass. The resulting trough is readily cleaned with detergent and wiped with tissues.

Type of group
Groups of two.

Instructions for the student, Results and Discussion
Cleaning up the cells
Out-dated human blood from a blood bank will be available. Each group should take about 40 ml of whole blood and wash the cells in 0·9% NaCl (which is roughly isotonic with serum). Dilute the serum with an equal volume of saline and divide into balanced portions in the centrifuge tubes. Centrifuge for 5 min at maximum revs. You should obtain a yellowish-red supernatant layer and a very deep red pellet. *Do not decant* the supernatant as the pellet is not solid enough. Suck off the supernatant at the filter-pumps, using a Pasteur pipette.

Wash the cells with further quantities of 0·9% NaCl in this way until the supernatant is only pinkish. The pellets will have a 'buffy layer' at the very top. This consists of leucocytes, platelets, and the rest, and should also be sucked off if possible.

After cleaning the cells, measure out about 10 ml of the very viscous pellet in a 10 ml graduated cylinder. It is important to have fully 10 ml. If you have less available, then see if another group has some to spare. Wash this into a 50 ml volumetric flask with 0·9% NaCl, and make up to volume.

Counting the cells
Immediately before removing aliquots from this volumetric flask it is *essential* to shake it, because the cells settle out very quickly. Remove 0·5 ml (accurately) with a pipette and transfer to a 10 ml volumetric flask. Make this up to the mark with 0·9% NaCl. This second suspension is dilute enough for counting.

Wipe the haemocytometers clean *and dry*. Mount them in the microscope and focus on the counting squares, with the cover slip in place. Add a drop of the suspension and wait for long enough for the cells to settle out. Decide on a criterion for deciding whether a cell is inside a square or not and count the number inside a small square. I have tried to arrange it so that about 20 cells per square are obtained. If this is not so, then dilute the contents of the 50 ml flask accordingly. If the number is much less than 10/square then *consult a demonstrator*.

The numbers inside the squares will, of course, vary considerably and this limits the accuracy with which the count can be made. It is found that if there are 100 cells altogether, then the expected accuracy is about $\pm \sqrt{100}$ or ± 10 in 100, i.e. 10% accuracy. Count enough squares to give about 400 cells altogether. This should give an error of $\sqrt{400}$, i.e. 20 in 400 or $\pm 5\%$.

Hence find the average number of cells per square. The volume above one square is $\frac{1}{400}mm^3$. Hence if n is the number per square, there are n in $\frac{1}{4} \times 10^{-3} mm^3$

or $\frac{1}{4} \times 10^{-6}$ cm^3 Hence in one cm^3 there are N, where

$$N = 4 \times 10^6 n.$$

Therefore the number of cells in *25 ml* of the larger flask, N_T, is (remembering 0·5 ml from it was diluted to 10 ml)

$$N_T = \text{total number in 25 ml} = 4 \times 10^6 n \times 10 \times \frac{25}{0·5}$$
$$= 2 \times 10^9 n$$

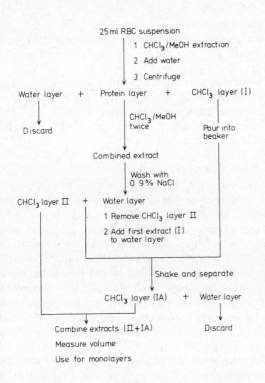

Figure 1 Summary of extraction procedure

Extraction of the lipids (See Fig. 1)
To sort out this maze in your minds try to remember where the lipids (which we are after) have got to; the lipids are always in the *heavier* fraction.

Pipette 25 ml of the suspension into the stoppered bottle. Add to it 50 ml of CH_3OH. Shake the mixture for 1 min. Now add 25 ml $CHCl_3$. Shake it again for 1 min. Finally, add a further 25 ml $CHCl_3$ followed by 25 ml of distilled water. At this stage two liquid phases will separate. Shake the flask for a final 15 s or so. Take enough centrifuge tubes to contain all of the liquid and pour all of the mixture, including the precipitated protein, into the tubes. Do not bother to wash any residue out of the flask; it will be re-extracted later. To ensure equal distribution of $CHCl_3$, etc., pour small amounts into each tube in rotation, swirling the bottle. Level up afterwards with a Pasteur pipette. Centrifuge the mixture by taking it up to maximum revs and then at once allowing it to stop.

A *lower* $CHCl_3$ phase, a compact red interface, and a clear supernatant aqueous phase is obtained. With the spatula, press the rubbery protein layer to one side (without breaking it) and pour both the liquids into a beaker. Pour the liquids into the separating funnel and allow them to settle out. (Make sure the tap is closed.) The *lower* has the lipid in it. Meanwhile, continue with the rest of this procedure. Now transfer the protein layer to the original shaking flask; this can be done in one piece balanced on the spatula. Add to them 50 ml of 1·1 $CHCl_3/CH_3OH$ and shake vigorously.

Prepare the Buchner funnel and flask by putting in a filter-paper and, with suction, make it fit firmly to the funnel by wetting it thoroughly with water and sucking dry. Empty the flask and filter some CH_3OH through the filter, emptying this out finally too. Now filter the protein suspension and press it dry on the pad. Do not discard the extract; it contains more lipid. Put the protein back in the flask and extract it once more with 50 ml of 1:1 $CHCl_3/CH_3OH$. Add this extract to the previous one, making 100 ml *in toto*.

The initial $CHCl_3$ extract in the separating funnel should now be poured off into a beaker. *Keep* the *lower* layer I and discard the rest. (But rather than throwing away any of the lower phase allow some aqueous phase to be added to it). Now pour the $CHCl_3/CH_3OH$ extract of the protein layer into the separating funnel and add to it 100 ml of 0·9% NaCl solution. Shake it thoroughly. The milky suspension so formed is best separated by brief centrifugation as before. Keep the lower phase in a beaker, II, and add the aqueous phase to the previous initial $CHCl_3$ extract I. Now put this I in the separating funnel and shake it with the aqueous phase. Allow the phases to separate and pour off the lower Ia into the same

99

beaker already containing the second and third $CHCl_3$ extracts II. Mix the solutions and pour into a measuring cylinder. Small amounts of water at the top should be ignored when the total volume is measured (approx. 80 ml).

Measurement of surface area

The best method is to use a Langmuir trough, but these are expensive, and comparatively difficult to use. A very simple alternative, and accurate enough for our purpose, is the following.

Teflon-coated Petri dishes with special flat edges are provided. Stand one in the tray provided and fill it with distilled water until the water level is well *above* the edge. Clean the surface with the 'teflon' barriers in the way demonstrated. After about four passages, arrange the light so that light is reflected from the clean surface. Then touch a drop of indicator oil on the surface at its geometrical centre. When you have stopped admiring the colours (see the appendix), allow the oil patch to settle down and, if it seems reasonably centrosymmetric and stable, proceed with the next step; if not, clean the surface and try again.

Fill the Hamilton syringes with the lipid solution and expel four or five times. Fill with solution and, drop-wise, add 50 μl to the *centre* of the oil film. The pressure exerted by the indicator oil is 16 dynes/cm (measured independently) and does not vary with its area. Hence it will compress any film contained in it with this pressure, which corresponds to the pressures on lipids observed in membranes. A dark patch forms at the centre and, with luck, it ought to be circular. The drops should be added at about 2 per second. If the patch ends up about circular, measure its diameter with the ruler. If an ellipse is formed, measure both axes and take the geometric mean ($\sqrt{ab}$) as the diameter. If the patch is very irregular, try again. Estimate the errors involved in making this measurement.

If D is the diameter of the lipid monolayer, then its area is $\frac{1}{4} \pi D^2$. If the total volume of lipid extract is V and 50 μl are added, then the total area of the lipids extracted from N_T cells when spread as a monolayer is given by:

$$\text{Total area } A_T = \frac{V \frac{1}{4} \pi D^2}{50 \times 10^{-3}}$$

The surface area of the erythrocyte

The total surface area is the surface area of one cell multiplied by N_T. We need, therefore, to measure the size and shape of an erythrocyte. The formula used by Gorter and Grendel was simply $2D^2$, where D is the diameter. More recent research, using micro-photography, has shown that $2 \cdot 1 D^2$ is a better empirical formula and accurate to about 3%.

An erythrocyte is a slightly collapsed sphere. The surface area of a sphere of radius D is πD^2. Hence this relation is equivalent to saying the area of an erythrocyte is 67% of the area of a sphere of the same diameter.

Measure the diameter of the erythrocytes in a wet smear under phase contrast and using oil immersion. Stage micrometers and eyepiece graticules are provided. Express your answer in both micrometers and centimeters. The answer is usually accepted as $8 \cdot 3 \mu m$.

If d is the diameter, then the total surface area of the erythrocytes in 25 ml suspension is

$$\text{Total cell area} = N_T 2 \cdot 1 D^2$$

The comparisons

Hence compare the total monolayer area of extracted lipid with the total cell area. Express your answer as a ratio.

When writing the experiment up, estimate the errors at each step, and say what final accuracy you think you answer would have. Discuss also the techniques being used and the effect any inadequacies would have on the final result.

Appendix on film colours

The colours are due to the wave nature of light and arise because light reflected off the top surface of the film *interferes* with light reflected off the bottom surface. If, the peak of one coincides with the peak of the other, then the two reinforce and light is seen. Otherwise they cancel and no light is observed. This happens when the thickness of the film, the angle of the ray and the wavelength of the light are such that the difference in path-length of the two rays is an integral number of wavelengths of light. Since there are complications due to the change in the velocity of light in the oil and to phase changes at boundaries, we will not go into the theory of this. Suffice it to say that, since light is cancelled, the colours seen are complementary to the one removed, and that the wavelength of the light removed is proportional, at a given angle of incidence, to the thickness of the film. When the film is too thin no interference occurs.

This can be demonstrated by sweeping most of the oil off and trapping a portion of the film between two barriers. By decreasing the area of the trapped film, it can be made to change its interference colour epeatedly throughout the entire (complementary) pectrum. You may want to try this.

urther reading
3ar, R.S., Cornwall, D.W. and Deamer, D.G.
   1966 Science, 153, 1010
ingleman, D.M.
   1969 Nature, 223, 1279

Hints for the teacher
The most difficult part is the measurement of the size of the erythrocyte. It is best to use the dimensions upplied, but to encourage the student to check up hat it is not wildly out. The difficulty is not only aking the measurement, but also in ensuring that the erythrocyte has not either collapsed of expanded in ome way. Clearly too the measurement of the film liameter is not very accurate, but ± 10% is perfectly ttainable, and the answer usually obtained is close to hat expected.

C.W.F. McCLARE
King's College
London.

2 Permeability of erythrocytes

Time required
3½–4 hr; 30 min for rapid method A.

Assumed knowledge
Use of spectrophotometer for method B.

Theoretical background
When an erythrocyte is placed in an isotonic solution containing a penetrating molecule, the penetrant enters the cell and this is followed by more water molecules until the cell bursts. In the erythrocyte this bursting is termed haemolysis and the release of haemoglobin into the solution can be monitored by the change of refractive index which occurs.

Necessary equipment
For each pair of students:
200 ml 0·17 M sodium chloride
100 ml 0·17 M ammonium chloride
100 ml 0·17 M ammonium acetate
100 ml 0·17 M sodium nitrite
100 ml 0·12 M sodium sulphate
100 ml 0·12 M ammonium oxalate
100 ml 0·32 M glucose
100 ml 0·32 M glycerol
100 ml 0·32 M ethyl alcohol
100 ml 0·32 M propyl alcohol
5 x 10 ml pipettes
3 x 1 ml pipettes
12 x test tubes and rack
Measuring cylinder filled with 0·17 M sodium chloride to take pipettes when not in use
Piece of newspaper, stopclock
5 ml horse blood (Wellcome Research Labs., oxalated, no preservative) or 5 ml sheep blood (Wellcome Research Labs., defibrinated, no preservative) (a smaller quantity required if only rapid method A)
Spectrophotometer or Eel colorimeter
5 x 50 ml beakers
2 x 50 ml measuring cylinders   Method B only
Graph paper

Type of group
Pairs of students, or singly if rapid method A only is used.

Instructions to the student
At all times keep your pipettes free from blood by placing them in a cylinder of saline.

Rapid method A
1. Dilute the horse blood taking 1 part blood to 10 parts isotonic saline (NaC1). This gives an opaque solution through which newsprint is unreadable. Use this as your stock solution.
   Now add 10 ml water to 1 ml diluted blood and watch the solution become clear; now you can read the newsprint. This gives you an endpoint, i.e. haemolysis has occurred, and you should use the same endpoint for each solution.
2. Add 10 ml of one of the isotonic solutions to 1 ml diluted blood in a labelled test tube, shake and note the time taken to haemolyse the cells. Work through the solutions provided, noting the time taken for the cells to haemolyse, and any colour changes.

Method B
1. *Calibration of spectrophotometer*
   Prepare solutions of 0% and 100% haemolysed blood in the following way:
   (a) Take a small volume of blood (say ½ ml) and totally haemolyse in smallest volume of distilled water, make up to about 30 ml with isotonic *saline;* this is your 100% haemolysed solution. Take an equivalent quantity of blood and dilute to 30 ml with isotonic saline; this is your 0% haemolysed solution. (The final concentrations of erythrocytes in these two solutions must be the same. Since the two solutions are going to be mixed it is essential to keep the quantity of distilled water used to a minimum or else haemolysis of the unhaemolysed blood will occur).
   (b) Using the 100% haemolysed solution as your standard, zero the spectrophotometer using either $OD_{455\,nm}$ or $OD_{600\,nm}$ (green filter for Eel colorimeter), and obtain a reading for 0% haemolysed.
   (c) Using varying quantities of these two solutions, obtain OD readings for 20%, 60% haemolysis etc. and plot these values to give a calibration curve for your instrument.
2. *Fragility of erythrocytes*
   Place small volumes of blood in varying dilutions

of isotonic saline, remembering to use the same final concentration of red cells as used in your calibration, i.e. ½ ml blood to 30 ml diluted saline. For each dilution obtain OD values. From your calibration read off the % haemolysis and then plot a second graph to show % haemolysis against concentration of saline.

3. *Permeability of erythrocytes*
   Selecting suitable solutions from the results of rapid method A, place small volumes of blood in each solution, and place in the spectrophotometer. Take readings of % haemolysis (i.e., OD readings) at known times. Plot graphs of % haemolysis against time for each solution. (Remember to keep the concentration of red cells constant)

Discussion and extensions
Which solutions contained non-penetrating molecules and what type of solutions were they? Can the ammonium salts be arranged in any order and can you account for the differences in permeability? Were there any colour changes in the red-cell solutions?

Experiments of this type have made classic contributions to the study of cell permeability and can be readily extended to examine species differences in permeability, for example, to glycerol or to studies using radioisotopes.

Further reading

General reading
Loewy, A.G. & Siekevitz, P.
    1969 'Cell Structure and Function', 2nd edn. Holt, Rinehart & Winston, New York
Davson, H.
    1964 'A Textbook of General Physiology', 3rd edn., J. & A. Churchill, London.

More advanced reading:
Davson, H. and Danielli, J.F.
    1952 'The Permeability of Natural Membranes' 2nd edn., Cambridge University Press
Stein, W.D.
    1967 'The Movement of Molecules Across Cell Membranes', Academic Press, New York.

Hints for the teacher
Fresh blood is available once a week from Wellcome Research Labs., and will keep in the refrigerator for a few days. Remember to allow time for the blood to come up to room temperature before the practical. Check that ½ ml blood to 30 ml diluting fluid gives

an adequate reading on the spectrophotometer, and alter quantities if necessary.

Part of this practical was adapted from one given by Dr. J.F. Danieli.

SHIRLEY E. HAWKINS
King's College
London

3 Narcosis in ciliates

Time required
2 hr. should be adequate.

Assumed knowledge
No particular skills are required.

Theoretical background
This experiment is designed to investigate the narcotic effects of a homologous series of alcohols on the ciliate *Tetrahymena pyriformis,* and is based on a practical of Dr J.F. Danielli. In a ciliate, narcosis is accompanied by loss of ciliary activity, i.e. no swimming cells can be observed. The endpoint of the experiment is the concentration of alcohol causing 80% narcosis.

Necessary equipment
Each student requires:
50 ml 5M methyl alcohol
50 ml 5M ethyl alcohol
50 ml 5M propyl alcohol
50 ml 0·8M butyl alcohol
50 ml 0·1M amyl alcohol
50 ml 0·025M hexyl alcohol
6 x solid watchglasses and lids
12 x test tubes and rack
6 x 1 ml pipettes and 1 x 10 ml pipette
Binocular microscope
Stopclock
Graph paper
Pasteur pipette and teat
10 ml suspension of *Tetrahymena pyriformis.*

Type of group
Students can work singly or in pairs, or groups, each student with one alcohol.

Instructions to the student
1 Determine the quantity of *Tetrahymena* suspension needed. Place a drop of the suspension into 1 ml of distilled water in a solid watchglass and observe under the binocular microscope. Are sufficient ciliates present to estimate 80%? If not, add another drop.
2 Dilute each alcohol with distilled water to give a series of concentrations ranging from 1 part

103

alcohol to 1 part water, to 1 part alcohol to 8 parts water.

3 Place 1 ml of the most dilute alcohol into a solid watchglass, add the determined quantity of *Tetrahymena* and cover with a lid.
4 After 2 min observe the % narcosis.
5 Repeat through graded series of alcohol until 80% narcosis is observed.

Recording results

Record your results and plot a graph of the log. concentration of alcohol giving 80% narcosis against the number of CH_2-groups.

Discussion and extensions

What conclusions can you draw from your results? Remember that the molecules are increasing in size as well as in the number of CH_2-groups, while the OH-group remains constant. Do the cells show any shape changes? If time is available, try to investigate reversibility of narcosis by adding distilled water to the narcotised cells.

Further reading

General reading
Davson, H.
    1964 'A Textbook of General Physiology', 3rd edn J, & A. Churchill, London

More advanced reading:
Davson, H. and Danielli, J.F.
    1952 'The Permeability of Natural Membranes', 2nd edn., Cambridge University Press
Stein, W.D.
    1967 'The Movement of Molecules Across Cell Membranes' Academic Press, New York.

Hints for the teacher

Tetrahymena pyriformis may be obtained from the Cambridge Culture Collection of Algae and Protozoa and can be grown in sterilised 1% proteose peptone (See Section 7.4). The cells should be cleaned by centrifugation and suspended in Chalkley's medium (See Section 7.4). One 250 ml culture grown in a 500 ml Erlenmeyer flask should provide enough ciliates for 25 students.

SHIRLEY E. HAWKINS
King's College
London.

4 Pinocytosis in Amoeba

Time required

The length of time spent on the practical depends on the extent to which the experiment is made quantitative and the number of solutions, etc., used. The time allowed can vary from 1 hr in introductory classes to 3—4 hr in more advanced classes.

Assumed knowledge

The only requirement for this practical is that the student should be able to use a phase-contrast microscope.

Theoretical background

The term *pinocytosis* was used by Lewis in 1931 to describe the uptake of fluid droplets from the surrounding medium by macrophages and was later used by Mast and Doyle in 1933 describing a similar process in amoebae. Later work indicated that uptake of solute was a more significant feature of the process, adsorption to the outer surface of the cell being the first phase of pinocytosis. Adsorption is followed by invagination of the membrane and formation of vesicles at the ends of narrow channels. These vesicles are then incorporated into the cell. In amoeba the cell rounds up, the granular cytoplasm contracting to the centre of the cell, and small blunt pseudopodia can be seen at the cell periphery. Each pseudopodium may possess a centrally placed narrow channel visible as a thin dark line when a salt solution is used as an inducing agent. After a period of about 15 minutes no more channels are seen and the amoeba cannot be induced to pinocytose again until after a period of membrane recruitment.

Necessary equipment

A culture of amoeba starved for 2 days.
Phase microscope
Teats
Pasteur pipettes (pulled a second time to make a fine pipette—number will depend on number of solutions to be attempted)
Slides
Cover slips
Piece of plasticine
Solid watchglasses(6)
Stopclock
3 x 2 ml graduated pipettes

0·125M NaCl	10 ml	adequate
1:10 000 Alcian Blue (Gurr)	10 ml	for a short
1:10 000 Toluidine Blue (Gurr)	10 ml	practical

Longer time period

Ribonuclease 1 x 10^4 (Koch- light) (adjust pH to 4·4)	10 ml
Bovine serum albumin 1% (BDH)	10 ml
Glass distilled water	50 ml
Ice cold 0·125 M NaCl	10 ml

Additional graduated pipettes, 3 x 2 ml, 1 x 10 ml

Type of group
Individual students, size of class does not matter.

Instructions to the student
1 Place 2 ml inducing solution, e.g. 0·125 M NaCl into a solid watchglass and, using a fine pipette, add amoebae in as little culture medium as possible.

2 After 2–5 min, pick up amoebae together with solution and place drop on a slide. Be sure to pick up as many cells as possible—gently swirling the watchglass tends to bring about a collection of amoebae in the centre of the vortex. Place a *small* piece of plasticine at each corner of the cover-slip and gently lower onto the slide. Be careful not to squash the cells.

3 Look at the shape of the amoebae—are they rounded up? Observe under phase-contrast (x 40) looking for small blunt pseudopodia lacking granules. When you find a blunt pseudopodium look at the central region and continually change the fine focus until you can see a narrow channel, observed as a thin, dark line. Try and count the number of channels in individual cells, and find out how long pinocytosis continues. Draw the cells.

4 Repeat with coloured solutions Alcian Blue and Toluidine Blue—do the channels look the same as in the salt solution? Draw the cells.

For longer periods of practical

Repeat above experiments with ribonuclease and bovine serum albumin. Try to make quantitative estimates of channel number in diluted salt solutions and in ice-cold salt solution. Do all amoebae respond and how variable is the channel number? Try a second immersion in inducing agent after the finish of the response.

Results
Make annotated drawings. If experiments with dilu-tion of NaCl solutions have been performed, plot mean channel number against molarity of NaCl.

Discussion and extensions
Did all amoebae respond to the inducing agent? What factors might influence the degree of response? What can you conclude about the nature of the inducing agent?

Pinocytosis is a membrane phenomenon and as such may be used to monitor such problems as membrane turnover, protein synthesis before and during channel formation, importance of nucleus and the effects of other agents on membrane binding sites. It can also be used as a means of transporting molecules into cells which are normally non-penetrants and this is important physiologically, for example, gastric intrinsic factor and entry of vitamin B_{12}.

Further reading
Brandt, P.W. and Pappas, G.D.
1960 'An Electron Microscopic Study of Pino-cytosis in Amoeba, I', J. Biophys. Biochem. Cytol., 8, 675

Brandt, P.W. and Pappas, G.D.
1962 'An Electron Microscopic study of pino-cytosis in amoeba, II', J. Cell Biol., 15, 55

Brownstone, Y.S. and Chapman-Andresen, C.
1971 'Degradation of Endocytosed Albumin by *Chaos chaos.*' Compt.Rend.Trav.Lab. Carlsberg, 38, 297

Chapman-Andresen, C.
1962 'Studies on Pinocytosis in Amoebae'. Compt. Rend.Trav.Lab.Carlsberg, 33, 73

Chapman-Andresen, C.
1964 'Measurement of Material Uptake by Cells', p.277 in 'Methods in Cell Physiology', Vol.I (ed. Prescott, D.M.) Academic Press, New York

Hints for the teacher
It is never possible to get enough amoebae in good condition from commercial suppliers. In fact, amoebae are relatively easy to grow and we use two methods routinely for research purposes. I am always willing to send starter cultures of amoebae together with food organisms for teaching purposes, providing advance warning is given and stamps are sent to cover postage.

Amoebae grow in Chalkley's medium in glass or plastic dishes and will tolerate temperatures from 17°C to 23°C; at lower temperatures growth is too slow, while at higher temperatures they tend to overgrow their food supply.

Chalkley's medium:

$$16 \text{ g NaCl}$$
$$0.8 \text{ g NaHCO}_3$$
$$0.4 \text{ g KCl}$$
$$0.2 \text{ g Na}_2\text{HPO}_4, 12\text{H}_2\text{O}$$

in 1 litre *glass* distilled water

Dilute 5 ml of concentrated solution to 1 litre glass distilled water for use.

Method A

About 300 amoebae together with food organisms (*Colpidium* and *Chilomonas*) are placed in Pyrex 'stewpot' lids or shallow glass containers containing 150 ml Chalkley's medium. Add 4 *boiled* wheat grains, cover with a lid and leave for a month. After this time the culture needs sub-culturing and cells are ready for experimental use.

Method B

This method is used for obtaining large numbers of cells for biochemical experiments and is less easy. Amoebae are placed in Chalkley's medium (to a depth of 2 cm) in plastic food containers. They are fed three times a week with *Tetrahymena pyriformis* (Cambridge Culture Collection). These ciliates are grown in sterilised 1% proteose peptone (1% proteose peptone; 0.25% yeast extract, sterilised at 15 lb/in^2 for 15 min), and washed free from growth medium. This is important because proteose peptone is toxic to amoebae. We clean our ciliates by placing them in a Buchner funnel 1 litre, (porosity 4, Griffin & George) and allowing Chalkley's medium to pass through; but repeated centrifugation and washing will also serve.

Prior to the practical the amoebae should be starved for 48 hr; pinocytosis is then much more vigorous.

SHIRLEY E. HAWKINS
King's College
London.

5 Aggregation of sponge cells to demonstrate features of cell adhesion

Time required
3 hr. (2 hr if parts of the practical are omitted).

Assumed knowledge
Elementary knowledge of cell biology at practical and theoretical levels.

Theoretical background
The experiments require the use of a cell suspension which is tested under various conditions to discover if it will aggregate. If aggregation occurs it is concluded that adhesion is permitted by those particular conditions. This experiment requires little in the way of advanced knowledge on the part of the student.

Necessary equipment
1 Sea-water aquarium (if marine sponges are used). This may be nothing more than a large plastic container filled with real or artificial sea water if the sponges can be collected or delivered within the two days prior to the practical. This aquarium should be kept (for British species) at temperatures between 8°C and 12°C.
Reciprocating or gyratory shaker to hold 10 ml or 25 ml conical flasks.
2 Biological materials.
Either marine or freshwater sponges may be used. Media outlined below are designed for marine sponges and will need modification for freshwater sponges. The following marine sponges give good aggregation: *Suberites ficus, Haliclona occulata, Halichondria panicea, Microciona sanguinea, Axinella spp., Petrosia spp. Pachymatisma spp.* Many other sponges aggregate well but *Grantia* and *Sycon* can present difficulties. They should be collected, preferably by diving, within the two days prior to the practical. If froth appears on the surface of the aquarium they are stored in, they should be discarded. Many sponges have an unpleasing smell even when alive so that this should not be used as a certain criterion of death. Many sponges carry a very large fauna of bacteria; if this is excessive

experiments may be spoilt. The species listed above are generally fairly free from bacteria.

The freshwater sponge *Ephydatia fluviatilis* may be grown axenically from gemmules in the laboratory following the method described by Rasmont. (Rasmont, R., 1961, Annal. Soc. Roy. Zool. Belg., 91, 147–56). It will provide small amounts of excellent material. Approximately 100 g of sponge (wet weight) will provide sufficient material for a class of 200 students.

3 Classroom equipment.

Small storage containers for sponges. If two species are used, store them separately.

Reciprocating or gyratory shaker. They usually have room for 25 flasks, which means that 5–6 students will be able to use one shaker. Shaker should be capable of being cooled to 2°C (Gallenkamp)

12 glass or plastic Petri dishes (60 mm dia.) per student.

4 Pasteur pipettes plus teats, per student.

7 stoppered 10 ml or 25 ml conical flasks, per student.

1 pair fine, 1 pair coarse forceps per student.

1 microscope with 10x and 20x objectives per two students (preferably phase-contrast).

1 haemocytometer per 5 students.

Tissues (paper).

4 Media.

Artificial sea water. Instant Ocean or sodium chloride 26·0 g, potassium chloride 0·7 g, calcium chloride 1·4 g, magnesium sulphate 3·3 g, tris 6·0 g; per litre, pH 7·8

CMF sea water. Sodium chloride 30·0 g, tris 6·0 g, potassium chloride 0·7 g, per litre, buffered to pH 7·8.

EDTA sea water. CMF sea water plus 0·74 g/litre di-sodium ethylene diamine tetra-acetate, pH 8·0.

Mg-artificial sea water. CMF sea water plus 1·2 g magnesium sulphate per litre, pH 7·8.

Ca-artificial sea water. CMF sea water plus 1·1 g calcium chloride (anhydrous) per litre, pH 7·8.

Sr-artificial sea water. CMF sea water plus 1·6 g strontium chloride per litre, pH 7·8.

These media need not be sterile but distilled water used in preparation must be free from heavy metals or pyrogens. Allow 2 litres artificial sea water and 1 litre of each of other media per 10 students.

5 Ancillary equipment: magic markers or chinagraph pencils.

Type of group
Individual students in large classes.

Instructions to the student
1 Preparation of cell suspension
Break off a piece of sponge about 1 cm by 1 cm by 1 cm or less. Place in a Petri dish with artificial sea water; remove any fragments of shell or sand and various commensal organisms with fine forceps. Place in CMF sea water (lacking Ca and Mg ions); after two minutes transfer sponge fragments to another Petri dish containing EDTA sea water (about 10 ml). After 5 min squeeze sponge fragments, keeping them under surface of Petri dish, with coarse forceps. Cells will emerge into suspension. A suitable suspension will be visibly opaque. If a haemocytometer is available the cell suspension should contain 2–6 million cells per ml.

Transfer cell suspension with a pipette to experiments in 2.

2 Conditions for cell adhesion
Two methods of achieving reaggregation are available.
(a) Pipette sufficient cell suspension into a Petri dish containing 10 ml artificial sea water (or other media, see (3) below) to make the medium opaque. Allow the cells to settle and take care not to disturb. Examine after 1 hr. Reaggregation takes place by cell movement. Cells adhere to Petri-dish surface as well as each other.
(b) Pipette sufficient cell suspension into a conical flask containing 5 ml artificial sea water to make the medium slightly opaque. Transfer flask to a shaker (shaker rate 60–100 impulses/min). Cells reaggregate passively due to shear of medium. Examine after 1 hr for aggregates (just visible to the naked eye).

3 Carry out the following experiments
(a) Will adhesion take place at low temperatures? Attempt reaggregation by methods (a) and (b) in artificial seawater in refrigerator or a cooled shaker respectively. Work at 1–2°C.
(b) Investigate how the ionic composition of the medium affects cell adhesion. Reaggregate cells by methods (a) and (b) in (i) CMF sea water (ii) Ca-sea water (iii) Mg-sea water (iv) Sr-sea water and (v) sea water. Carry out these experiments at 10–25°C.
(c) It has been claimed that when sponge cells of two species are aggregated together in

suspension each individual aggregate that forms is composed of the cells of one species alone. If sponges of two very differently coloured species are available, this experiment can be re-examined in the laboratory. Prepare suspensions from two species, examine the suspensions to discover whether you can detect visible colouration in a large proportion of the cells. Aggregate in the shaker for two hours, then examine in a Petri dish and score the colour of the aggregates. Do you regard aggregate colour as a clear criterion of the species purity of each aggregate? Does specific adhesion appear to occur? How would you improve the test if you had the technical means?

The quantitative assessment of the degree of aggregation may present some difficulty. Most simply, all that you are asked to do is to observe whether aggregates have visibly formed. This will probably fail as a method of assessing how much aggregation has occurred. One suggestion is that you measure the population density of un-aggregated cells at various times during the process. If this is done it is desirable to arrange for all suspensions which you have under comparison to be made up the same initial population density, for the collision rate is roughly proportional to the square of the density. Thus if you do this, aggregation will be inversely proportional to the density of unaggregated cells. Are there any assumptions in this statement? If you think there are, identify them.

Recording results
Results should be recorded in a standard manner.

Discussion
Consider the results on adhesion at low temperatures. Is there any difference between the two systems of aggregation? If there is, advance an explanation. Do either of the systems require any cellular property other than adhesiveness to produce aggregation? Does your evidence suggest that adhesion is dependent on metabolism? If you think that it does, consider whether there might be other explanations. If the evidence suggests otherwise, does this give you any clues as to the mechanism of adhesion?

Do the experiments on ionic effects on adhesion indicate any difference between the two methods of aggregation? If they do, suggest explanations for this.

What role in the mechanism of adhesion can yo suggest for divalent cations? If there any evidence o cation specificity?

The questions that are relevant to the experimen on specific sorting out are set out at the end of th 'Instructions' section.

Extensions and applications
These techniques for investigating cell adhesion ar widely used in research on cell adhesion. The que: tion of whether specific adhesion occurs is still partially unsolved problem (see 'Further Reading').

Further reading
Specific adhesion:

Humphreys, T.
    1963 'Chemical Dissolution and *in vitro* Recon struction of Sponge Cell Adhesion', Devl. Biol., 8 27–47.
Curtis, A.S.G.
    1970 'Problems and Some solutions in the Study of Cellular Aggregation', Sym. Zool. Soc. Lond. 25, 335–52.
General:

Curtis, A.S.G.
    1973 'Cell adhesion', Prog.Biophys.Mol.Biol., 27 315–86.

Hints for the teacher
There are no special problems other than the collec tion of healthy sponges.

A.S.G. CURTIS
Glasgow

5 Agglutination of cultured animal cells by concanavalin A

Time required

3 hr. (This assumes that students are given explicit information about concentrations of con A and glycosides. It may be more valuable to withhold this, in which case 2 sessions of 3 hr would probably be required.)

Assumed knowledge

Familiarity with the use of a haemocytometer would be an advantage if this counting technique is to be used.

Theoretical background

Phenomenon

Concanavalin A (con A) is a protein from jack bean which has binding sites for certain carbohydrate residues. Since con A is multivalent, and its specificity includes residues found at the surfaces of animal cells, it causes aggregation of such cells in suspension (agglutination). This effect is of particular interest at present because transformation of cultured cell lines by tumour viruses (probably both DNA and RNA) enhances the agglutinability of the cells by con A, and also by another plant protein from wheat germ. Non-transformed cells are rendered agglutinable by very gentle treatments with proteolytic enzymes. The mechanism of this change, and of that caused by transformation, is controversial.

Agglutination by con A can be reversed by certain low molecular carbohydrates, which presumably act as analogues of the cell-surface carbohydrate to which con A binds.

Technique

Agglutination of cells requires that they be brought into collision with one another. In this experiment, collisions are brought about in a reproducible manner by gently shaking the cells in suspension. Keeping cell suspensions stationary, however, does not completely suppress collisions and consequent agglutination, but fortunately con A-induced agglutination is negligible at $0-2°C$, so it can be started promptly by transferring cells from ice to a shaker bath at $37°C$. Under these conditions, both the rate and the extent of cell clusters is a reproducible function of the con A concentration.

Agglutination is often scored subjectively, simply by examining cell suspensions under the microscope. In this experiment, it is measured by using a Coulter Counter to count electronically the concentration of clusters (N) in the suspension. N_t is the concentration at time t of single cells, plus clusters of all sizes counted as one each. Thus if the initial cell suspension is completely single, N_0 equals the total single cell concentration, and N_0/N_t measures the mean number of cells per cluster at time t. As agglutination proceeds, N_t decreases

Necessary equipment

Equipment per student

12 x 10 ml conical flasks, with stoppers. These should all be approximately the same shape.

1 x 0.1 ml pipette of the type which fills to a constriction. The tip should not be too fine, and may have to be cut and heated to leave an aperture of about 0.25 mm.

6 x 2 ml graduated pipettes

30 universal bottles, each containing 20 ml 0.9% sodium chloride

2 Haemocytometers, e.g. Fuchs—Rosenthal type (Gallenkamp) with cover slip

6 Pasteur pipettes

1 Bucket of crushed ice

Solutions per student or pair

50 ml cells suspended in a balanced saline solution (BSS), at a density $0.5-1.0 \times 10^6$ per ml

5 ml con A, 0.4 mg/ml in BSS

2 ml each, α—methyl glucoside, α—methyl mannoside (Sigma), 0.2M in BSS,

20 ml BSS (*Calcium* and *magnesium containing* saline, such as Hank's, preferably buffered, for example with Tris or HEPES. Flow Labs.)

Equipment per 6 students or per 6 pairs of students.

1 (minimum) phase-microscope with a condenser of working distance long enough to permit phase viewing of cells in haemocytometer

1 (minimum) Coulter counter with 200 μm aperture, Celloscope, or other electronic particle counter

3 thermostatted water baths with reciprocating shaking tables equipped with clips to hold 10ml flasks. Set to $37°C$.

Type of group

Pairs of students, number likely to be restricted by availability of counting instruments. Could be readily adapted to dispense entirely with electronic counting, using haemocytometers exclusively, to measure N_t. In this case, a minimum of 3 haemocytometers per pair would be required.

Instructions to the student

The aim of this experiment (which exploits a new technique for quantitating agglutination) is to determine:

1 The dependence of agglutination on con A concentration.
2 The relative affinities for con A of the competing glycosides α—methyl glucoside and α—methyl mannoside.

Make additions of cells, con A and glycosides to 10 ml flasks, as Table 1 below. It will probably prove convenient to incubate 6 flasks at a time. Keep the stock cell suspension and flasks cold by immersing them in crushed ice. It is extremely important when pipetting animal cells to swirl the stock immediately before removing each aliquot, to avoid settling. For the same reason, use a 2 ml pipette times, not different segments of a larger pipette When the additions are complete, take a Coulter count from each (see below). Values should deviate less than 5% from the mean.

Transfer a small sample from one flask to haemocytometer to compare with agglutinated samples. Start agglutination by transferring flasks to the (moving) shaker bath at (say) ½ min interval. Take a Coulter count for each again after 30—45 min. Examine the initial sample and a sample from an agglutinated suspension by phase microscopy. Does the reduction in the total number of clusters (including single cells as 'clusters') agree with the value obtained using the Coulter counter? You should count sufficient squares of the haemocytometer grid to

Table 1 Concentration dependence

	Flask No.					
	1	2	3	4	5	6
Con A 0·4 mg/ml	—	1·0	0·2	—	—	—
Con A as above, diluted 1/10in BSS	—	—	—	1·0	0·5	0·2
BSS	1·0	—	0·8	—	0·5	0·8
Cells	3·0	3·0	3·0	3·0	3·0	3·0

Table 2 Effect of glycosides

	Flask No.					
	1	2	3	4	5	6
Con A 0·4 mg/ml	0·1	0·1	0·1	0·1	0·1	0·1
α—methyl mannoside 0·2M in BSS	—	—	1·0	—	—	—
α—methyl mannoside 0·02M in BSS	0·1	1·0	—	—	—	—
α—methyl glucoside 0·2M in BSS	—	—	—	—	—	1·0
α—methyl mannoside 0·02M in BSS	—	—	—	0·1	1·0	—
BSS	0·9	—	—	0·9	—	—
Cells	3·0	3·0	3·0	3·0	3·0	3·0

cumulate at least 300 counts.

Sampling for the Coulter Counter

mple 0·1 ml of suspension (immediately after
irling) with a coarse-tipped constriction pipette.
ansfer the sample to 20 ml ice-cold 0·9% saline, in a
iversal bottle, screw on the cap, and mix. These
luted samples can be stored at least 30 min before
ounting, but must be re-mixed by inversion before
ounting. Vigorous shaking should be avoided, since
ibbles may be generated and counted. (You can
eck this with a blank saline bottle). If the Coulter
ounter is set to count 2 ml samples, it should be
ossible to obtain 3 readings from each 20 ml diluted
mple. Since the 0·1 ml pipette is to be used
peatedly, it should be rinsed with saline between
ch sampling.

ecording results

lot histograms of the reduction in Coulter count
nean of the three counts) against con A and
ycoside concentrations. Convert the counts to
ounts/ml, taking into account the sampling dilution
id volume setting on the manometer of the counter.
here is a coincidence error in the count (about 10%,
hen count is 10^6 per ml of suspension, and a 200
m aperture is used) which may be neglected.

iscussion

1 Why does the highest concentration of con A not
   show the most agglutination?
2 Which glycoside has the highest affinity for con
   A? Does a competitive experiment such as this
   identify the carbohydrate residue at the cell
   surface with which con A is interacting?
3 What do you suppose limits the precision of the
   Coulter count?
   The haemocytometry count?

Extensions and applications, and Further reading

Method

This experiment exploits the principle of the Coulter
ounter in a very simple and restricted way. The
nstrument is capable of giving much more informa-
ion about the sizes of suspended particles, and an
xtensive bibliography of its use for a range of
iological applications is supplied to users. For an
ccount of its application to measuring the state of
ggregation of suspended animal cells, see the chapter
Cell Adhesion', by J.G. Edwards, in 'New Techniques
n Biophysics and Cell Biology' Eds. R. H. Pain and
3. Smith, Vol. 1, p. 1 (1973) John Wiley, London.

Lectins

For a review on the properties and uses of cell-
agglutinating proteins such as con A, see Sharon, N.
and Lis, H. (1972) Science, 177, 949.

Hints for the teacher

Preparation of cells

We have used BHK21 cells, which are a robust line,
easy to grow and readily available. Other cell types
could be used, but dose response to con A may be
different. BHK cells 'agglutinate' spontaneously, unless a rather high level of trypsin is used to prepare
the cell suspension—a suitable procedure follows:

Confluent cells on a $120 \, cm^2$ glass surface are
washed twice with BSS, then 10 ml BSS containing
200 μg/ml crystalline trypsin is added. The cells are
incubated at 37° for 15 minutes. They are then
transferred to a universal bottle, and dispersed by
aspiration in a Pasteur pipette. 1 ml of DNAse I
(Sigma) 100 μg/ml is added, and the incubation
continued for a further 15 min. 2 ml calf serum is
then added, and the cells are sedimented by centrifugation and resuspended in BSS. They are washed by
a second centrifugation and resuspension in BSS. One
such culture may easily yield 30×10^6 cells, but at
higher densities, poorer suspensions are generally
obtained.

Setting the Coulter counter

This use of the instrument is very undemanding.
Simply arrange amplification and aperture current so
that most pulses reach about $\frac{2}{3}$ up the CRT display,
and use a lower cut off at about 1/10 this level to
remove instrument background and counts due to cell
debris. No upper cut off is used. If settings are
appropriate and the cell suspension reasonably clean,
the count will be insensitive to small changes in the
lower cut off, and will agree within 10% with the
haemocytometer count.

J.G. EDWARDS
Glasgow

Section 8

Cell Motility

1 Down to the molecular level: Brownian motion, particle size and algal movement

Time required
Two sessions of about 3 hours.

Assumed knowledge
An ability to use a phase-contrast microscope is a advantage, although we have usually found th enough can be taught actually in the practical.

Theoretical background
The point of the practical comes across better if th student studies the theory *after* the experiments.

Molecular biology is about the molecules in livir organisms. But molecules are so much outside o' ordinary experience that we have no feeling for wh they are like.

The purpose of this practical is to give son imaginative grasp of the size, environment and tim scale appropriate for molecules; to impart some ide of how they can be thought about—and also ho they cannot. For this reason there is no point i hurrying. Use your imagination.

The data to be collected need only be *ver approximate*. Visual estimations are made which ca only be rough. This means that the calculations nee only be made in your head. One year we ha calculations carried out using five-figure logarithm This was ludicrous. Quite apart from being far mor work than necessary, it is very misleading as to th accuracy of the results obtained. In this practic; therefore we also consider errors and approximation Hence *read the appropriate section under 'Errors* Particularly the first sentence.

The completely open mind is a myth; it impossible. Even Bacon argued that anyone with a unprejudiced mind could *see* that the sun goes roun the earth. Every baby knows from experience that h is at the centre of the universe. To a large extent w only see what we expect to see from previou experience. When we look down a microscope there fore we naturally assume that things will behave th same way as they always have behaved in ou ordinary experience; but they do not. These experiments are really about the prejudices we carry around with us. In them you will be asked to make some estimates of certain quantities and then to compare those estimates with values obtained from theory—unless you make informed guesses you will certainly be wrong. But the idea is *not* to catch you out but to confront your natural expectations with what, from totally different evidence, we know to be the case. This teaches two things: first, only by making precise, written down, estimates can we know whether they are wrong (and therefore whether there was anything amiss with our assumptions) and, second, sense data are only reliable in situations for which the senses were specifically evolved. (Eyes are useful for swinging through trees, but only very recently started looking down microscopes.)

Molecules. When thinking about molecules it very easy to overwhelm yourself with numbers. For example, one mole of hydrogen (2 g) contains 6.0232×10^{23} or, for our present purposes, 6×10^{23} molecules. Just how big a number is that? The radius of the earth is 7×10^6 m which gives it an area of 5×10^{14} m^2. How many molecules would there be per square metre if they were spread evenly over the earth's surface? The answer is 10^9. This means that one molecule would occupy an area bounded by a square of sixe 3×10^{-5} m or about 30 μm (10^{-6} m). (Note the approximations here). The same would of course be true of 100 kg of a molecule of molecular weight 100000.

We *ought* to wonder therefore what justicication there is for believing in molecules, let alone supposing that that we can think about them in the same way we think about tables and chairs. We must remember that even when we study physical properties we are usually studying vast numbers of molecules. Yet in a living cell sometimes a single molecule, all by itself, is extremely important. For example. a bacterium has just one DNA molecule for most of its life-cycle. How do we know that odd things don't start happening at the molecular level? This is not an easy question, and one still being thought about deeply.

What we want to do is to make these numbers seem more manageable and also to bring some of the problems into focus. *Please do not read ahead in this text.* The experiments may make more impact if you do them in sequence, for they may contain some surprises.

Necessary equipment
Good phase-contrast microscopes, capable of about 400 x magnification. Oil immersion is not required

(and in fact a nuisance in this case).
Stage micrometers
Eyepiece graticule
Simple ruler
Slides with a well
Cover-slips

Suspensions of particles: the first unknown, the second *Chlamydomonas reinhardii* (alive). Ideally an electron microscope suitable for demonstrations. A projection screen is perfect, but few laboratories will have such equipment. Some of the immediacy of the experiment is lost of course, but photographs of the resulting structures have to be used and the actual demonstration could be dispensed with altogether. Suitable photographs can be found in Hopkins, J. (1970), *J. Cell. Sci.* 7, 823–39, where the technique for preparing the grids can also be found.

Type of group
Individual students, but they are very much encouraged to discuss their findings with each other.

Instructions to the student
Writing up this practical
You will miss the whole point if your observations are not only written down but also *worked out as you go along.* That is, we want you to think about the implications of what you see. We will ask what answers you get to each question during the practical itself. Hence write down your observations and the answers, together with any immediate implications that you can see. A fair copy, with further thought, can be made later for handing in.

The scale of living cells

If you have not used a phase-contrast microscope before, or if in any doubt, consult a demonstrator.

Each microscope has an eyepiece graticule. There are also a number of cm scales on microscope slides available (2 cm divided into 200 divisions or 1 cm into 100).

The idea is to use these scales to measure the size of something in the field of view. For this purpose we need to work out how many μm the divisions in the eyepiece graticule correspond to. For example: length of object = 2·3 eyepiece divisions; 1 division = 3·2 μm, hence object is 2·3 x 3·2 = 7·1 μm long. For our purposes the most powerful objective which does not use oil immersion will be suitable. Hence *measure* how many μm the eyepiece divisions correspond to.

To grasp the significance of the magnification *measure* the thickness of a human hair and then estimate its apparent size by placing a ruler on the microscope stage and viewing it with one eye and the hair with the other. The ratio of the two measurements gives the magnification (i.e. apparent size = 3 cm; actual size = 60 μm; magnification = $3 \times 10^{-2}/60 \times 10^{-6} = 500$).

Typically a living cell would be from 0.2 (Mycoplasma) to 100 μm (some animal cells) across. They are found in many different shapes. A common shape for bacteria is coccoid, which means spherical. Take a drop of the suspension provided, put it on to a cavity microscopy-slide under a cover glass (it is quite harmless) and look at it under the microscope.

Are the particles you can see alive or dead? Try to devise tests which would answer this question. One obvious criterion of life is movement, hence it may be natural to assume that they are alive. But is this sufficient? If not what else could explain the movement?

Brownian motion

Whether or not you were taken in by the ambiguous language above, you should now be convinced that the particles were not alive in the first place. In fact, the particles are made of polystyrene latex. Take another look at them (or as many looks as you like), and try to appreciate the significance of what you see. *They are not alive* and yet the motion is incessant. It is hardly surprising that Robert Brown, who first observed this phenomenon in 1827, thought at first that he had come upon the 'primitive molecule' of life. Only when he found that undoubtedly dead things were also in incessant motion, was he convinced otherwise. What is it due to? Remarkably enough no explanation of this *Brownian motion* was put forward for 40 years and not generally accepted until Einstein accounted for it quantitatively in 1905.

The water molecules are in incessant, chaotic motion. The latex particles are so small that when unequal numbers of molecules strike them on opposite sides they are moved. Thus Brownian motion is simply due to heat; it is thermal motion. By observing this phenomenon it is possible to measure the size of molecules directly. This was a great surprise in 1905. (The theory of Brownian motion is given under 'Results' for those who are interested. You could estimate N, Avogadro's number, using your observations).

Measure the diameter of the particles. This is not easy because they are moving. Occasionally they get stuck to the glass; these are easier to measure particularly if you can find a string of them. Then *estimate* the distance in μm moved by the particles in one second, but be *quite clear* what is required here. What we want is the *average distance* one would expect the particle to be away from its original position after one second. That is; if two snap-shots were taken one second apart the quantity required, $\bar{x}$, is the average distance between the particles in the superimposed snap-shots:

$$\bar{x} = (x_1 + x_2 + \dots x_n)/n$$

Position at $t = 0$ •
Position at $t = 1$ s ⊕

Figure 1

Perhaps the easiest way to estimate x is to count off seconds and see how many of its own diameters the particle moves, on average, in this time. A satisfactory estimate could only be expected to be + 20% here. Does the motion appear to be consistent in direction at all? (It may be; if so, why?) Do you think (and this is a difficult question) the particle would move, on average, twice as far in twice the time? If not, why not? Take a fixed point and *sketch out* what the path of an individual particle seems to be.

A very illuminating consideration is the following: the particle is continually changing its direction, hence its instantaneous velocity must be greater than its average drift velocity. First *guess*, just by looking, at what this instantaneous velocity might be. (Unless it is an informed guess you are going to be wrong, but the object of the exercise is to understand *why* you are wrong.) Call your guess V'.

Then *calculate* the kinetic energy of the particle at V'. If the radius of the particle is r μm, and the density of the material is ρ kg/m^3, then the mass of the particle is

$$m = \frac{4}{3}\pi\rho(r \times 10^{-6})^3 \text{ kg/particle} \qquad (1)$$

The density of the polystyrene is 1.1 g/cm^3 which of course is 1.1 x 10^3 kg/m^3. (Don't forget that $\pi = 3$ for our purposes.) Therefore the kinetic energy of he particle is E', where

$$E' = \tfrac{1}{2}m(V')^2 . \qquad (2)$$

f m is in kg, V' in m/s, then E' will be obtained in J/particle.

But the motion you have observed is due to *heat*. In fact it *is* heat. (What would happen to x if the temperature were higher?) Therefore *in fact* the instantaneous kinetic energy is, on average, equal to the average heat energy which is $2 \cdot 2 \times 10^{-21}$ J/degree of freedom. *Compare E'* with this quantity; was your estimate of E too large or too small? Hence *calculate* what V, the average instantaneous velocity, should have been. How quickly would the particle cover the field of view of the microscope at this velocity in a straight line?

But why does the instantaneous velocity look so much less than this? The answer to this question gives considerable insight into the main difference between molecules and the large particles in our immediate experience. Suppose you fired the latex particles into water so that they hit it at V m/s, how quickly would the friction due to motion through the water slow them down to $V/3$ m/s? Again first make a *guess*, and then work it out. (This time your guess ought to be influenced by what you have just learned).

The calculation is easy. Suppose the particle is travelling at V m/s. It will then experience a force slowing it down due to the viscosity of the water, F, where

$$F = 6\pi\eta\, r\, V; \qquad (3)$$

η = the viscosity of water, 10^{-3} N s/m^2. Substituting in (3) the force is obtained in Newtons. (A Newton is that force which would cause 1 kg to accelerate at 1 m/s^2.) Hence the acceleration of the particle is given by

$$\frac{\mathrm{d}V}{\mathrm{d}t} = \frac{-6\pi\eta r\, V}{m}$$

or

$$\frac{\mathrm{d}V}{V} = \frac{-6\pi\eta r}{m}\mathrm{d}t$$

Therefore, integrating, we obtain

$$\log_e \frac{V_O}{V} = \frac{6\pi\eta r t}{m}$$

that is

$$V = V_O e^{-\frac{6\pi\eta r t}{m}}$$

That is, the velocity decreases to $1/e$ ($\doteqdot 1/3$) of its previous velocity each time t increases by $t_{1/3}$ where

$$t_{1/3} = \frac{m}{6\pi\eta r}$$

Calculate $t_{1/3}$. Even allowing for your slightly informed guesswork, the result may be another surprise.

The idea of these guesses is to confront our expectations for the microscopic world with what we can calculate and measure must be the case. This should make the point forcibly: *things happen very much quicker than we expect*. The latex particle can be thought of perfectly legitimately as just a very large molecule; with more usual molecules the time-scale is *even shorter*.

Diffusion. The movement of the particles is in fact *diffusion;* the forces which make the particles migrate also make any other molecule migrate. For example ATP. The diffusion of small molecules is important because we would like to know, for example, whether ATP made in one part of a cell can simply diffuse to another part to be used or whether it needs to be transported there in some way. We have $\bar{x}$ for our particles; we would like to have a relation between the size of a molecule and its own particular x. It is shown in the section in Brownian motion under 'Results' (you are not expected to be able to derive the equation unless it interests you) that if two spherical molecules of the same density have molecular weights M' and M'' then

$$x'(M')^{1/6} = x''(M'')^{1/6}. \qquad (5)$$

In other words if one particle has a molecular weight 10^6 times the other then it will diffuse $1/10$ as far in an equal time.

Work out the molecular weight of the particles by calculating the weight in grams of 6×10^{23} of them. The molecular weight of ATP is about 600. Therefore, making the crude assumption that it is spherical and has the same density as the particles (which assumption is good enough for our purposes), *calculate* what x should be for one second for ATP.

$$\bar{x}_{ATP} = \bar{x}_{\text{ particle}} \left[\frac{\text{M.W. particle}}{\text{M.W. ATP}} \right]^{1/6}$$

115

Your drawing of the path of an individual particle above should have made you realise (if it wasn't already obvious) that the particle diffuses less than twice as far in twice the time. In fact it is easy to show (see 'Brownian motion' under 'Results' on average, twice as far in four times the time. That is

$$\bar{x}^2 \propto t \qquad (6)$$

Hence *calculate* how long it takes ATP to diffuse a biologically relevant distance—say, 1 μm.

As an illustration of how important these considerations are, remember that a muscle takes about 1 ms to 'turn on' after the nervous impulse has reached it. It is turned on by the release of Ca ions from organelles and the ions have to diffuse some 1–2 μm to where they are needed. Are these observations compatible?

Chlamydomonas reinhardii

Chlamydomonas reinhardii is a green alga. It is motile and has two flagella. *Observe* (low power) a drop of the culture of *Chlamydomonas* on a covered slide. You should have little difficulty in deciding whether the particles are alive this time. The organisms are inclined to lose their flagella and, in any case, many seem to stick to the glass. Find a non-motile one and *measure* its diameter. Also measure the length of the flagella and, if you can, measure, or make a guess at, their thickness. We will measure it exactly later in the electron microscope. *Time* how quickly the organisms can cross the field of view and hence *calculate* their velocity. As before, the viscous drag is given by

$$F = 6\pi\eta rv$$

for a sphere of radius r moving at velocity v through a medium of viscosity η. Hence, given that the viscosity of water is 10^{-3} Ns/m^2, *calculate* first the viscous drag on the organism, assuming it to be a sphere, and secondly *calculate* the rate of energy expenditure of these organisms in swimming. The energy released from one molecule of ATP on being hydrolysed to ADP, and usable for work, is about 40 kJ/mole, which works out as 7×10^{-20} J/molecule. Hence *calculate* how many molecules of ATP are hydrolysed per second in one organism.

The flagellum does not hydrolyse one molecule at a time. It has been concluded that all along the flagellum there are sites at which both the hydrolysis and the action take place. These have been identified, by electron-microscope observation and reconstitution experiments.

The electron microscope

For the present purpose the electron microscope will be treated merely as a rather better microscope which can give considerably better magnification than a light microscope. (Later in the course, the principle of this instrument will be explained to you). Specimens in electron microscopy must be stained with a heavy-element stain to give contrast, and must be supported on a film. For this purpose a grid of copper which has been covered by a very thin film of plastic or carbon is used. Take one of the grids provided and have a look at it in the light microscope. *Measure* the size of a typical square on the grid. You can use that to give you a rough idea of the scale in the electron microscope. Better ways are given at the end of this script.

The simplest technique of staining is negative staining in which the relatively transparent object is dried down into a pool of electron-dense material (such as uranyl acetate) and can be observed as a light patch in a dark background.

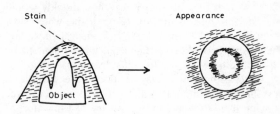

Figure 2

Imagine a perspex object in water (hardly visible) and in ink; the object became detectable by contrast. Small groups will be taken to an electron microscope and will be shown what the resulting pictures look like. First, a magnification at which one square just fits inside the field of view will be demonstrated (1000x), followed by another magnification which is sufficient to resolve large molecules (40 000x). In this way some of the remarkable molecular structure of the flagella will be visible.

The interpretation of this structure is work which is still proceeding (in this department and elsewhere and not agreed about as yet, but pictures will be on display.

In outline: thin sections of the flagella show '9 + 2' structure in which a central 2 fibres are surrounded by 9 pairs of fibres. These outer pairs are joined firmly along one side so that in section they form a 'figure 8' with the long axis inclined slightly to the tangential direction. On the end of the figure 8

116

losest to the centre, a pair of 'arms' have been etected. The protein which makes these up, called ynein, is an ATPase. You may also be able to see pokes'.

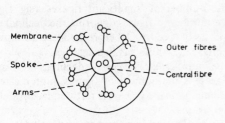

Figure 3

Recently it has become clearer what these structures are for. Gibbons (Proc. Natl. Acad. Sci. U.S.A. 8, 3092, 1972) has shown that sperms treated with etergent lose both heads and membrane but will still wim when ATP is added. If trypsin is added briefly, owever, the spokes are broken and, on adding ATP, he outer 9 fibres are *extruded* from the cilium. Thus TP seems to supply energy for a relative displacement etween the outer fibres. Normally the spokes stop rge-scale movement so that only bending can result. ossibly the 'arms' ('legs' might be better) walk up he adjacent fibre. The central fibres may be involved a coordinating the movement.

Photographs of negatively stained fibres are proided, with the magnification at which they were aken indicated. Identify the outer 9 and the entral 2 fibres. Repeating structures are very clear on he central 2 (function not known) but much more ague on the outer 9. Longitudinal striations are isible on the micrographs, and these may well epresent the packing of the individual sub-unit roteins of which the fibres are composed. We are own to the molecular level.

Measure the distance between the 'arms'. This is ifficult to do exactly, because it is by no means clear vhat to count, but our previous measurements have nly been rough so that we do not require great recision. But this is no excuse for carelessness; be uite clear what you have done and explain it well. rom this figure, and the length of the flagellum, btained from the light-microscope measurements, alculate the number of ATPase sites per flagellum nd hence get an idea of the number of ATP nolecules hydrolysed per site per second. In this way e have inferred what is happening at the molecular evel.

The most misleading thing, of course, about these electron-microscope pictures is their complete lack of motion: how quickly, for example, would the fuel, ATP, be able to diffuse from the body of the organism right up the flagellum? *Make an estimate* of this (from your previous results) and try to decide whether diffusion alone is sufficient to keep the flagella active.

Magnification in Electron Microscopy
Using the grid itself as a scale, as we did, is extremely crude and better methods of estimating the magnification have been devised. The latex particles used earlier are made for this purpose and have accurately known diameters ($1 \cdot 099 \pm 0 \cdot 001 \ \mu m$). Smaller ones are also made. The lines on the finest optical diffraction gratings are about $1 \ \mu m$ apart, and may similarly be used as length standards. An excellent scale, which is applicable right down to molecular dimensions, is a crystal of the enzyme catalase. This is a protein of molecular weight 250 000 which produces crystals of great regularity. The spacings can be independently measured by X-ray crystallography, and then used to calibrate distances in the electron microscope.

The appearance of these crystals in the electron microscope is truly remarkable, and if time permits, students will be taken to see this. Again, individual molecules can be readily detected. The diameter of these particles is 9 nm. Given that their density is $1 \cdot 3 \ g/cm^3$, *calculate* their mass. Hence obtain their molecular weight by multiplying by N_0. Does your answer agree with the quoted molecular weight? This, of course, was precisely what we did with the latex particles.

The significance of Brownian motion
This is a subject which is well worth pondering over. Democritus realised that the characteristic of all phenomena is *change*, and, wondering how change is possible, realised that matter must be divisible and also must be in motion. It would be difficult to find a more profound conclusion from such a simple argument. By observing Brownian motion we are looking at the first direct evidence for Democritus' conclusions and this was put forward 2200 years afterwards.

So in the living organism Brownian motion is indeed useful—it is necessary for diffusion and also for reactions to take place, but another interesting question is: Can the organism use the energy of Brownian motion to do *work*? That is, the *Chlamydomonas* is continually being pushed in all directions—is there any way for it to *select* one particular direction

117

for it to be moved so that it would need no other source of energy? The accepted answer is *no*, but do not accept it, instead try to think of ways for the organism to do this; this will give you a better understanding of why they cannot! If, after some thought, you can think of no solution, comfort yourself with the fact that there would be such an enormous advantage for an organism to be able to dispense with food that, were is possible, natural selection would certainly come up with some way of doing it. And, since all living things require food (even after 2–3×10^9 years of trying to do without it) we can be fairly certain that Nature hasn't thought of a way to do it either, and it must be impossible.

Results
(See also previous section)

On Large Numbers, Approximations and Errors

1 *Large numbers* If we want to work out $\dfrac{33\,000 \times 5\,670}{0.008}$ it is only too easy to lose a few noughts in the process. With very large numbers, such as those used in this practical, it is even easier. It is convenient to re-write the fraction above as:

$$F = \frac{3.3 \times 10^4 \, \times 5.67 \times 10^3}{8 \times 10^{-3}}$$

or

$$F = \frac{3.3 \times 5.67}{8} \times 10^{(4+3)-(-3)} = 2.3 \times 10^{10}.$$

This makes errors easier to avoid.

$$N.B. \frac{1}{10^{-n}} = 10^n.$$

$$\sqrt{(3 \times 10^5)} = \sqrt{(30 \times 10^6)}$$

$$= (\sqrt{30}) \times 10^{-3}.$$

We apologise for such elementary remarks, but one year we had errors of factors of 10^6 or more.

2 *Approximations* Suppose we know that the radius of a sphere is somewhere between 1.5 and 2 mm. Its volume, $\frac{4}{3}\pi r^3$, will therefore be between 14 and 33 mm^3. Now π can be evaluated as accurately as we wish, e.g., $\pi = 3.14115926536$, but if the volume might be anywhere within a factor of two there is no point in using any more than the first figure, i.e. $\pi \approx 3, V \approx 4\,r^3$.

In the practical we are interested only in orders of magnitude, i.e. factors of 10, and few measurements will be more accurate than $\pm 20\%$. Hence 1, or at most 2, significant figures is all that need be given. (One year we had the volumes worked out using five-figure logs and quoted to five figures.)

Notes

(a) The number of 'significant figures' (sig. figs.) is the number of figures, ignoring the decimal point and zeros occurring before or after all non-zero digits, e.g. 0.0068 has two sig. figs.

(b) The number of 'places of decimals' (or simply places) is the number of figures quoted after the decimal point, e.g. 3.14 is to two places (and to sig. figs).

(c) The convention is that when a quantity is quoted to n places, if the $(n+1)$th figure after the decimal point is greater than or equal to 5, then 1 is added to the nth figure, but, if not, then the nth figure is left alone. e.g.: π to 4 places is 3.1416; to 2 places it is 3.1.
Note that $x = 3.0$ contains more information (i.e., $2.95 \leqslant x < 3.05$) than $x = 3$.

(d) Factors like $\frac{1}{2}$ are like π; they are as accurate as necessary. Hence $x = 0.53$ is *not* $\frac{1}{2}$ to 1 place of decimals, but 0.5.

3 *Errors* No measurement should *ever* be quoted without *some* indication of the errors involved. While a measurement given as 3.065 g strictly should imply that it is within 3.0645 g and 3.0655 g, very often it does not and, to avoid ambiguity, an error, e.g. 3.0650 ± 0.0005 should also be given. Any measurement will contain a *random* error and may also contain a *systematic* error. Random errors are due to uncontrollable factors and cannot be eliminated because they are an inherent property of the measuring system. Systematic errors can be either eliminated altogether, with care, or may be reduced to a point when they have less effect than the random error (and then don't matter).

(a) *Random error*
Suppose we need to know the delivery volume of a pipette. For this purpose we could weigh aliquots of pure water delivered from the pipette. Suppose we do this and obtain $m = 0.2493$ g. How do we know this is correct? We do not, because slight variations in manipulation will affect the answer. The measurement must be repeated.

Suppose that 5 readings are obtained: 0.2493, 0.2496, 0.2488, 0.2495, 0.2496 g. We take the mean 0.24936 g. But this is misleading because the measurements differed in the third and fourth place. So we must express the mean to show the expected variation., i.e. $\overline{m} = 0.2494 \pm$ error.

(i) We could simply show the maximum *spread* of the results by quoting the largest difference, i.e.: $m = 0.2494 \pm 0.0006$. But this is unfair because most of the readings agreed better than this.

(ii) Better, we could quote the *mean deviation*, ignoring the sign of the deviation, i.e., taking the modulus. That is; if m is a reading and $\overline{m}$ the mean then, with n readings.

$$\text{Mean deviation} = \frac{1}{n} \Sigma \left| \overline{m} - m \right| ,$$

where Σ means 'the sum of all' and $| \; |$ 'the modulus of'.
In this case we would obtain $m = 0.2494 \pm 0.0003$.

(iii) Best of all, we could calculate the *standard deviation of the mean*.
If we take n measurements of a quantity $m_1, m_2, m_3, \ldots m_i, \ldots m_n$ then the *mean*, m, is

$$\overline{m} = (\tfrac{1}{n}) \, \Sigma \, m_i;$$

the *standard deviation* σ, is

$$\sigma = \surd (\tfrac{1}{n} \Sigma \, (m_i \overline{m})^2)$$

and the *standard deviation of the mean*, σ_m, is

$$\sigma_m = \surd (\tfrac{1}{n^2} \Sigma \, (m_i - \overline{m})^2).$$

σ_m is the best measure of the *precision* of a set of measurements; it estimates how much we would expect the 'true' value to differ from the mean actually obtained by measurement.

This figure, however, is only of interest where *large numbers of measurements* are taken and where statistical comparisons are being made. For the purposes of giving the accuracy of a measurement, the mean error is good enough. (It gives the same answer in this case anyway.)
Thus we can say weight of aliquot $= 0.2494 \pm 0.0003$ g. That is; the reproducibility of the measurement is just under 0.1% (actually obtained using a Lang pipette).

(b) *Systematic error*

We want the volume, however. Suppose that, by mistake, we had used an aqueous salt solution of specific gravity 1.01, thinking it was water, then in calculating the volume we would make a systematic error of 1%. If we ignored the temperature we might make another, even with pure water. There would be *systematic* errors due to faults in the method.

(c) *Estimating errors*

Sometimes it is not practical to take enough readings to obtain the random error, in which case an estimate—a reasonable guess—should *always* be made. At the same time other sources of error should be carefully considered. In the present example we should consider whether the temperature of the water makes any difference (at a 0.1% accuracy a change of 1°C is sufficient to make a real difference. At ± 1% it could be ignored); whether the evaporation of the water whilst being weighed introduces systematic error (it does, a stoppered bottle is used); whether the pipette will drain the same way with other solutions (it usually does), and so on.

A competent experimenter always considers such points and estimates how much they affect the final result.

No practical write-up can be considered complete without a discussion of errors. It is sufficient to discuss the error of the *final* results rather than of each measurement separately (which gets tedious for everyone). Errors are applied common sense; ask yourself 'How far can I trust this result?' The answer lies in your discussion of errors.

Brownian motion

These remarkably simple derivations are due to Einstein.

(a) *The average diffusion distance and the diffusion coefficient*

Consider a suspension of particles in a tube of unit cross-sectional area. Let the concentration, c, of these particles increase in the x-direction. Therefore there will be a net diffusion in the opposite direction, and if we consider a plane at x we may define D, the diffusion coefficient, by

$$\frac{\mathrm{d}n}{\mathrm{d}t} = -D \frac{\mathrm{d}c}{\mathrm{d}x} \qquad (1)$$

Note that $\mathrm{d}n/\mathrm{d}t$ is negative, because the movement is in the opposite direction to the increase of concentration. We want to derive the relation between D and x, the average distance moved by an individual particle in a time t.

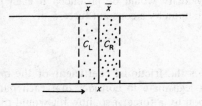

Figure 4

Consider therefore the volume enclosed by x and $x + \bar{x}$. Let the concentration in this volume be C_R. Then there will be $C_R x$ particles in it.

The particles are moving at random. Hence, on average, in the time t, half will move $\bar{x}$ in the positive direction and half $\bar{x}$ in the negative direction. (This is what we mean by $\bar{x}$.) Hence the number crossing the plane at x from the right, in the time t, will be

$$n_- = \tfrac{1}{2} C_R \bar{x},$$

and similarly

$$n_+ = \tfrac{1}{2} C_L \bar{x}$$

from the left.

Thus the net flow across x will be

flow across x in positive direction $= \tfrac{1}{2} (C_L - C_R) \bar{x}$.

$$\therefore \frac{dn}{dt} = \tfrac{1}{2} (C_L - C_R) \bar{x}$$

But $\dfrac{C_R - C_L}{\bar{x}} = \dfrac{dc}{dx}$

$$\therefore \frac{dn}{dt} = -\frac{\bar{x}^2}{2t} \frac{dc}{dx}$$

$$\therefore D = \frac{\bar{x}^2}{2t} \tag{2}$$

Thus the distance diffused is on average proportional to the square root of the time, i.e. it takes four times as long to migrate twice the distance. (This is so because the particles as often reverse their direction as continue it).

(b) *Osmotic pressure and diffusion*

This flow rate $\dfrac{dn}{dt}$ is the same as if each particle had been given a small velocity v, where

$$\frac{dn}{dt} = Cv. \tag{3}$$

Such a velocity would be produced if each particle was acted upon by a constant force F, if

$$F = fv, \tag{4}$$

where f is the frictional coefficient of the medium. But is it legitimate to consider that each particle is acted upon by a force; if so, how big would the force be?

120

Suppose that the plane at x was a semi-permeable membrane. Then an osmotic pressure would be developed at x. Imagine that the semi-permeable membrane can be moved by this pressure. Then the membrane will be displaced until the concentration of the particles is uniform. But this is the point where diffusion stops also. Hence the osmotic force and the diffusion force are identical.

If C is in particles/unit volume, then a mole of particles are found in a volume N_0/C, where N_0 is the Avogadro number.

Therefore the osmotic pressure π at x is

$$\pi = \frac{C}{N_0} RT,$$

where R is the gas constant and T the absolute temperature. What we are interested in however is the pressure causing the flow, i.e. how it varies as the concentration varies. But differentiating

$$d\pi = \frac{RT}{N_0} dC.$$

Thus we can imagine that the flow is caused by a pressure $d\pi$ acting on all the particles together within the volume between x and $(x + dx)$. In this volume there are $C\,dx$ particles.

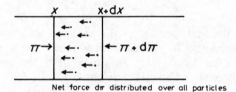

Net force $d\pi$ distributed over all particles

Figure 5

$\therefore$ The force on each particle, F, is

$$F = -\frac{RT}{N_0 C} \frac{1}{dx} \frac{dc}{dx}$$

Note that f is negative, because it is a *back*-pressure

$\therefore$ From (4), $v = -\dfrac{1}{C} \dfrac{RT}{N_0 f} \dfrac{dC}{dx}$

$\therefore$ From (3) $\dfrac{dn}{dt} = -\dfrac{RT}{N_0 f} \dfrac{dC}{dx}$

$$\therefore D = \frac{RT}{N_0 F} \tag{5}$$

ence we have a relation between x, the distance
migrated in t, and F, which may be calculated from
the molecular parameters, thus:

$$\bar{x}^2 = \frac{2RT}{N_0 F} \, t \qquad (6)$$

ence if x is measured and the particle is a sphere of
radius r, then $F = 6\pi\eta r$. Therefore we can measure N_0
R and T are known.

$$R = 8\cdot3 \text{ J/deg. mole.}$$
$$T = 300^{\circ}\text{K.}$$
$$\eta = 10^{-3} \text{ N s m}^{-2}$$

stimate N_0 from your measured values of r and x.
) *The relation between molecular weight and x for*
spherical molecules
et molecular weight of molecule $= M$.

Mass of particle $= M/N_0$

et ρ = density of particle.

Volume $= \dfrac{M}{N_0 \rho}$

r = radius of particle, volume $= \dfrac{4 \pi r^3}{3}$

$$\therefore r = \left[\frac{3M}{4\pi\rho N_0}\right]^{1/3}$$

$$\therefore F = 6 \pi\eta \left[\frac{3M}{4\pi\rho N_0}\right]^{1/3}$$

ut
$$\bar{x}^2 = \frac{2RT}{N_0 F} t \qquad (6)$$

$$\therefore \bar{x}^2 = \frac{2RT}{6\pi N_0 \eta} \left[\frac{4\pi\rho N_0}{3M}\right]^{1/3}$$

$$\therefore x \propto \frac{1}{M^{1/6}}$$

hat is, if a molecule has molecular weight 2×10^6,
hen it will diffuse only $1/10$ as far as H_2 in the same
me.

xtensions
he approach is invaluable in all research. To be
ware that you *may* be making fundamental errors,
ven when taking extreme care, simply because
omething is so obvious that you have not questioned
, is an important thing to realise. It prevents a
ogmatic attitude; it leaves the mind flexible enough

to contemplate changes. Another useful lesson perhaps is that ultra accuracy just for its own sake is
pointless; one only needs to be accurate enough for a
particular purpose, namely, to test a particular idea.

Further reading
(1) Explanation of Brownian Motion.
    Einstein, A.
        1908 Zeit. fur Electrochemie 14,235
        1910 'Investigations on the theory of
        Brownian Motion' (English translation),
        Dover books, New York
(2) More on the philosophy of science which shows
    how unconscious prejudice meets things it can-
    not explain and thus discovers problems; makes
    guesses at the answer and then tests the guess:
    Popper, K.R.
        1963 'Conjectures and Refutations', Rout-
        ledge and Kegan Paul, London
(3) A paper which has attempted to wrestle with the
    problem of whether Brownian Motion can be
    used to do useful work:
    McClare, C.W.F.
        1971 'Chemical Machines, Maxwell's Demon
        and Living Organisms', J. Theor. Biol., 30,
        1−34

Hints for the teacher:
We usually find that it is best to aim at getting as far
as measuring the diameter and the speed of swimming
of the *Chlamydomonas* by the end of the first
session. The students should have obtained rough
answers for the various quantities by the beginning of
the next. It is best to conduct the next session as a
seminar−which can be pretty lively−while groups go
away to be shown the electron microscope and the
structures visible in it. They can do the measuring up
of the photographs during this time. Latex particles
(spheres) can be obtained from: Polaron Equipment
Ltd., 60 Greenhill Crescent, Watford, Herts.

C.W.F. McCLARE
King's College
London.

2 Cytoplasmic streaming in *Nitella*, *Physarum* and *Amoeba*

Time required
Variable: this can be as little as 2 hr, but 3—4 hr is preferable.

Assumed knowledge
Use of microscope (phase-contrast if available) and bench centrifuge.

Theoretical background
This practical is an investigation of three types of cytoplasmic streaming. Movement of the cytoplasm without change in cell shape can be seen in plant cells (cyclosis) where cytoplasm circulates in a continuous band close to the cell wall. Gradual progression can be seen in the plasmodia of myxomycetes (slime moulds), e.g. *Physarum*. In these organisms cytoplasmic flow occurs in well-marked 'channels' and is characterised by rapid reversal of direction. Amoeboid movement is a term given to locomotion in a variety of cells which involves change of cell shape and flow of cytoplasm. In *Amoeba* movement is in the direction of the extending pseudopodium, while in many tissue cells movement involves extension and retraction of pseudopodia, the cell being drawn to the point where the tip of the retracting pseudopodium attaches itself to the substratum.

Necessary equipment
Each student should have:
A culture of *Nitella* (or *Chara*), *Physarum polycephalum* and *Amoeba*
Binocular microscope
Microscope with good standard optics (or phase-contrast)
Slides and cover slips (about 8)
Length of cotton
Stopwatch or clock
Graduated eyepiece and micrometer slide (several for the class)
Solid watchglasses (2)
Pasteur pipettes, pulled a second time to make a finer pipette, and teats

Use of bench centrifuge and plastic 15 ml centrifuge tube
10 ml Chalkley's medium
10 ml Chalkley's medium + fine suspension of carmine particles (BDH)
10 ml Chalkley's medium + 0·01% Neutral Red (BDH)

Type of group
Individual students, number in the class does not matter.

Instructions to the student
Nitella (or Chara)
Make a preparation of a small piece of algal material on a slide. Support the cover slip by small pieces of plasticine at the corners and mount in culture medium.
1 Using a microscope, look for cyclosis and try and estimate the rate of flow of the cytoplasm (make use of graduated eyepiece and stopwatch).
2 Centrifuge some material so that you have vacuolate and non-vacuolate halves. Ligature the mid-line with cotton and observe the cytoplasm in the non-vacuolate portion. Is there any streaming?
3 Try centrifuging at a higher speed—can you remove the chloroplasts? If so, does streaming continue?

Physarum polycephalum
The plasmodium is growing over agar; remove the lid of the Petri dish and observe *Physarum* under a binocular microscope. Make some sketches indicating direction of flow in the channels.
(a) Obtain some information of the rate of flow of cytoplasm in the channels.
(b) Record the number of reversals of flow in a given time.
(c) Try 'prodding' the plasmodium or banging hard on the bench (remember this may affect your neighbour's material!). How soon does streaming recommence?

Amoeba
(Adapted from a practical of Dr L.G.E. Bell).
Using a fine pipette, place some amoebae in a drop of culture medium on a slide and cover with a cover slip bearing a small piece of plasticine at each corner (for support).
Using a binocular microscope:
1 Do the amoebae move before attaching to the glass?
2 Select a field of view with several cells; draw the

field and every five minutes indicate the new positions of the amoebae. (In this way you can draw the tracks of the amoebae.)

Observe another preparation under a microscope, using a minimum amount of light.

3 Draw several amoebae in outline—is the shape constant, or is any one shape preferred? Try and estimate rate of flow of cytoplasm.

4 Is there a front and rear?

5 Draw the advancing pseudopodium—indicate clear and granular cytoplasm and direction of movement.

6 What happens in a retreating pseudopodium?

7. What happens in strong light?

Make a preparation of amoebae in a fine suspension of carmine particles. Some particles will stick to the surface which will enable you to look at surface movement.

8 Do carmine particles move at the same rate as the cytoplasmic particles, and in which direction?

9 What happens when carmine particles reach the tip of the pseudopodium? Place several amoebae in a watchglass and add 0·01% Neutral Red in Chalkley's medium. At varying intervals make preparations and look at the internal distribution of the dye—do you find that it accumulates in the tail region?

Discussion

With approximate rates of flow available, which of the three organisms showed the fastest flow rate? What other questions might be asked about the three types of streaming, e.g. what about use of ATP or chemical inhibitors? Try and list requirements of this type of locomotion, e.g. forces involved, control mechanisms, etc.

Extensions and applications

The general approach used in this practical was to indicate the importance of observation; any theory ...ust be supported by fact, observational or experimental. Nevertheless, these experiments have been on apparently structureless material and it is for other techniques both electron microscopical and biochemical to throw more light on the problem of the structures involved.

Further reading

Kamiya, N.
1960 'Physics and Chemistry of Protoplasmic Streaming', Ann. Rev. Plant Physiol., 11, 323
Wolpert, L.
1965 'Cytoplasmic Streaming and Amoeboid Movement', Symp. Soc. Gen. Microbiol., 15, 270
Wolpert, L. and Gingell, D.
1965 'Cell Surface Membrane and Amoeboid Movement', Symp. Soc. Exp. Biol., 22, 169
Pollard, T.D. and Ito, S.
1970 'Cytoplasmic Filaments of Amoeba proteus, I', J. Cell Biol., 46, 267
Pollard, T.D. and Korn, E.D.
1971 'Filaments of Amoeba proteus, II', J. Cell Biol., 48, 216
Spooner, B.S., Yamada, K.M. and Wessells, N.K.
1971 'Microfilaments and Cell Locomotion', J. Cell Biol., 49, 595
Williamson, R.E.
1972 'A Light Microscope Study of the Action of Cytochalasin B on the Cells and Isolated Cytoplasm of the Characeae', J. Cell Sci., 10, 811

Hints for the teacher

The living material can be obtained as follows:

1 *Nitella* (or *Chara*) from Gerrard and Haig Ltd., Gerrard House, Worthing Road, East Preston, Nr. Littlehampton, Sussex BN16 1AS.

2 *Physarum polycephalum* Vigorous plasmodia are required for observation and Dr M. Carlile, Department of Biochemistry, Imperial College London, has kindly agreed to supply this myxomycete to people wishing to put on the practical. Since plasmodia cannot be sent by post, Dr Carlile will supply *only personal callers*, provided a few day's notice is given. These plasmodia can then be transferred from their nutrient agar onto sterile non-nutrient agar for the practical by cutting out squares of nutrient agar and plasmodium using a sterilised scalpel. The small piece of *Physarum* will migrate across the non-nutrient agar and will spread over the plate within 18—24 hr. Students may like to keep their plasmodium after the practical, and should add a little moistened porage oats to the agar plate.

A supplier of *Physarum* resistant stage or sclerotia is Harris Biological Supplies Ltd., Oldmixon, Weston-super-Mare, Somerset.

3 Amoebae can be obtained from Cambridge Culture Collection, 36 Storey's Way, Cambridge or from myself, if advance warning is given and stamps are sent to cover postage. Amoebae are relatively easy to grow in Chalkley's medium. Two methods are given in the 'Hints for Teachers' (Section 7.4)in the Pinocytosis practical. Chalkley's medium consists of the following:

16 g NaCl
0·8 g NaHCO$_3$
0·4 g KCl
0·2 g Na$_2$HPO$_4$.12 H$_2$O
} in 1 litre *glass* distilled water

Dilute 5 ml of conc. solution to 1 litre glass distilled water for use.

SHIRLEY E. HAWKINS
King's College
London

3 The Regeneration of Cilia in *Tetrahymena*

Time required
This study of regeneration can be completed in 3 hr, but 4 hr should be allowed to investigate the effects of inhibitors in more detail.

Assumed knowledge
The student should be able to use a phase-contrast microscope and a haemocytometer. Some facility with sampling techniques is useful, as is general laboratory 'awareness', so that the experiment is more suitable for advanced students.

Theoretical background
These experiments are designed to investigate the regeneration of cilia in *Tetrahymena pyriformis* and were based on the work of Rosenbaum and Carlson (1969). Earlier methods for the detachment of cilia involved lysis of the cells. Modification of these methods by lowering pH, omitting ethanol and using mechanical shearing in the presence of calcium, detaches cilia at the level of the pellicle, leaving immobile although swollen cells. The rate of ciliary regeneration of cells grown at 25°C can be compared with that of cells grown at 17°C. The synthesis and assembly of microtubule sub-units during regeneration can be investigated using different inhibitors of known activity, e.g. cycloheximide (an inhibitor of protein synthesis on 80s ribosomes), neomycin (an inhibitor of protein synthesis on 70s ribosomes) and colchicine (an inhibitor of microtubule assembly).

Necessary equipment
The experiment should be run by individual students, who each require the following:
1 x 250 ml culture of *Tetrahymena pyriformis* in 1% proteose peptone and grown in a 500 ml Erlenmeyer flask at 25°C.
1 x 250 ml culture of *Tetrahymena pyriformis* in 1% proteose peptone and grown in a 500 ml Erlenmeyer flask at 17°C.
2 x 50 ml Erlenmeyer flasks each containing 15 ml 1% sterilised proteose peptone
3 x 100 ml Erlenmeyer flasks each containing 20 ml 1% sterilised proteose peptone

1 x 100 ml Erlenmeyer flask containing 20 ml 1% sterilised proteose peptone plus 10 μg/ml cycloheximide (Sigma)

1 x 100 ml Erlenmeyer flask containing 20 ml 1% sterilised proteose peptone plus 50 μg/ml neomycin (Sigma)

1 x 100 ml Erlenmeyer flask containing 20 ml 1% sterilised proteose peptone plus 2 mg/ml colchicine (Sigma)

Ice bucket containing ice together with
  25 ml medium A (10 mM EDTA.2Na; 50 mM sodium acetate; pH 6·0)
  20 ml distilled water
  5 ml calcium chloride 0·2 M

Bench centrifuge and plastic tubes

3 large centrifuge tubes or boiling tubes, capacity at least 15 ml

Graduated pipettes, 5 x 5 ml and 1 x 1 ml

Stopclock or watch

Water bath running at 25°C and rack for 100 ml flasks

2 x 5 ml syringes fitted with 19 gauge needle (Gillette Surgical)

5 Pasteur pipettes with teats or constant-volume pipettes with wide bore

Phase-contrast microscope

Haemocytometer

Slides and cover slips

Solid watchglasses

Parafilm

Graph paper

Marking pencil

Type of group
Individual students

Instructions to the student

1 *Tetrahymena*, grown at 25°C, should be concentrated by centrifugation (800 g for 10 mins) and resuspended in fresh proteose peptone. Resuspend in 15 mls proteose peptone in a labelled 50 ml Erlenmeyer flask. This should be repeated for *Tetrahymena* grown at 17°C—remember to mark flask with growth temperature.

2 The next steps must be followed carefully, and all are *on ice*:
   (a) At zero time place 2·5 ml concentrated *Tetrahymena* suspension (grown at 25°C) into 5 ml medium A on ice, using a large tube—mix by swirling.
   (b) At zero + 30 s add 2·5 ml cold distilled water.

   (c) At zero + 90 s add 0·25 ml cold 0·2M $CaCl_2$, cover with parafilm and mix by inversion.
   (d) At 3 min 30 s shear in a syringe fitted with a 19 gauge needle (i.e. draw up syringe and expel into cold tube). Quickly remove a sample, examine in a watchglass and check that all are immobile.
   (e) Immediately after deciliation place 1 ml deciliate cells into a 100 ml Erlenmeyer flask containing 20 ml proteose peptone, previously placed to warm in water bath at 25°C. Incubate at 25°C and commence experiment.

3 Assay % motility with time by withdrawing a constant-volume sample every 10 min. Use a Pasteur pipette and mark volume. Remember to agitate the flask before sampling since deciliate cells are immobile. Examine in a haemocytometer, or a watchglass and estimate numbers of motile and non-motile cells. Make quick preparations and observe the condition of the cells under phase contrast, for example are they swollen? Can you see any new growing cilia?

4 Deciliate *Tetrahymena* grown at 17°C and compare the regeneration time with those grown at 25°C (run experiment at 25°C).

5 Prepare some fresh deciliate cells from concentrated *Tetrahymena* suspension grown at 25°C. Add 1 ml of deciliate cells to each of the following flasks:
   (i) proteose peptone + 10 μg/ml cycloheximide
   (ii) proteose peptone + 50 μg/ml neomycin
   (iii) proteose peptone + 2 mg/ml colchicine
   (iv) proteose peptone only as a control
Sample flasks every 10 min. Keep experiment running for as long a time as possible to see whether their response alters.

Recording results
Present your results as a graph of % motility against time in minutes for (1) *Tetrahymena* grown at 25°C and 17°C and (2) the effects of inhibitors on the regeneration.

Discussion and extensions
The time taken for the regeneration of cilia is the resultant of many processes including synthesis and assembly of microtubule protein. Why would the previous temperature at which the cultures were grown influence the regeneration rate? Using results from inhibitors, can we determine how far the regeneration depends on the size of the existing precursor pool and on new protein synthesis during

the experiment? What can be learnt about protein synthesis in one lower eukaryote? As an extension to these experiments, further inhibitors might be used to indicate whether production of new informational molecules was required, or whether the stage in the cell cycle was important. The inhibitors might be investigated for reversibility, since after a period of time cells overcome the assembly block imposed by colchicine. These experiments look at a system of regeneration where the endpoint is readily recognisable, and have provided information on the synthesis and assembly of microtubular structures such as cilia and flagella.

Further reading

Rosenbaum, J.L. and Carlson, K.
  1968 'Cilia Regeneration in *Tetrahymena* and its Inhibition by Colchicine', J. Cell Biol., 40, 415
Rosenbaum, J.L. and Child, F.M.
  1967 'Flagellar Regeneration in Protozoan Flagellates', J. Cell Biol., 34, 345
Rosenbaum, J.L., Moulder, J.E. and Ringo, D.L.
  1969 'Flagellar Elongation and Shortening in *Chlamydomonas*', J. Cell Biol, 41, 600
Coyne, B.U. and Rosenbaum, J.L.
  1970 'Flagellar Elongation and Shortening in *Chlamydomonas*, II', J. Cell Biol, 47, 777

Hints for the teacher

1 *Tetrahymena pyriformis* can be obtained in pure culture from Cambridge Culture Collection of Algae and Protozoa, 36 Storey's Way, Cambridge and then grown to produce the number of cultures required. They are grown in 1% proteose peptone [1% proteose peptone (Difco); 0·25% yeast extract (Difco)] which should be sterilised at 15 lb/in^2 for 15 min. Sterile techniques must be used to inoculate cultures and throughout growth, but there is no need to run the class experiment using sterile sampling techniques. A 500 ml Erlenmeyer flask containing 250 ml proteose peptone will take 1–2 days at 25°C and 2–3 days at 17°C to reach numbers required experimentally. A culture of this size will serve to inoculate 25 new, 250 ml cultures. I have found that the growth temperature is important and that rate of regeneration is considerably slowed unless the *Tetrahymena* have been grown at 25°C for at least 6 days. Two different growth temperatures thus form the basis of part of the experiment.

2 Centrifugation of *Tetrahymena* requires care, because the cells tend to 'swirl up' during

deceleration. If a lot of students are involved, or they are inexperienced, it is better for a technician to centrifuge the cells just prior to the practical. Generally 800 g will suffice, but up to 1 000 g can be used.

SHIRLEY E. HAWKINS
King's College, London

4 The structure of myofibrils and the mechanism of contraction

Time required

Minimum of 3·5 hr if the demonstrator prepares the fibrils.

Assumed knowledge

No knowledge assumed, but the students must be familiar with the use of the phase-contrast microscope.

Theoretical background

Myofibrils are the contractile elements in a muscle cell. They are made up of two kinds of filaments which can be resolved only by electron microscopy. In cross-striated muscles, however, the filaments are arranged in separate assemblies and these can be distinguished in the light microscope, so that it is possible to observe their positions relative to one another and to study how this varies with the length of the fibril. In practice, because the cross-striated fibril is a linear series of identical units (sarcomeres), we need only study how the positions of the filament assemblies vary in a single sarcomere. Using a phase-contrast microscope, we can study 'living' fibrils while they are contracting, and deduce how the filaments move. The sliding-filament theory of contraction, proposed in 1954, was based on the results of observations and experiments very much like the ones suggested in this article.

The purpose of the exercises suggested here is to give you the opportunity to test for yourselves some generally accepted fundamental concepts about the mechanism of contraction. I will not spoil the fun by describing these concepts now. I must, however, explain why the experiments are to be done on fibrils initially in the rigor state, and why contraction is induced by giving them ATP.

The chain of processes coupling the nervous excitation of the muscle cell (fibre) with the contraction of its fibrils goes as follows:

1 Changes in the muscle cell membrane, elicited by excitation, spread rapidly over the whole surface (action potential) and also into the interior along numerous transverse tubules which are invaginations of the surface membrane. In this way the stimulus is conducted to the vicinity of all the fibrils.

2 The transverse tubules make numerous contacts with the sarcoplasmic reticulum (cf. endoplasmic reticulum) that envelops all the fibrils. The stimulus transmitted to the reticulum causes it to release Ca^{++} ions which diffuse into the fibrils, raising the concentration around the filaments to about 10^{-6} M. (While the muscle is at rest, the activity of the Ca^{++} pump in the membrane of the reticulum keeps the concentration of Ca^{++} ions in the fibrils at less than about 10^{-8} M).

3 The Ca^{++} ions that enter the fibril are bound by a protein called troponin C which is situated together with actin in the thinner of the two kinds of filaments. Whilst the muscle is relaxed, actin is kept in an inhibited state by the co-operative action of troponin C and the other proteins of the regulatory system in the thin filaments. This inhibition is released when troponin C binds Ca^{++}. Actin is now able to interact with the myosin molecules in the thick filaments to convert the energy stored in ATP into mechanical work.

In summary, the contractility of the fibrils in the intact muscle is controlled by the local Ca^{++} concentration. When Ca^{++} is withdrawn from the fibrils they relax, provided sufficient ATP and Mg^{++} are present. When Ca^{++} enters the fibrils they contract, using the ATP. Normally, the muscle cell can keep its fibrils well supplied with ATP, but if the level of ATP falls too far, the fibrils go into a state of rigor.

Fibrils that are initially in the rigor state are recommended for the following experiments. But why not start with a preparation of fibrils in the relaxed state? In practice, it is not very easy to prevent the fibrils from contracting while separating them from the muscle, or to relax them again once they have contracted violently. On the other hand, it is not difficult to make good preparations of muscle in the rigor state, to store them in the deep-freeze, and to obtain large quantities of fibrils when needed. Much of the work on which the sliding-filament theory of contraction was originally based was done on fibrils prepared in this way. A very useful so-called 'glycerination' procedure, introduced by Szent-Györgyi, yields fibrils whose ability to contract in a normal manner has not been impared. They have lost none of the 10 or so different proteins that compose the contractile apparatus of the intact muscle. The only sign of damage is that glycerinated fibrils sometimes fail to relax properly: their actin molecules cannot be 'switched off' completely, presumably because one or more of the regulatory proteins has been damaged.

The term 'glycerinated', though conventional, is not very informative. Muscle fibres, held so that they cannot shorten, are placed in an aqueous solution of glycerol. This destroys membranes and soluble molecules diffuse out. The glycerol allows the preparation to be stored at a low temperature. After a few weeks the only organelles that remain functional are the myofibrils, and these are in rigor, but this state is reversible.

Necessary equipment

For each student
A phase-contrast microscope with a mechanical stage, a high-resolution oil immersion objective and a medium-power objective. An eyepiece with a measuring scale and, if possible, another one without a scale.

Immersion oil and lens tissue

20 slides and cover slips (of a suitable thickness for the oil immersion objective)

A few sheets of Kleenex

A few sheets of absorbent filter paper

20 Pasteur pipettes and 5 teats

10 small test tubes (e.g. 10ml) and a rack to hold them

Pen that will write on glass

For each group of 10 students
Two glycerinated psoas muscle strips, one long, one short

250 ml of standard salt solution (0·1M KCl, 1mM $MgCl_2$, 6mM phosphate buffer, pH 7·0)

25 ml of a 1mM solution of ATP in standard salt solution

50 ml of a solution containing 1·0M KCl, 10mM sodium pyrophosphate, 1mM $MgCl_2$, buffered with phosphate to pH 7·0

Insulated bowl of crushed ice (for the stock solutions) or easy access to a refrigerator

Extra slides, cover slips, test tubes, Pasteur pipettes, Kleenex, filter paper

Optionally, a micrometer slide

For the whole class
A blendor with a 25—50 ml cup (e.g. MSE or Virtis) and enough standard salt solution to prepare all the fibril suspensions needed

Pair of small scissors, two pairs of fine forceps

Optionally, a polarising microscope capable of resolving the A— and I—bands in fibres or bundles of fibrils

Type of group
Each student will do all the experiments himself. group of about 10 students can conveniently shar the stock suspensions of fibrils and the various stoc solutions.

Instructions to the student, Recording results an Discussion
Preparation and examination of fibrils
Each group of students is provided with two glycerin ated strips from the psoas muscle of a rabbit. Eac strip has been tied to a plastic rod when it was cu out of the carcass. One strip goes into rigor while hel at a length somewhat greater than the maximur attainable in the body. The other strip is slack whe tied, and it shortens while going into rigor. Thes strips are taken out of the deep-freeze a few hour before the experiment. The glycerol solution replaced by cold 'standard salt solution' (0·1M KCl 1mM $MgCl_2$, 6mM phosphate buffer, pH 7·0) an the strips divided into narrow bundles of fibres. Th demonstrator will show you how to obtain a suspen sion of fibrils by mechanically disintegrating th fibres. The stock suspension of fibrils must be kep cold (0—4°C), and it is desirable to keep all the othe stock solutions on ice or in the refrigerator. Eacl student in the group will take away small samples ir test tubes. Do not forget to label the tubes.

The stock suspension of fibrils is likely to be to concentrated to allow an uncluttered view, so dilute small sample with the standard salt solution. Using Pasteur pipette, place a small drop on a clean slid and lower a clean cover slip onto the drop withou entrapping air bubbles. Examine, using moderat magnification. Identify the myofibrils and try to fine an unbroken fibre (usually a few are left); notice th nuclei.

Before using the oil-immersion objective, mak sure that only a thin film of liquid is present unde the cover slip. Use filter paper to suck away any excess. To do this competently, tear a disc of filte paper to give you a triangular piece, and place on point of the triangle very near one edge of th cover slip making contact with the liquid under th slip, but without touching the slip. In the irrigatio experiments suggested below, you will need to b able to draw experimental solutions through th preparation while observing a myofibril at high magnification, and if you jar the cover slip with th filter paper you may lose the fibril.

Using the oil-immersion objective, examine fibrils from both long and short muscles. Identify the repeating units in the pattern of cross-striation. Most of the fibrils from the long muscle will look alike and you will be able to identify the Z-discs (narrow dense lines) delimiting the sarcomeres. The long bands of high optical density are called the A-bands and the long bands of low optical density, bisected by Z-discs, are called the I-bands. If a polarising microscope is available, confirm that the A-bands show anisotropy (the reason for naming them anisotropic bands) and that the I-bands are relatively isotropic. Identify also the H-zone, a short band of relatively low optical density mid-way along the A–band.

The fibrils from the shorter muscle will show a variety of band-patterns. Make a preliminary attempt to classify the main types of band-pattern you observe in the preparations from the long and short muscles, and correlate the length of the repeating unit with the type of band-pattern. Come back to this problem later after you have watched individual fibrils contracting; you will then find it easier to make sense of your observations.

General technique for experiments on fibrils
Make a thin preparation from a sample of a fibril suspension diluted so that only a few fibrils appear in the field of view at high magnification. Select a good-looking one lying just under the cover slip (where the best images are obtained with a phase-contrast microscope). Fill a Pasteur pipette with some of the experimental solution you are going to apply to the fibril. Place a piece of filter paper at one side of the preparation, bending it so that it stays in position without your having to hold it. Check that it is beginning to draw liquid through the preparation. Check that your chosen fibril is still in view. Now apply a small drop of the experimental solution (see below) to the opposite side of the preparation, placing it on the slide so that it makes contact with the liquid under the cover slip but does not spread over the top of the slip. Quickly transfer your attention to the fibril and keep it in focus as the height of the cover slip changes.

Contraction
Use fibrils from the long muscle for this experiment. Irrigate with a solution of ATP made by diluting the stock solution with standard salt solution. There is no need to add Ca^{++} ions because sufficient (above the threshold of about 10^{-6} M) are present as contaminants in the ATP solution and on the glass of the slides, pipettes, etc.

You will have to find the best concentration of ATP by trial. I suggest you start with 0·1 mM and then try 0·01 mM, etc. You need not make the dilutions very accurately (note the order-of-magnitude steps in the sequence I suggest), so use Pasteur pipettes, counting drops; for example, dilute one drop of 0·1 mM ATP with 9 drops of standard salt solution. Be as economical as possible: ATP is extremely costly. If the ATP solution is too concentrated, the fibrils will contract so rapidly that you will not be able to watch what happens. I suggest you go on diluting the ATP until you find that the fibril you are watching contracts a little and then stops. Now stop the irrigation (by withdrawing the filter paper) and look at other fibrils elsewhere in the preparation. Some will have contracted to short lengths, some not at all. The chances are that on repeating the experiment several times with this dilution of ATP you will get a good view of what happens during contraction; at least you will be able to deduce what happened by studying the population of fibrils at the end of irrigation.

Your record of the results of your observations on contracting fibrils, and on the band-patterns of fibrils obtained from long and short muscles, should be illustrated with drawings of sarcomeres, alongside which you note the relative sarcomere lengths, expressed in units of eye-piece scale divisions. Using a micrometer slide, and given the magnification, you can calibrate the eye-piece scale, but if it is not feasible to do this, make use of the fact that the A–band in the rabbit psoas muscle is 1·6 μm long.

Extraction of myosin
The myosin molecules in the fibril separate from one another as the ionic strength of the medium is raised, and provided they are prevented from linking on to the actin molecules, they can be extracted from the fibril. In the muscle, the actin–myosin cross-links are broken by ATP (in the presence of Mg^{++}). Inorganic pyrophosphate also breaks cross-links (if Mg^{++} is present). What are the similarities and differences between ATP and inorganic pyrophosphate?

The extraction solution you will now use contains 1·0 M KCl, 10 mM sodium pyrophosphate, 1 mM $MgCl_2$ and is buffered with phosphate at pH 7·0. Start with fibrils from the long muscle, and when you are satisfied that your results are well-defined and reproducible, apply the extraction solution to fibrils from the short muscle.

Plan carefully (with proper controls) and carry out an experiment to test whether or not fibrils that have been thoroughly extracted with this solution are capable of contracting.

1 Do the lengths of the A-band, I-band and H-zone stay constant or vary with the length of the sarcomere? Does the distance from the Z-disc to the edge of the H-zone vary with sarcomere length? If your observations suggest that any of these parameters stays constant, how accurately do you think you have shown this? Given that the filaments consist of regularly packed molecules whose arrangement repeats many times along the length of each filament, what physical method could be used to investigate whether or not the filaments in the intact muscle change their lengths according to the length of the muscle?

2 The long muscle strip had been moderately extended after it was cut out of the carcass. What would the band-pattern of the sarcomeres have looked like if the muscle had been extended even further?

3 What was the maximal amount of shortening you observed when you added ATP? Could the fibrils shorten as much in *situ*?

4 Assuming that you succeeded in extracting all the protein of the thick filaments, and given that the thin filaments remained after extraction, what can you deduce about the lengths and positions of the thin filaments from your observations on sarcomeres of different lengths after extraction? For instance, do the two assemblies of thin filaments (one attached to each Z-disc) abut or overlap in short sarcomeres?

5 Do your experiments suggest that myosin plays an essential role in contraction? Where, according to the results of your experiments, do you think myosin is situated in the intact sarcomere?

6 If you extracted all components of the thick filaments, and if the thin ones extend only from the Z-disc to the edge of the H-zone (as the literature has it), would you not expect the extracted fibril to disperse into short pieces? Does this happen to fibrils not stuck to cover slip or slide?

7 According to the sliding filament theory of contraction, the contractile machine in the cells of a muscle consists of discontinuous filaments of two kinds which change their positions relative to one another, but do not change their lengths, when the muscle contracts or is extended. Furthermore, the motion of the filaments during contraction is thought to be caused by the

interaction of myosin molecules in the thick filaments with actin molecules in the thin filaments; during this interaction, ATP is hydrolysed and the energy so obtained is used to drive the filaments past one another. You have tried to repeat some of the original experiments on which this theory is based. Are your results consistent with the theory?

8 There are many muscles that are not cross-striated, but it is known that they consist of two kinds of discontinuous filaments and contract by a sliding-filament mechanism. Can you suggest why such muscles appears unstriated (smooth) in the light microscope?

Extensions and applications

Phase-contrast optics allow us to observe many details of structure within living cells or in separated organelles still capable of functioning. It is obvious that the irrigation methods you have used can be applied generally in cell biology.

Using similar methods, there are many other interesting experiments that can be done on myo-fibrils. For instance, we can investigate whether or not a fibril that has contracted down to a short length will extend when exposed to a solution that causes it to relax (when, in the presence of ATP and Mg^{++}, the concentration of Ca^{++} is reduced to less than about 10^{-8} M with ethanedioxybis (ethylamine) tetra-acetic acid (EGTA)). We can do selective extraction experiments designed to discover the location of other proteins besides myosin. When such extraction experiments give clear-cut results in the light microscope, it is worth while to fix the extracted fibrils and examine them at higher resolution in the electron microscope, and the experiments can be repeated on bulk suspensions of fibrils, using gel electrophoretic methods to determine which proteins have been extracted and how much of each has been removed. Probably it will have occurred to you to ask if, once certain protein molecules have been extracted from a fibril, they can be put back again and, if so, whether the contractility or some other lost function is restored. You could also try to investigate if a fibril is capable of contracting once it has been stretched to such an extent that the thick and thin filaments no longer overlap one another. A more ambitious project, currently being tackled in some industrial laboratories, was started when it was noticed that the sarcomeres are exceptionally long in the psoas muscles of beef carcasses hung in the conventional manner. Is the tenderness of fillet steak (psoas) accounted for by the relatively small extent of overlap and cross-linkage between the filament assemblies? Could other muscles be made as tender by stretching them before rigor set in?

Further reading

These experiments were done originally by J. Hanson and H.E. Huxley and reported first in *Nature,* 172, 530, (1953); 173, 973 (1954) and then in more detail in *Symp. Soc. exp. Biol.,* 9, 228 (1955). Huxley wrote an article on the sliding-filament theory in the *Scientific American* (November 1958) and a subsequent article about the cross-bridge mechanism (*Sci. Amer.,* December 1965). For a detailed up-to-date account of the mechanism of contraction, written for students, see the chapter by G.W. Offer in *Companion to Biochemistry* ed. A.T. Bull, J.R. Lagnado, U.O. Thomas and K.F. Tipton, published by Longmans in 1974. If this book is not available, see the earlier but still useful short review by Huxley in *Science,* 164, 1356 (1969).

Hints for the teacher

Glycerinated psoas strips are prepared as follows. A rabbit is anaesthetised with Expiral (Agro-Vet Division) (up to 5 ml injected into the ear) and then killed by cutting the jugular veins (a licence is required). The abdomen is opened and the gut removed to expose the two psoas muscles which lie alongside the backbone. Strips that are several cm long and a few mm in diameter are peeled off and tied firmly to Perspex sticks, using thick cotton thread. If the muscle is dissected within about an hour after the animal's death, the fibres are extensible. The length of the long strips should be about 125% of that in *situ*. The short strips should be tied so that they are slack and able to shorten to about 75% of the *in situ* length. The strips are placed in cold glycerol solution (equal volumes of glycerol and of 6 mM phosphate buffer, pH 7·0). They are left at 0–4°C for 48 hr, during which time the glycerol solution is changed twice. Then they are put in a deep-freeze (−15 to −20°C) for at least 3 weeks. They remain usable for several years.

Expiral is manufactured and sold by Agro-Vet Division, Abbot Laboratories Ltd., Queensborough, Kent. It contains 200 mg/ml Nembutal (pentabarbitone).

To prepare the fibrils, the strip is cut off the stick and placed in cold standard salt solution in a large Petri dish. With two pairs of fine forceps it is pulled apart lengthwise into two equal pieces, and each of these is again divided lengthwise, and so on. Continue until the fragments are 1 mm thick. Pick up bundles of fragments and cut them into short lengths while dropping them into the blender cup. (If not cut they become wrapped round the blades.) Fill the cup with cold standard salt solution and cool on ice; also cool the blade assembly. Blend until most of the fibres have been broken up into fibrils. Do not attempt to disrupt the remaining fibres, or the fibrils already released will be damaged. If an MSE homogeniser is used at top speed, the process is complete in about 1 min and the temperature does not rise too far if the cup is surrounded by ice water.

Dissolve ATP in distilled water and adjust the pH to 7·0 using KOH. Store frozen in small polythene bottles.

E. JEAN HANSON
King's College, London

Section 9

Subcellular Organelles

1 Methods of disrupting cells

Time required
4–5 hr.

Assumed knowledge
For second-year students familiar with microscopes, spectrophotometers and basic biochemical techniques.

Theoretical background
Many biochemical procedures require the preparation of a cell-free extract of an organism. There are many different methods available for the preparation of such extracts, the relative efficiency of these methods depending on (1) the nature of the cell, particularly whether it has a cell wall or not, and (2) the experiment to be performed; some procedures for disruption inhibit or destroy some enzymic activities.

Ultrasonic radiation causes cavitation in liquids. The effect of this is to cause localised and transient regions of high vacuum in the liquid. A cell finding itself on the borders of such a region of high vacuum will burst. The French pressure cell and the homogeniser are both devices for producing shear forces. The French cell generates much larger shear forces than the homogeniser if used carefully. Grinding with sand, of course, is a way of generating very slight shear forces.

As well as physical methods of breaking cells chemical and enzymic methods have been used. Lysozyme hydrolyses murein, the molecule which forms the sacculus of a bacterium, and sodium dodecylsulphate (SDS) dissolves phospholipid membranes.

Necessary equipment
Required per pair of students
Bacillus megaterium suspension 15 mg/ml 50 mM Na/K PO_4, pH 7 (25 ml)
Washed bakers yeast 30 mg/ml 50 mM Na/K PO_4, pH 7 (25 ml)
Myxamoebae of *Dictyostelium discoideum* 3×10^8 cells/ml pH7 (25 ml)
50 mM Na/K PO_4, pH7 (1 Litre)

Lysozyme solution (BDH) 1 mg/ml in phosphate
buffer
37°C water bath
Microscope (sufficiently good to see *B. megaterium*
cells clearly)
Slides and cover slips
1% (w/v) sodium dodecyl sulphate (sodium lauryl
sulphate) (BDH)
MSE Minor bench centrifuge
Bovine serum albumin solution 10 mg/ml
(50 ml) (BDH)
NaCl 0·9M (200 ml)
Bonner's salts solution (100 ml)

Required per group of students
Ultrasonic disintegrator (Dawe Soniprobe)
French pressure cell and hydraulic press (Aminco,
from V.A. Howe)
Homogeniser and drive motor (Gallenkamp, Jencons)
Washed sand (50 g) (BDH)
Spectrophotometer, (Unicam SP 600) and cells
Biuret reagent (BDH) (500 ml)

Type of group
Pairs of students. It is often convenient to make Part
1 a class experiment so that every student uses every
piece of equipment once only, and the results of the
group are just pooled to obtain the necessary data
needed for calculation of the percentage efficiency
of breakage. This saves a lot of time and material
without detracting from the point of the experiment.

Instructions to the student
You are provided with suspensions of three different
cells:
  (a) *Bacillus megaterium*—15 mg/ml 0·05M Na/
      KPO$_4$ pH 7
  (b) Washed baker's yeast—30 mg/ml 0·05M Na/
      KPO$_4$ pH 7
  (c) Cellular slime mould amoebae—3 x 10^8
      cells/ml 0·05M Na/KPO$_4$ pH7
Bacteria and yeast both have a cell wall, whereas the
slime-mould amoebae merely have a cell membrane.
1 Examine drops of the suspensions under the
   microscope. Then treat *each* suspension with the
   following procedures:
   (a) Expose 5 ml of each suspension to ultrasonic
       radiation for 3 min.
   (b) Pass 10 ml of each suspension through the
       French pressure cell once.
   (c) Homogenise 5 ml of each suspension (10
       strokes, maximum speed).
2 Grind together 1g of bakers yeast and 2g of sand

in a pestle and mortar for approximately 15 min.
Then add 5 ml of 0·05M Na/KPO$_4$ pH 7 and mix
well. Allow the sand to settle and decant the
supernatant into a clean tube.
3 Dilute the *B. megaterium* suspension with phosphate buffer and 1·0M NaCl until it has a
   reading (Unicam at 625 nm) of about 0·7, a
   conc. of 0·2M and is 1mM in EDTA. Add 500
   μg of lysozyme to 5·1 ml of this suspension.
   Incubate at 37°C and take readings at intervals.
   Remember to zero the instrument using a water
   blank.
4 Place a drop of the slime mould amoebae suspension diluted with a drop of Bonner's salt solution,
   if necessary, on a clean microscope slide and a
   drop of 1% SDS close by it. Place a cover slip on
   the two drops so that they mix and examine
   under a microscope. Repeat with the yeast and
   B. megaterium suspensions.
   After procedures 1 (a), (b), (c) and 2 you will
   have a mixture of cells and possibly cell fragments. Examine the suspensions under a microscope, then centrifuge in the bench MSE for
   10 min. Determine the protein content of the
   supernatants using the biuret reagent. Also
   measure the protein content of the original cell
   suspension and thus calculate the % release of
   cellular protein and hence the relative efficiences
   of the various methods you have used.
5 Estimate the protein with biuret reagent as
   follows.
   To 1·0 ml of solution add 4·0 ml Biuret reagent,
   and mix. Read the OD at 550 nm after at least
   30 min (the colour is stable) and compare with a
   standard curve prepared using bovine serum
   albumin (1—10 mg).

Recording Results
Sketch the appearance of the cell suspensions and
make a note of the magnification. Score the apparent
cell breakage on an arbitrary scale.

Plot OD at 625 nm against time of incubation for
the lysozyme digestion.

Tabulate the protein release data so that calculation of the efficiency of breakage is straightforward.

Discussion
Do your microscopic observations correlate with your
protein determinations? Would you expect them to?
The *Bacilli* form spores. Did you see any? Would
spores behave differently from vegetative cells in this
experiment? Yeast cells when starving undergo autolysis. Did any autolysis occur? How would you

133

measure the extent of autolysis?

Extensions and applications

All the procedures that you have used have been used in research. The choice of method depends on the kind of cell used and the use which is to be made of the cell extract.

Further reading

'Methods in Enzymology', 'Methods in Microbiology', (Academic Press) and a number of other such compilations describe in greater detail these kinds of techniques.

Hints for the teacher

1 Growth of *Bacillus megaterium*—medium:
  0·7% sodium acetate
  5 ml/1 trace element solution
  12 ml/1 1·3% nutrient broth (Oxoid, Difco)
  1 in 2 dilution of Bacillus salts medium
  Autoclave inoculate and grow at $37°C$.

2 Trace elements solution
  Dissolve in 200ml of distilled water
    8 g $MgSO_4$ $7H_2O$
    0·4 g NaCl
    0·4 g $FeSO_4$ $7H_2O$
    0·4 $MnSO_4$ $4H_2O$
    80 g Ammomium molybdate
    0·8 ml Conc. HCl

3 *Bacillus* salts solution
  Dissolve in 1 litre of water
    4 g NH_4Cl
    9·3 g K_2HPO_4
    1·8 g KH_2PO_4

4 Washed baker's yeast is obtained by suspending a block of commercially available (Distillers Co. Ltd.) yeast. Most large bakers have or can obtain such blocks. Suspend 1 g of yeast in 50 ml phosphate buffer pH 7 and centrifuge. Wash pellet once or twice and finally resuspend in phosphate buffer.

5 Myxamoebae of *D. discoideum*. We use this organism because it is readily available at Leicester or Essex, but a suspension of any cell lacking of a cell wall would do as well. *D. discoideum* myxamoebae are grown in the medium prepared as follows.
  Dissolve in 1 litre
    14·3 g bacteriological peptone
    7·15 g Oxoid yeast extract
    15·4 g D-glucose
    1·28 g $Na_2HPO_4 . 12H_2O$
    0·486 g KH_2PO_4

Final pH 6·5—6·7; autoclave; inoculate with spores or myxamoebae of *D. discoideum* strain Ax–2 (Watts and Ashworth, Biochem. J. (1970) 119, 171—174). *D. discoideum* Ax-2 can be obtained from the American Type Culture Collection (ATCC 24397)

J.M. ASHWORTH
Essex.

2 Isolation and identification of subcellular components of rat liver

Time required
About 3½ hr. If suitable a lunch break during the 45 min of centrifugation is convenient.

Assumed knowledge
Familiarity with density-gradient centrifugation, phase microscopy and histochemistry. These techniques can be well illustrated in this experiment.

Theoretical background
Density-gradient centrifugation is widely used for the separation and isolation of subcellular components of many types of cells. The gradient can be either continuous, which requires special apparatus, or discontinuous, which calls for simple techniques. The size, density and shape of the particles determines their movement in the gradient under the influence of centrifugation which, together with the concentrations of sucrose used and the length of centrifugation, determines the degree of separation of the various subcellular organelles. A more detailed discussion of the technique is given by Packer.[1]

The histochemical techniques are quite straightforward and are described in detail by Pearse.[2] The student must not be sloppy in performing the staining steps otherwise no clear-cut staining will result.

Equipment required
Items required per pair of students

Gradient sucrose solutions
1 2·2M sucrose + 2 mM MgCl$_2$ → 10 ml
2 1·8M sucrose + 2 mM MgCl$_2$ → 10 ml
3 1·6M sucrose + 2 mM MgCl$_2$ → 10 ml
4 1·3M sucrose + 2 mM MgCl$_2$ → 10 ml
5 0·7M sucrose + 2 mM MgCl$_2$ → 10 ml

Staining set
95% alcohol	100 ml
66% alcohol	100 ml
33% alcohol	100 ml
H$_2$O	100 ml

Methyl green pyronin	100 ml

Methyl green pyronin is prepared as follows:

A { 5% aqueous pyronin (BDH) 17·5 ml
    2% methyl green (BDH) (chloroform washed) 10 ml made up to 50 ml with H$_2$O

B   0·2M acetate buffer pH 4·8

For use mix A and B in equal volumes (not stable for long)

Water	100 ml
Acetone	100 ml
Acetone + xylol 1 : 1	100 ml
Xylol	100 ml

Janus green B (BDH)
1 : 10 000 Janus B in 0·15M NaCl 25 ml

Aceto-orcein (Gurr)
1% w/v orcein dye in 45% v/v acetic acid 25 ml

Glass ware and others
1 dozen glass slides
1 diamond pencil
1 centrifuge tube (clear), 15 ml capacity (MSE)
10 Pasteur pipettes + teats
6 clear test tubes
1 phase microscope
High-speed centrifuge (MSE 25)(swing out rotor)

To prepare rat homogenate
2 rats
2 homogenisers
4 layers of muslin
Dissecting instruments
Extracting soln. 500 ml of 0·25M sucrose + 2mM MgCl$_2$

Type of group
Students work singly or in pairs. Space in the centrifuge rotor may be the main determinant.

Instructions to the student
The fractionation of rat liver homogenate will be accomplished by a sucrose density-gradient centrifugation. This provides macroscopic amounts of various cell constituents in a relatively pure form which can then be identified histochemically.

Preparation of rat liver homogenate
Remove the livers from 2 rats using dissecting boards and instruments. Weigh, and transfer to a beaker containing 200 ml cold 0·25 M sucrose + 2mM MgCl$_2$.

Cut up the livers into small pieces with pointed scissors, decant carefully, wash again with 100 ml cold sucrose, and then add the sucrose medium to a final volume of 60 ml. Homogenise in a glass homogeniser with a teflon plunger until the tissue is seen to be well homogenised. Pass the homogenate through 4 layers of muslin and measure the homogenate volume.

Density-gradient centrifugation
Layer successively from bottom to top in a 15 ml clear centrifuge tube 2 ml of each of the following sucrose solutions containing 2 mM $MgCl_2$: 2·2M, 1·8M, 1·6M, 1·3M, and 0·7M. Then layer on top 2 ml of homogenate (in 0·25M sucrose + 2mM $MgCl_2$). Mark the boundary layers of the sucrose solutions.

Centrifuge at 50 000 g in the MSE '25' for 40 min, using the horizontal swing-out rotor (6 x 15 ml). Balance the buckets carefully before fixing to the rotor.

After centrifugation is completed, carefully remove the centrifuge tube and make a drawing of the appearance of the layers. Carefully, using a Pasteur pipette with a rubber bulb, remove each layer for identification. Report the approximate densities of each cellular fraction according to its position in the centrifuge tube.

Caution The gradients should be made very carefully and slowly before starting with the liver extraction. Keep all the solutions and gradients at ice temperatures throughout in order to minimise diffusion problems.

Microscopy
Examine each fraction and the whole homogenate under phase-contrast.

Staining procedures
Place a drop of each fraction, *including the whole homogenate,* onto slides and smear it. Allow to dry about 1 min. Examine under phase-contrast.

1 *Methyl green—pyronin.* This combination of dyes results in the simultaneous detection of chromatin DNA (stained green by methyl green) and nucleolar and cytoplasmic RNA (stained red by pyronin).
  (a) Fix smears in 95% alcohol for 5 min; then rehydrate through successively lower concentrations of alcohol.
  (b) Stain in dye solution for 10 min.
  (c) Rinse in water for only a few seconds.
  (d) Blot dry.

  (e) Dehydrate rapidly in absolute acetone.
  (f) Rinse briefly in equal parts of acetone and xylene.
  (g) Clear with xylol, one minute.
2 *Janus green B.* This is a redox dye which is colourless in the reduced form and coloured when oxidised. Mitochondria show a high specificity toward Janus green B stainability. Add the dye to unfixed preparations.
3 *Aceto-orcein.* A chromosome stain used as 1% (w/v) orcein dye solution in 45% (v/v) acetic acid. Add to unfixed preparations.

Recording results
Drawings, both descriptive and detailed, should be made of the gradient, the various unstained fractions and the stained preparations. Tabulating the results of the staining is advisable.

Discussion
The student should determine the density of the various subcellular fractions from a knowledge of the density of the sucrose layers at 0°C (Handbook of Chemistry and Physics; Handbook of Biochemistry and Molecular Biology).
Why is the calculated density different from that determined by other techniques (look up the values)? Is any damage done to the organelles during the gradient centrifugation? Explain. Could the isolated organelles carry out their usual biochemical functions and how could this be determined?

Extensions and applications
The experiment could be designed to test additional histochemical assays, e.g. Feulgen, tetrazolium, and to determine the biochemical properties of the organelles. Different tissues, e.g. kidney, brain, leaves, could also be used. Density-gradient centrifugation is now extensively used to isolate purified preparations of DNA and RNA viruses, cell walls, membranes, mitochondria, chloroplasts, nuclei, etc. A continuous system called zonal ultracentrifugation using specially designed rotors is now in fairly general use for the large-scale preparation of larger quantities of subcellular components.

Further reading
Packer, L.
  1968 'Methods in Cell Physiology', Academic Press, New York p. 32
Pearse, A.G.E.
  1968 & 1972 'Histochemistry—Theoretical and

Applied Vol. I and II', Churchill Livingstone, London
Birnie, G.D. and Fox, S.M.
1969 'Subcellular Components'. Butterworths, London.

Hints for the teacher

Preliminary trials should be carried out to determine the centrifugation speed and time required for optimum separation of the fractions. This will vary depending on the centrifuge and size and shape of the tubes. To kill rats outright with no experimentation no animal licence is required.

D.O. HALL
King's College
London.

3 Succinic dehydrogenase activity in rat liver fractions

Time required
About 3½ hr.

Assumed knowledge
Use of simple spectrophotometer.

Theoretical background
The enzyme succinic dehydrogenase is localised in the mitochondria of the cell and its activity can be measured by following the reduction of the dye, triphenyl tetrazolium chloride (TCC). The rat liver homogenate is fractionated by differential centrifugation and the degree of purity (or contamination) of the fractions is estimated from their succinic dehydrogenase activity.

Necessary equipment
1 Rat liver fractions are prepared by the technician; usually 3 large male rats are required for a class of 40 students.
2 Solutions required per pair of students:
   (a) 0·1M phosphate buffer pH 7·4 → 20 ml
   (b) 0·2M sodium succinate (BDH) → 20 ml
   (c) 0·1% triphenyl tetrazolium chloride (TTC) (BDH) → 20 ml (freshly made up)
   (d) 0·25M sucrose → 25 ml
   (e) Acetone → 150 ml
   (f) Graded solutions of TTC
       (i) 50 μg/ml → 5 ml   to be prepared
       (ii) 100 μg/ml → 5 ml   on the day of
       (iii) 150 μg/ml → 5 ml   practical and
       (iv) 200 μg/ml → 5 ml   kept in dark
       (v) 250 μg/ml → 5 ml
       (vi) 300 μg/ml → 5 ml
   (g) Sodium hydrosulphite crystals (BDH)
3 Glassware:
   (a) 1 x 10 ml graduated pipette
   (b) 10 x 1 ml graduated pipette
   (c) 12 x glass centrifuge tubes.
   (d) 12 x Spectrophotometer tubes.
4 Apparatus and others
   (a) Water bath at 37°C
   (b) Timer

(c) Spectrophotometer
(d) Parafilm and paper tissues
(e) Centrifuge, e.g. MSE 18.

Type of group
Single students or pairs.

Instructions to the student
You are provided with fractions obtained by centrifugation of rat liver homogenate. These fractions will be on ice and must be kept in this way until used for enzyme assay. The fractions have been obtained as follows.

Livers are removed from freshly killed rats and homogenised in 0·25M sucrose at 0°C. A sample of whole homogenate is retained for assay. The homogenate is then centrifuged at 600g (the actual rev/min will depend on the centrifuge used) at 0°C for 10 min. At this speed nuclei and cell debris are brought down. The supernatant is set aside and the nuclear fraction is washed by resuspending in 0·25M sucrose and spun again at 600g for 10 min. This supernatant is discarded, the pellet resuspended and retained for assay. The supernatant, which now lacks nuclei, is spun again at 9 000 g for 10 minutes at 0°C to bring down the mitochondrial fraction. This pellet is resuspended and respun at 9 000g for 10 min, then resuspended again and assayed for enzyme activity. At the same time the material which remains, i.e. ribosomal and soluble fraction of the cell, must also be assayed.

Thus you will be provided with a sample (on ice) of (1) whole homogenate, (2) nuclear fraction, (3) mitochondrial fraction, (4) supernatant fraction containing ribosomes, etc.

Estimation of enzyme activity
The method used is that of Kun and Abood, Science, *109*, 144 (1949). The estimation uses TTC as an indicator of succinic dehydrogenase activity in the presence of succinate. The dye is reduced to a red, water-insoluble formazan. This is easily dissolved in acetone, which by precipitating tissue proteins leaves a clear supernatant ready for colorimetric estimation.

1 *Calibration curve.* There is a linear relationship between the number of micrograms of dye reduced and the optical density. To obtain a calibration curve tubes are set up as follows:
Into centrifuge tube pipette
0·5 ml phosphate buffer (0·1M, pH 7·4)
0·5 ml sodium succinate (0·2M)
1 ml sucrose (0·25 M)
1 ml of a variable TTC solution

lastly
a *few* crystals of sodium hydrosulphite.
Tubes containing 0, 50, 100, 150, 200, 250 and 300 μg of TTC are set up, reduced by the addition of a few crystals of sodium hydrosulphite, shaken vigorously and incubated for 5 min at 37°C. After removal from the water bath, 7 ml acetone is added, tubes are shaken and centrifuged. The optical density of the clear supernatant is determined in the spectrophotometer at 480nm. The tube containing *no* dye is used as a blank. A calibration curve is then obtained plotting OD_{480} against number of micrograms of TTC.

2 *Estimation of enzyme activity.* It is necessary to measure enzyme activity in the four fractions provided. Each reaction tube should be duplicated and various blanks should be set up as controls.
Thus for each fraction you will need 3 tubes as follows (all reaction tubes must be cooled on ice):
2 reaction tubes containing
  0·5 ml phosphate buffer
  0·5 ml sodium succinate
  1 ml tissue homogenate
  1 ml 0·1% TTC dye
1 tissue blank—substitute 0·5 ml water for sodium succinate.
Cover the tubes and incubate at 38°C for 15 min. Then add acetone, centrifuge and measure the optical density of the clear supernatant. Other controls may be devised and measured.
All the tubes are read against water and the readings for the tissue blank and the reagent blank are subtracted from the test reaction. Using the calibration curve, it is now possible to read off the number of micrograms of dye reduced by 1 ml of tissue homogenate in 15 min. You will be given the dry weight estimations so that you can give your results as micrograms dye reduced in 15 min by 1 g dry weight of tissue. The weights will be determined by pipetting 1 ml of the fraction into a weighed tube, adding 5 ml 80% ethanol and allowing to stand for 10 min. The tube is then centrifuged, supernatant removed and precipitate dried at 110°C for 2 hr before cooling and weighing. Express your results on a dry weight and a volume basis.

Discussion
The student should understand clearly the units used in expressing enzyme activities and discuss how better standard units could be used in this experiment. What is the succinic dehydrogenase activity of a purified rat liver mitochondrial preparation? Why are rat livers used so frequently in biological experimentation and

would other animals and/or tissues be preferable? What other mitochondrial or nuclear enzyme activities could be used to test a tissue fractionation procedure?

Extension and applications

Colorimetric enzyme assay procedures are the easiest systems available, especially if they can be recorded continuously for kinetic studies—where initial rates of reactions are so important. Reduction of dyes in oxidation/reduction reactions are widely used in histochemical assays of many properties of organelles within cells—both normal and malignant—since rapid results can be obtained once the procedure is well established.

Further reading

Roodyn, D.B. (Ed.)
    1967 'Enzyme Cytology', Academic Press, London
Bernath, P. and Singer, T.P.
    1962 in 'Methods in Enzymology', Vol. 5, Academic Press, New York
Pearse, A.G.E.
    1968 & 1972 'Histochemistry—Theoretical and Applied Vol I and II', Churchill Livingston, London

Hints for the teacher

The students could fractionate the liver if time and centrifuges were available. It is a useful experience in tissue fractionation but requires about 1½ hr. If this is done then standard curves may be provided and/or fewer assays performed. It is important to include the necessary controls and blanks.

D.O. HALL
King's College
London.

4 Electron transport in rat liver mitochondria

1 Isolation of rat liver mitochondria
2 Utilisation of substrates and effect of selected inhibitors

Time required

(1) 1½–2 hr (2) 2 hr.
If carried out as a single experiment should be complete in 3 hr.

Assumed knowledge

Basic schemes of oxidative phosphorylation and electron transport as presented in most biochemistry textbooks. It would also be an advantage to have had previous experience in the operation of an oxygen electrode. See Section 10.2.

Theoretical background

The preparation of isolated mitochondria introduces the student to the use of differential centrifugation as a method for isolating subcellular organelles in which the g (relative centrifugal force) x minutes (time of centrifugation) is used to cause the fractionation of the homogenate. A low value causes the separation of nuclei and higher values separate mitochondria from microsomes.

It also involves the use of the oxygen electrode, which is a specific application of polarographic analysis designed to measure the concentration of oxygen dissolved in aqueous solutions. The principle has been summarized in several reviews, notably:

Davies, P.N.
    1962 in 'Physical Techniques in Biological Research', 4, 137
Delieu, T. and Walker, D.A.
    1972 New Phytologist, 71, 201

In brief the apparatus consists of a thin platinum wire sealed in glass or plastic to act as a micro-cathode with a silver/silver chloride anode. A polarising voltage of between 0·6 and 0·7 v is then applied across the electrodes and oxygen in the vicinity of the platinum cathode undergoes electrolytic reduction. Under these conditions the current which flows is directly proportional to the concentration of oxygen in solution, because the polarising voltage is sufficiently high to reduce all of the oxygen in the vicinity of the micro-cathode. Thus the current which flows is limited by the rate at which unreduced oxygen in the bulk of the solution reaches the cathode surface, through the layer of reduced oxygen; this is governed by the laws of diffusion and is directly related to the concentration of oxygen in solution.

Necessary equipment

Each pair of students will need:

1 male (e.g., Sprague-Dawley) rat (200 g)
1 dissecting-board and instruments
1 refrigerated high-speed centrifuge with a rotor capable of spinning 200 ml at 10 000 g (MSE 18)
1 loose-fitting glass—teflon homogeniser (100 ml capacity) Wesley Coe, Cambridge.
Several Pasteur pipettes and bulbs
1 ice bath
1 oxygen electrode and control box. A suitable 'Clarke' type electrode is supplied by Rank Bros. Bottisham, Cambridge. The $\frac{3}{8}$ inch internal diameter electrode, not fitted with a rubber '0' ring on the control stopper, is most suitable for mitochondrial experiments. This electrode operates best with 1·0 ml of reaction medium
1 Recorder, preferably one with 5 and 10 mV full scale deflection. The Servoscribe (1s) is very suitable.
1 Magnetic stirrer (Gallemkamp)
3 x 25 μl syringes; 3 x 100 μl syringes (SGE)
Several 5 and 1 ml pipettes.

Solutions required

Each pair of students will need:

300 ml of extracting solution consisting of 0·25M sucrose; 0·1mM EDTA (BDH); 0·01M tris-(hydroxymethyl)-methylamine (BDH) pH 7·6

10 ml of 10% sodium deoxycholate (BDH)

2 ml of bovine serum albumin standard (10 mg/ml) (BDH)

50 ml Biuret reagent (BDH)

100 ml of reaction mixture consisting of 0·1M KCl; 0·01M MgCl$_2$; 0·015M KH$_2$PO$_4$; 0·001M EDTA; 0·25 M sucrose; 0·0025M ADP (pH 6·8)

5 ml succinate; 1·0M Na salt pH 7·0

5 ml β-hydroxybutyrate; 1·0M Na salt pH 7·0

2 ml ascorbate; 0·5M Na salt pH 7·0

2 ml tetramethylphenylenediamine (TMPD) (BDH) 0·02M

1 ml rotenone (BDH); 1 mg/10 ml ethanol
1 ml potassium cyanide; 30 mM
1 ml Antimycin A; 1 mg/10 ml ethanol.

Type of group

Both experiments are best carried out in pairs.

Instructions to the student

Isolation of rat liver mitochondria

Kill the rat by a sharp blow to the back of the neck. The rat should bleed through the nose after the blow; if it does not, quickly sever the arteries in the neck as rapid bleeding is necessary or blood will collect in the liver and be difficult to remove during the isolation procedure. If in any doubt get a technician skilled in the procedure to kill the rat. Quickly remove the liver, make sure you separate it from the gall bladder, and drop it into 30 ml of ice cold extracting solution; this cools the tissue and removes any blood. Then cut the liver into small cubes and drop them into 100 ml of extracting medium in the homogenizer and grind the tissue by hand, taking care to keep the homogenizer tube cool. The liver tissue should be well ground after passing the pestle up and down the tube about 10 times. The pestle should be revolved slowly during this process. Transfer the homogenized liver into cooled centrifuge tubes; balance the weight of the tubes and spin at 600 g for 10 min. This removes the cell debris, red blood cells and nuclei; carefully remove and retain the supernatant. Take great care not to remove the cell debris and red blood cells which are very loosely packed at the bottom of the tube; this separation is best achieved by pipetting the supernatant off rather than attempting to pour. Re-centrifuge the supernatant at 8 500 g for 10 min, which will precipitate the mitochondria. Remove any lipid which may be floating on the surface of the supernatant, using a paper tissue twisted into a point, then pour off the supernatant and retain the mitochondrial pellet. If the surface of the pellet is very light in colour and loosely packed this is composed of damaged and swollen mitochondria and is known as the 'fluffy' layer; it should be discarded along with the supernatant by washing the surface of the pellet with a small volume of extracting medium. Well-prepared mitochondria do not have much of the 'fluffy' layer and are themselves fairly loosely packed, and it is possible for an inexperienced operator to throw away the complete preparation at this point, so take great care; if in any doubt retain all the pellet. Suspend the pellet in 10 ml of cold extracting solution by sucking it up and then expelling it through a Pasteur pipette or a 10 m graduated pipette with a large orifice. Do not suck up into a small pipette because the shearing forces can damage the organelles. Recombine all the resuspended pellets and make up to 100 ml with extracting solution and then centrifuge this solution at 8 500 g for 10 min, then pour off the supernatant, add 2 ml of extraction solution and suspend as before. Place in a test tube in an ice bath, and the preparation is now ready for use. Remember to always keep all apparatus and solutions cold during this preparation—this includes the rotor during centrifugation—also avoid excessive frothing during resuspension if you require high-quality mitochondria. The biochemical activity of this preparation can now be measured as described in the following section.

The concentration of protein, if required, can be measured by adding 0·1 or 0·2 ml of the preparation to 0·4 ml of 10% sodium deoxycholate solution and then made up to 1 ml with water. Add 4 ml of biuret reagent, mix well, and stand for 15 min, measure the optical density at 540 nm. The method should be calibrated against a bovine serum albumin standard containing between 1−5 mg per aliquot.

Assay of rat liver mitochondria

The volumes quoted here are suitable for use with the *standard* $\frac{5}{8}$ diameter Rank electrode which can hold between 1·5 to 5·0 ml of reaction medium and a mitochondrial preparation containing approximately 10 mg protein per ml. Start by adding 1·6 ml of reaction mixture into the electrode vessel and insert the stopper, screwing it down until the solution enters the narrow hole to act as a seal between the bulk of the solution and the atmosphere. Then adjust the sensitivity control so that the recorder gives a full-scale deflection. When using the Rank control boxes it is often necessary to use the 5 mV setting on the recorder. After ensuring that the 100% reading is steady, add 2−3 cystals of dithionite; the pen should now drop to within 10% of the zero line and represents the anaerobic position, which should be noted for future reference. Once this has been done remove the solution from the electrode and wash it out well; the instrument is now ready for use.

Next check that the mitochondria are capable of respiration in the following manner. Add 1·6 ml of reaction mixture and adjust the sensitivity control to obtain a full-scale deflection on the recorder, activate the chart drive set at 30 mm per minute, add 0·2 ml of mitochondria, insert and adjust the stopper to seal the electrode. Good mitochondria are anaerobic, so a downward deflection of the pen should occur; once this has ceased add 20 μl of succinate and wait

until the pen reaches the bottom of the chart. The slope of the line is proportional to the rate of oxygen consumption; assuming the concentration of dissolved oxygen to be 0·240 μmoles per ml, calculate the rate of oxygen uptake as n moles/minute/mg protein. Repeat the experiment with fresh mitochondria, using either 50 μl of β-hydroxybutyrate or 0·1 ml of ascorbate followed by 50 μl of TMPD to act as a redox mediator.

It is now possible to investigate the influence of various inhibitors. This is a complex experiment and requires that the operator knows exactly what he is going to do and has all the substrates and inhibitors next to the electrode, preferably already loaded into the syringes before starting the run. It is also essential to make the additions fairly rapidly or the solution will become anaerobic before the final additions can be made; it is only necessary to have about one inch of trace in order to calculate the rate of oxygen uptake. Add 1·6 ml of reaction medium and 0·2 ml of mitochondria into the electrode and insert the stopper, and as soon as the trace is linear add 50 μl of β-hydroxybutyrate. After about 30 s, when the trace is steady, add 10 μl of rotenone; when you are sure of the result add 20 μl of succinate, followed 20 s later by 5 μl of antimycin A, and a further 45 s later by 0·1 ml of ascorbate and 50 μl of TMPD. Wait until the rate is steady and then add 20 μl of KCN. Finally investigate the influence of antimycin A and KCN separately on the rate of electron flow from β-hydroxybutyrate and the effect of KCN on the rate of electron flow from succinate.

Recording results

All the rates should be calculated as n moles of oxygen consumed per mg protein per min and recorded in a tabular form as shown in the table.

Discussion

It is a useful exercise to attempt to interpret your results without reference to the scheme of electron transport in the text books. Can you locate the points at which the various substrate electrons enter into the respiratory chain in relation to the points at which the inhibitors appear to inhibit? Draw up the most efficient sequence of events that your data suggests, and then compare it with the accepted scheme. They should be similar because the experiment you have just carried out forms the basis for accepting the published scheme.

Further reading

Basic background information can be obtained from:

Treatment	Rate of oxygen consumed n moles/min/mg protein
β-hydroxybutyrate alone	
Succinate alone	
Ascorbate + TMPD alone	
β-hydroxybutyrate alone	
β-hydroxybutyrate + rotenone	
β-hydroxybutyrate + rotenone + succinate	
β-HOB + rotenone + succinate + antimycin A	
β-HOB + rot. + succ. + anti/A + ascorbate/TMPD	
β-HOB + rot. + succ. + anti/A + asc./TMPD + KCN	
β-HOB + antimycin A	
β-HOB + KCN	
Succ + KCN	

Lehninger, A.L.
  1964 'The mitochondrion', Benjamin, New York
Lehninger, A.L.
  1971 'Biochemistry', Worth, New York
Palmer, J. M. and Hall, D.O.
  1972 'The Mitochondrial Membrane System' in 'Progress in Biophysics and Molecular Biology', 24, 127.
Hall, D.O. and Palmer, J.M.
  1969 Nature, 221, 717

More advanced and detailed information can be obtained from:
W.W. Wainio
  1970 'The Mammalian Mitochondrial Respiratory Chain', Academic Press, New York

Hints for the teacher
It is essential that all the reagents should be readily available and cold before the start of the experiment. The students must know what they are going to do *before* commencing the experiment, since it is essential that they do not take longer than the stipulated times or the mitochondria will suffer. It is also important to remember that rotenone, antimycin A and potassium cyanide are poisons so they must *never* be pipetted by mouth. It is advisable to have the antidote to KCN available. Solution A: 158g $FeSO_4$. $7H_2O$ + 3g citric acid to one litre water (inspect regularly). Solution B: 60g anhydrous $NaHCO_3$ per litre. When required mix A:B as 50:50. SWALLOW. If antimycin A is not available then 2 heptyl-4-hydroxyquinoline N-oxide can be substituted, using the same concentration.

It is also very useful to have all the oxygen electrodes set up and tested by a technician before the students arrive. Should the electrodes fail to respond towards the end of the experiments, this can often be corrected by washing the electrode vessel with ethanol, which removes any traces of inhibitors bound to the walls and membrane.

J.M. PALMER
Imperial College
London

5 Electron transport in plant mitochondria

1 Isolation of wheat seedling mitochondria
2 Operation and location of various dehydrogenases within the organelle

Time required
1 1–1½ hr.
2 1½–2 hr.

Assumed knowledge
The same as in the previous experiment.

Theoretical background
Although plant mitochondria are similar to animal mitochondria in several respects, there are many aspects in which they show marked differences. Two well-known differences are in their ability to oxidise exogenous NADH and malate, neither of which substrates are oxidised by mammalian mitochondria. Malate is not oxidised in rat liver mitochondria unless a system to remove oxalacetate is added, because the equilibrium is very much in favour of converting oxalacetate to malate; NADH is not oxidised because the inner membrane is impermeable to the co-enzyme. Recent research has shown that plant mito-chondria oxidise malate using an NAD^+-linked malic enzyme (E.C. 1.1.1.39) rather than the malic de-hydrogenase (E.C. 1.1.1.37), the equilibrium of which is in favour of pyruvate production and malate oxidation. Plant mitochondria oxidise exogenous NADH using a NADH dehydrogenase located on the outer surface of the inner mitochondrial membrane; this enzyme is not found in mammalian mito-chondria. The location of enzymes with respect to the faces of the inner membrane can be carried out using potassium ferricyanide as a non-penetrating electron acceptor. Potassium ferricyanide is a redox reagent with $E_0{}' = +360$ mV and is therefore capable of accepting electrons from any component in the electron transport chain providing it has access to the active site. Ferricyanide, however, is an ion and thus cannot cross hydrophobic biological membranes, and is therefore unable to interact with redox systems located on the inside of these membranes. In the mammalian system the only component which we know to be located on the outside of the inner membrane in a configuration which allows ferricya-nide to interact with it is cytochrome c. We know this because electron flow from succinate or pyruvate dehydrogenase to ferricyanide is inhibited by anti-mycin A, showing that the electrons need to pass from cytochrome b to c before they reach ferricya-nide on the outside. Thus it is assumed that succinic dehydrogenase is located on the matrix side of the inner membrane and unable to interact with the ferricyanide. You will observe the antimycin A sensi-tivity of ferricyanide reduction by NADH or succinate in order to establish their relative location on the inner membrane.

Necessary equipment
1 Supply of wheat seeds with a high germination rate
 2 seeds trays with covers approximately 15in x 9in x 2in per pair of students
 Incubator for germination at 27°C (dark)
 Pestle and mortar (200 ml capacity)
 Muslin for filtration
 Refrigerated centrifuge (MSE 18)
 Rotor with capacity of 200 ml and speeds up to 18 000 rev/min
 Pasteur pipettes,
 Homogenizers (glass/teflon) 20 ml capacity e.g., Jencons, Wesley Coe
 10% sodium hypochlorite (BDH)
 Extracting solution (250 ml per group) consisting of 0·5m sucrose, 30mM 3-(N-morpholino) propanesulphonic acids (MOPS) (BDH); 1mM ethylenegly-col-bis-(β-amino-ethylether) N, N-tetra-acetic acid (EGTA) (BDH); 0·1% (w/v) bovine serum albumin at pH 7·5 (ice cold)
2 Oxygen electrode (for details of this see Section 9.4 and 10.2)
 Recorder
 Magnetic stirrer
 Spectrophotometer, e.g. Spectronic 20 (Bausch and Lomb) is suitable
 3 x 25 μl syringes (SGE)
 3 x 50 μl syringes (SGE)
 Various 1 and 5 ml pipettes
 50 ml Assay medium consisting of 0·3M sucrose, 5mM MOPS buffer, 5mM KH_2PO_4, 10mM $MgCl_2$ at pH 7·2 (room temperature)
 10 ml succinate 1·0M Na salt pH 7·2
 10 ml malate 1·0M Na salt pH 7·2
 1 ml NADH 0·1M
 1 ml potassium ferricyanide 0·1M (make up fresh each day)
 1 ml potassium cyanide 50 mM

1 ml antimycin A (Sigma) 1 mg/10 ml ethanol
1 ml ADP (Sigma) 25mM pH 6·8

Type of group
This experiment is best done in groups of two. It is not easy to do single-handed.

Instructions to the student
Growth of wheat seedling
Wash 500g of dry seeds in tap water to remove any seed dressing (this is important if you are to obtain high-quality mitochondria), then soak the seeds in 10% (v/v) hypochlorite. Wash well in tap water to remove hypochlorite, and transfer the seeds into two seed trays lined with 6 thicknesses of paper tissue saturated with distilled water; cover and place in a dark incubator for 3–4 days at 27°C.

Isolation of mitochondria
After 3–4 days of growth, the etiolated shoots should be approximately 1–1½ in long.

Harvest 50 g of these shoots and grind fairly gently in 200 ml of the isolating medium in an ice-cold pestle and mortar for about 1–1½ min. Filter the homogenate through two layers of muslin and transfer to suitable centrifuge tubes, balance and spin at 2 500 g for 2 min. Then decant the supernatant and re-centrifuge at 4 000 g for a further 3 min. Suspend the resulting pellet in approximately 10 ml of extracting medium by drawing up and then expelling through a Pasteur pipette; if this does not work, transfer to a loose-fitting glass-teflon homogeniser and homogenise carefully, since this can lead to damage. Dilute the resulting suspension to 40 ml and re-centrifuge at 40 000 g for 3 min. Finally suspend the pellet in 2–3 ml of the extracting solution and homogenise in a glass-teflon homogenizer. Transfer to a test tube kept in an ice bath. The mitochondria, which should be quite yellow in colour, are now ready for assay.

Assay for oxygen uptake
Place 1·6 ml of assay medium in the electrode vessel, set the 100% deflection on the recorder (see details for the experiment in Section 9.4). Add 0·2 ml of the wheat mitochondria and adjust the stopper. To start the oxidation add 20μl of either succinate or NADH or 100 μl of malate. The ADP:O ratio can be obtained by adding 5 μl of ADP at this point. The sensitivity to respiratory inhibitors can be tested by following the instructions given in Section 9.4. You should be able to show that while the oxidation of malate is partially sensitive to rotenone

the oxidation of NADH is not effected. The ADP:O ratio accompanying NADH oxidation is also lower than that found for malate.

Reduction of ferricyanide
1 *Using succinate.* Place 2·9 ml of assay medium in a spectrophotometer tube, add 0·1 ml of KCN, 0·1 ml ADP and 0·1 ml of mitochondria. Mix well and place in the spectrophotometer, wavelength set at 420 nm, and adjust the zero optical density (100% transmission), then add 25 μl of potassium ferricyanide. This should give a reading of approximately 0·7 OD which should remain relatively steady. Take reading every 5 s. Then after 30–60 s remove the tube and add 30 μl of succinate; this should initiate the reduction of ferricyanide and the OD should decrease. Take readings every 10 s until you have established the rate of reduction, then add 20 μl of antimycin A. If the mitochondria are intact this should result in an inhibition of the rate of ferricyanide reduction.

2 *Using NADH.* This is basically similar to the methods for succinate, but reduce the volume of mitochondria to 30 μl and initiate the reaction by adding 20 μl of NADH.

Recording results
Calculate all the results into acceptable units, i.e. in n moles oxygen consumed per min per mg protein, or n moles of ferricyanide reduced per min per mg of protein; for this latter calculation take the millimolar extinction coefficient for ferricyanide to be $E_{420} = 1·03 \, mM^{-1} cm^{-1}$ Before the rate of ferricyanide reduction can be obtained you will need to plot the rate of reduction against the time and measure the slope of the tangent to the line. Antimycin A often takes a little time to react, so always measure the slope about 30 s after adding the inhibitor.

Discussion
Compare your results with the predictions made in the theoretical background section. Do they agree with theory? Also compare the electron flow rates from NADH and succinate to both oxygen and ferricyanide. Which is the most rapid? Can you tell anything about where the rate-limiting step in normal electron flow may be situated?

Further reading
Many of the references mentioned in Section 9.4 are useful.
Metabolism of plant mitochondria is well reviewed by

Hanson, J.B. and Hodges, T.K.
    1967 In 'Current Topics in Bioenergetics', ed.
    Sanadi, D.R., vol 2, pp. 65–98, Academic Press,
    New York

More detailed information concerning the use of
potassium ferricyanide to locate the various dehydro-
genases is given in:
Von Jagow, and Klingenberg, M.
    1970 Eur. J. Biochem., 9, 519
Brunton, C.J. and Palmer, J.M.
    1973 Eur. J. Biochem., 39, 283
Douce, R. and Bonner, W.D.
    1973 Biochem. Biophys. Acta., 292, 105

Hints to the teacher
This is a straightforward practical, although it
requires more theoretical background than experimental background. The only problem in the preparation of the mitochondria is to obtain the right
amount of grinding in the pestle and mortar. If in
doubt undergrind rather than overgrind, so that the
homogenate will have some large pieces of tissue still
in it after grinding. Remember to keep the temperature low at all the steps of preparation.

J.M. PALMER
Imperial College
London

6 The preparation isolated rat liver and the determin the amount of DNA per average nucleus

Time required
6 hr.

Assumed knowledge
Use of the Potter-Elvehjem homogeniser, haemocytometry, spectrophotometric analysis.

Theoretical background
The cells of rodent liver may be disrupted by
homogenisation in citric acid solution. The latter
stabilises the nuclei and dissolves cytoplasm. The
number of nuclei per aliquot of culture may then be
determined by haemocytometry and all the nuclei
then sedimented by centrifugation. This nuclear
deposit may then be successively extracted with cold
trichloroacetic acid, ethanol and ethanol-ether, and
hot trichloroacetic acid to remove the acid-soluble
components, the lipids and the nucleic acids, respectively. The nucleic acid fraction contains both
deoxyribonucleic acid (DNA) and ribonucleic acid
(RNA) as soluble products of hydrolysis. The DNA
may then be determined using a colour reaction
which is specific for deoxyribose.

Necessary equipment
Rat liver (about 10 g per animal depending upon
    size).
Ice bath and 50 ml beaker
Chilled saline (0·9% NaCl)
Dissecting instruments
Filter papers
Balance
Chilled 2% citric acid
Potter-Elvehjem   tissue homogeniser and motor
    (Gallenkamp, Jencons)
25 ml stoppered measuring cylinder
3 test tubes and rack
0·01M citric acid containing 0·5% brilliant cresyl blue
    (BDH)

_auer haemocytometer and coverslips (Gallen-kamp)
Pasteur pipette and teat
Microscope with mechanical stage
2 x 1 ml pipettes
1 x 2 ml pipette
3 x 4 ml pipettes
5 x 5 ml graduated pipettes
2 x 10 ml pipettes
1 x 10 ml graduated pipette
Pumpette (Pipette filler, Gallenkamp)
2 x 15 ml conical centrifuge tubes
Refrigerated swing-out centrifuge (M.S.E. 6L)
2 x tapered glass rods
1 x 10 ml spouted measuring cylinder
2 x 10 ml stoppered measuring cylinders
Water bath at 90°C
DNA solution (667 μg/ml, dissolved with the aid of NaOH) (BDH)
5% trichloroacetic acid
Chilled 10% trichloroacetic acid
20% trichloroacetic acid
Boiling-water bath
Fume cupboard
Timing clock
Spectrophotometer and cuvettes
8 boiling tubes
Diphenylamine reagent: 1 g diphenylamine (BDH) dissolved in 100 ml glacial acetic acid, plus 2·75 ml of concentrated sulphuric acid.

Type of group
Intended to be performed by individuals.

Instructions to the student
Preparation of homogenate
The livers from freshly killed rats are rapidly removed and plunged into chilled saline in an ice bath. Pieces of the chilled liver are later drained and blotted dry on filter paper. Then 2·5 g is weighed out, cut into smaller pieces, and homogenised in chilled 2% citric acid (after Fakuda and Sibatani, 1953) using a Potter-Elvehjem tissue homogeniser. The homogenate is diluted to 25 ml volume with 2% citric acid and stored in an ice bath until required.

Enumeration of nuclei
Mix the liver homogenate thoroughly and dilute a small sample 1 in 3 with 0·01M citric acid containing 0·5% brilliant cresyl blue. Enumerate the nuclei (in duplicate) in this diluted sample using a Neubauer haemocytometer as described in Section 13.7. Calculate the number of nuclei in 10 ml of undiluted homogenate.

Isolation of nuclei
Pipette 10 ml of well-mixed liver homogenate into each of two conical centrifuge tubes and centrifuge in a refrigerated centrifuge at about 1400 g (average) until all the nuclei have been sedimented (about 5 min). Decant the supernatants and examine microscopically. These will contain some refractile lipid droplets of similar size to nuclei (of which there are two main sizes). The deposits, which should have been placed in an ice bath, are now each resuspended in 5 ml of chilled 2% citric acid and recentrifuged as before. The supernatants, which should contain no nuclei, are discarded. The deposits, containing known numbers of nuclei, are then treated to extract the nucleic acids.

Extraction of nucleic acids
This is carried out on each nuclear deposit.
1 Resuspend the deposit in 1 ml of water by stirring with a tapered glass rod and treat with 2·5 ml of cold 10% trichloroacetic acid. Centrifuge and discard the supernatant. Repeat the process and discard the supernatant (containing acid-soluble products).
2 Remove the lipids by resuspending the deposits in 1 ml of water and mixing with 4 ml of 95% ethanol. Centrifuge and discard the supernatant. Extract the deposit successively with 3 x 5 ml of alcohol: ether 3:1 (v/v) (use measuring cylinder), discarding the supernatant after each centrifuging and resuspending at the beginning of each operation. It is advisable to carry out this procedure in a fume cupboard and *observe appropriate fire precautions*.
3 The deposit is resuspended in 1·2 ml of water, mixed with 1·3 ml of 10% trichloroacetic acid and centrifuged. Ensure that the duplicate tubes are individually labelled and decant the clear supernatant from each tube into a labelled 10 ml graduated measuring cylinder. The deposit is then resuspended in 5 ml of 5% trichloroacetic acid and heated in a water bath in a fume-cupboard at 90°C for 15 min with occasional stirring. This hydrolyses the nucleic acids to soluble products. The tube is then cooled in a water bath and centrifuged. The supernatant is then decanted into the appropriate stoppered measuring cylinder. The deposit is resuspended in 2·5 ml of 5% trichloroacetic acid and recentrifuged, the clear supernatant again being added to the appropriate measuring cylinder. The nucleic acid extract in the stoppered measuring cylinder is then made up to 10 ml volume with 5%

trichloroacetic acid. (The extraction procedure is that of W.C. Schneider, J. Biol. Chem., 161, 293 (1945).)

Estimation of DNA in nucleic acid extracts
This is determined using a specific colour for the deoxyribose of the DNA and is modified from the method of Dische. (1930) Mikrochemie, *8*, 4.

1 *Calibration curve.* To a series of solutions of DNA (in the range 0–2000 μg DNA per 4 ml) in a final volume of 4 ml (which also includes 1 ml of 20% trichloroacetic acid), add 8 ml of freshly prepared diphenylamine reagent, using a pumpette, and immerse for exactly 15 min in a vigorously boiling water bath in a fume cupboard. Also treat 4 ml of 5% trichloroacetic acid in the same way, as a control. It is advisable to add the tubes to the boiling water bath at timed intervals so that the bath continues boiling vigorously during the additions. After exactly 15 min, remove each tube and immerse in running cold water to cool. The colours, which are stable, may then be estimated by determination of their optical densities at 600 nm, reading against water. The cuvettes should be dry and should not be washed with water, as this causes precipitation. Subtract the blank value from the test values and plot optical density against concentration.

2 *Estimation of DNA.* Repeat this assay procedure using 4 ml aliquots of each nucleic acid extract. These may be boiled at the same time as the tubes for the calibration curve.

Results
Use mean values for the number of cells harvested and for the DNA content of the extracts. Calculate the amount of DNA which has been extracted from the number of nuclei you originally sedimented and divide one value by the other to give the DNA content per average nucleus.

Discussion
Try to estimate the accuracy of your result. Which parts of the procedure are likely to introduce the greatest errors? Why are the nuclei in two size populations?

Extension and applications
A knowledge of the average nuclear DNA content can be used to make a comparison of the number of cells present in a tissue. (Knowing the tissue DNA content and provided that the mitotic rates are low). Knowing these two values it is possible to compare other analytical results with them in order to estimate the cellular concentrations of the other components. This may enable changes in cellular concentrations, as opposed to increases in the number of cells, to be detected. This differentiation cannot be achieved simply by percentage composition comparisons.

Further reading
Fakuda, M. and Sibatani, A.J.
    1953 J. Biochem. (Tokyo), 40, 95
Davidson, J.N.
    1969 'The Biochemistry of the Nucleic Acids', 6th edn. Methuen, London
Davidson, J.N.
    1954 Proc. Nutr. Soc., 13, 112

Hints for the teacher
1 In the isolation of the nuclei an ordinary bench centrifuge may be used for sedimentation without too great an error, provided that the tubes are only removed from the ice bath immediately prior to sedimentation and are cooled again rapidly after this.

2 The trichloroacetic acid and diphenylamine reagent are quite corrosive.

3 It is also possible to analyse the nucleic acid extracts for RNA using the orcinol reaction (BDH) (and correcting for the colour due to the DNA in this reaction).
    Other methods for the isolation of the nucleic acids are possible and some of these result in an extract containing only DNA. The DNA can be estimated in such solutions by simply measuring their ultraviolet absorption at an appropriate wavelength.

4 Because of cost it is probably better to use an inexpensive DNA sample for the calibration curve. It should however be realised that this is not ideal and it should be possible to give the students the slope of a calibration curve prepared using highly polymerised DNA, so that they can calculate accurately their true average nuclear DNA contents.

T.R. RICKETTS
Nottingham.

Isolation of chloroplasts from higher plants

Time required
1 hr for each preparation; 2 hr if all three preparation techniques are done in one class.

Assumed knowledge
Use of centrifuges and microscopes including phase contrast.

Theoretical background
The first photosynthetically active chloroplast preparations were obtained by Hill; these preparations were active only in oxygen production coupled to the reduction of nonphysiological electron acceptors. Other photosynthetic activities of chloroplasts were not discovered until the development by Arnon and Whatley of isolation techniques using isotonic media; they isolated *whole* chloroplasts in isotonic sodium chloride. Broken chloroplasts which exhibit photophosphorylation and photosynthetic electron transport were obtained simply by diluting the *whole* chloroplast preparation. Sugars were not used in these isolation media as it was thought that they might act as substrates for oxidative reactions which would confuse the results obtained. More recently Good and his co-workers have developed methods of isolation using sucrose as osmotic *buffer*. These chloroplasts have a higher ATP/2e ratio than those prepared by the method of Arnon and Whatley. Although whole chloroplasts prepared by the procedure of Arnon and Whatley contain all the enzymes of the reductive pentose phosphate cycle, they are able to fix CO_2 only at extremely low rates (1–2% of that in whole plants). Examination of these chloroplasts (which appear whole under the normal microscope) either by phase-contrast microscopy or electron microscopy, however, indicates that they have lost their outer membranes, and are *naked lamellar systems* or *class 2* chloroplasts. Chloroplasts prepared by Good's method are similar. Recently Walker has developed techniques for the isolation of whole, intact chloroplasts, which gives preparations in which the majority of the chloroplasts retain their membranes and are *class 1* chloroplasts. Walker's procedures have been modified by Bassham and Jensen and by Kalberer *et al*. Chloroplasts prepared by these groups fix CO_2 at rates between 30 and 75% of those of whole leaves. The procedure of Kalberer *et al* which uses the simplest media is described.

Necessary equipment
1 *Shared:*
  (a) Atomix (MSE) or equivalent blendors with approximately 200 ml capacity. At least one for every 3 groups. Keep on ice until required.
  (b) MSE super minor or equivalent bench centrifuge with head for 100 ml tubes. One for every two groups. The head should be refrigerated until required. Any refrigerated centrifuge may be used instead.
  (c) Microscopes with phase-contrast.
  (d) Narrow bandwidth spectrophotometer with cuvettes.
2 For each group:
  (a) 1 ice bucket
  (b) 4 x 100 ml centrifuge tubes
  (c) Test tube rack with 12 test tubes
  (d) 1 glass rod
  (e) 2 x 100 ml measuring cylinders
  (f) 1 x 25 ml measuring cylinder
  (g) 1 small filter funnel with Whatman No. 1 filter paper
  (h) 1 x 4in filter funnel
  (i) 1 bag made of a double layer of 3 600 holes/cm^2 terylene fabric, available from J.R. Carmichael Ltd., Liverpool; or enough cheesecloth to make a filter bag with four layers in the 4in funnel. The cheesecloth should be discarded after use; the nylon bag can be washed and reused.
  (j) 50–100 g of fresh or refrigerator-stored (4°C) spinach
  (k) Absorbent cotton wool
3 Solutions required for class of 20–25 students all doing preparation 1 and half doing 2 and half 3:
  (a) 0·35M NaCl   1 litre
  (b) 0·2M tris-HCl pH 8·0   250 ml
  (c) 0·04M TES (BDH; buffer)   ⎫
       0·35M NaCl   ⎬ pH 7·3 250 ml
       0·001M $CaCl_2$ ⎪
       0·001M EDTA   ⎭
  (d) 0·03M tricine (BDH; buffer) ⎫
       0·02M sucrose   ⎬ pH 7·3 250 ml
       0·005M $CaCl_2$ ⎭
  (e) 0·35M sorbitol
       0·025M HEPES (BDH;buffer) ⎫ pH 7·6 250 ml
       0·002M EDTA   ⎬
       0·002M ascorbic acid   ⎭

(f) 0·35M sorbitol
0·05M HEPES
0·004M Na pyrophosphate $\Big\}$ pH 7·6 250 ml
(g) 80% acetone in H_2O 1 litre
All solutions except the acetone should be kept
in the cold room.
Solutions (e) and (f) must be made on the day
they are used.

Type of group
Pairs or larger groups.

Instructions to the student
Chloroplast Preparation—General
These notes are *important to all* preparations.
1 Keep all solutions and apparatus used on ice. This is particularly important if the centrifuges used are not refrigerated.
2 Treat the chloroplasts gently. They have thin skins.
3 Resuspend chloroplasts using a *small* piece of absorbent cotton wool wrapped round the end of a glass rod. Wet with the correct medium before using it.
4 Check all stages of the procedure and make up reaction mixtures *before* starting the chloroplast preparation.
5 Carry out the preparation as quickly as possible. Use the chloroplasts as soon as they are made.

Chloroplast preparation 1
Based on the procedure of Whatley, F.R. and Arnon, D.I. 'Methods in Enzymology', 6, 308 (1963).

Grinding medium (solutions required):
0·35M NaCl
0·2M Tris/HCl buffer pH 8·0

Tear 25 g of spinach leaves into small pieces about 2–3cm² Place in the Atomix with 100 ml 0·35M NaCl and 10 ml 0·2M tris buffer pH 8·0. Blend for 10 s at low and 20 s at high speed. Filter the homogenate through the nylon bag into a centrifuge tube placed in an ice bucket. Centrifuge 4 min at 2 000 g.

Resuspend the sedimented chloroplasts in a small volume (2–3 ml) of 0·35M NaCl. Dilute to 40 ml with 0·35M NaCl with gentle stirring and centrifuge 4 min at 2 000 g. Resuspend the pellet in 2 ml 0·35M NaCl. This preparation consists of whole chloroplasts. Broken chloroplasts are prepared by diluting the suspension with 10 volumes of water.

Chloroplast preparation 2
Based on the procedure of Horton, A.A. and Hall, D.O. Nature, 218, 386 (1968), modified from Izawa, S. and Good, N.E.
Grinding medium:
0·04M TES
0·35M NaCl
0·001M $CaCl_2$
0·001M EDTA $\Big\}$ pH 7·3
Resuspending medium:
0·03M Tricine
0·2M Sucrose
0·005M $CaCl_2$ $\Big\}$ pH 7·3

Cut up finely 25 g of washed spinach leaves. Blend them in an Atomix with 50 ml of grinding medium for 5 s at half speed and 10 s at full speed. Squeeze the resulting slurry through a nylon bag into a centrifuge tube and centrifuge the filtrate at 2 500 g for 5 min. Discard the supernatant and resuspend the pellet of chloroplasts in a few ml of resuspending medium. Make up the suspension to 50 ml with resuspending medium and centrifuge at 2 500 g for 5 min. Discard the supernatant and resuspend the pellet as before in approximately 5 ml resuspending medium.

Both the grinding and resuspending media should be at about 4°C and the whole procedure should be carried out in an ice bucket.

Chloroplast preparation 3
Based on the procedure of Kalberer, P.P., Buchanan, B.B. and Arnon, D.I. Proc. Natl. Acad. Sci. (US) 57, 1542 (1967)

Grinding medium:
Sorbitol	0·35M	
Na HEPES	0·025M	
EDTA	0·002M	pH 7·6
Ascorbic Acid	0·002M	

Reaction medium:
Sorbitol	0·35M	
Na HEPES	0·05M	pH 7·6
Na pyrophosphate	0·004M	

Free 10 g of spinach leaves of midribs and tear them into strips about 1 cm wide. Mix the strips with 30 ml of grinding medium and disrupt in the Atomix blender for 5 s on full power. Then squeeze the slurry through the nylon bag to remove large particles, unbroken cells and nuclei. Centrifuge the filtrate for

1 min at 2 500 g. Resuspend the pellet in a small volume of reaction medium.

Chlorophyll determination

Measure the amount of chlorophyll in each of your preparations by the following method.

Add 0·1 ml of chloroplast suspension to 20 ml of 80% acetone. Filter. Read the optical density at 652 nm. The chlorophyll concentration of the suspension can be calculated as follows:

OD_{652} x 5·8 = mg Chlorophyll/ml of original suspension.

From method of D.I. Arnon, Plant. Physiol., 24, 1 (1949).

Recording results

Examine the chloroplasts under the microscope, using normal and phase-contrast illumination. Note the shiny appearance of intact chloroplasts which have retained their outer membrane under phase-contrast, and record the percentage of intact chloroplasts in each preparation. Make drawings and notes on the shape, and structure of the different types of chloroplasts. Measure the amount of chlorophyll in each preparation.

Discussion

See later experiments using the chloroplasts.

Further reading

San Pietro, A. Ed.
   1972 'Methods in Enzymology', Vol. XXIII Section III. Academic Press, New York.

Hints for the teacher

The chloroplasts in this experiment may be used in the subsequent experiments on chloroplast electron transport. The activity of the chloroplasts can most easily be determined by measuring O_2 evolution coupled to dye reduction. The experiments were developed using spinach and the best results will be obtained with it. It can be bought from greengrocers in the October—December term and again in April—June. Greengrocers often sell spinach beet as spinach; it tastes much the same but you cannot make chloroplasts easily from it, so check the identity of the plant carefully. If spinach is not available, pea seedlings can be used. These can be grown by planting pea seeds in vermiculite or sand and growing them for 3/4 weeks. Pea chloroplasts can be used for all experiments except the assay of ferredoxin. The ferredoxin-NADP reductase of pea chloroplasts is soluble so they will not catalyse NADP reduction.

Lettuce and swiss chard have also been used for chloroplast isolation.

M.C.W. EVANS
University College
London

8 Photosynthetic oxygen evolution and the relationship between electron transport and ATP synthesis in spinach chloroplasts

Time required
3–4 hr.

Assumed knowledge
Use of oxygen electrode and ancillary equipment. A good understanding of electron transport systems and photosynthesis.

Theoretical background
Higher plants use light to provide the energy required to fix carbon dioxide using water as an electron donor, and oxygen is evolved as a waste product of water oxidation. CO_2 fixation is not directly coupled to the light absorption processes, but uses ATP and $NADPH_2$ in the reductive pentose phosphate cycle. ATP and $NADPH_2$ are made as a direct result of light-activated electron transport in the chloroplast lamellae. All of the components, except ferredoxin, required for the absorption of light and the use of the absorbed energy to oxidise water and to transport the electron to NADP with the coupled synthesis of ATP are bound to the thylakoid membrane of the spinach chloroplast. Chloroplasts prepared by any of the three procedures described in Section 9.7 can be used to demonstrate the requirements for photosynthetic electron transport and associated ATP synthesis. Oxygen evolution coupled to the reduction of a suitable electron acceptor will be used as a measure of light-dependent electron transport, and the effect of phosphorylating conditions and different types of inhibitor on the rate of electron transport investigated.

The natural electron acceptor in photosynthesis is NADP

$$2NADP + 2H_2O + 4ADP + 4Pi$$
$$\text{light} \downarrow \text{chloroplasts}$$
$$2NADPH_2 + O_2 + 4ATP$$

However, NADP is expensive, and also ferredoxin which is not commercially available must be added to the reaction mixture. It is therefore more convenient to use potassium ferricyanide as an artificial acceptor.

$$4\,Fe\,(CN)_6{}^{3-} + 2H_2O \xrightarrow[\text{chloroplasts}]{\text{light}} 4\,Fe\,(CN)_6{}^{4-} + O_2 + 4H^+$$

If the oxygen electrode available to you does not have a back-off system to allow you to use it to measure oxygen evolution, a different acceptor, methyl viologen, may be used. This is an acceptor which is autoxidisable, reacting with oxygen to form hydrogen peroxide. The overall effect observed in this system is an uptake of oxygen.

1 Photosynthesis with methyl viologen as acceptor:
$$4MV_{ox} + 2H_2O \xrightarrow[\text{chloroplasts}]{\text{light}} 4MV_{red} + 4H^+ + O_2$$

2 Autoxidation of methyl viologen:
$$4MV_{red} + 4H^+ + 2O_2 \rightarrow 4MV_{ox} + H_2O_2$$

Net result observed
$$2H_2O + O_2 \rightarrow 2H_2O_2$$

The H_2O_2 accumulates if no catalase is present (or is inhibited by azide) and O_2 uptake is then measured.

The rate of electron transport in chloroplasts is stimulated by addition of ADP and phosphate; if a limited amount of ADP is added, the electron transport rate slows down again when all the ADP has been phosphorylated to ATP. These effects are called photosynthetic control. The photosynthetic control ratio, which is the phosphorylating electron transport rate divided by the rate measured after all the ADP has been converted to ATP, is a measure of effectiveness of coupling of phosphorylation and electron transport.

Necessary equipment
1 For each group:
  (a) 1 oxygen electrode
  (b) 1 light source e.g., 300W projector
  (c) Pipettes, 0·5 ml, 1 ml and 2 ml
  (d) Microliter syringes, 10 and 25 μl (Hamilton, S.G.E.)
  If the students are to prepare their own chloroplasts the apparatus described in Section 9.7 is also required.
2 Solutions for a class of 20–25 students
  (a) 100 ml 1M Sorbitol
  (b) 100 ml 1M-Tris HC1 pH 8·0
  (c) 100 ml 0·1M $MgCl_2$
  (d) 100 ml 0·5M $K_3\,Fe\,(CN)_6$

(e) 100 ml *0·002M Methyl viologen (BDH)

(f) 100 ml *0·015M NaN$_3$ (BDH)

(g) 50 ml *0·01M Dichlorophenol indophenol (BDH)

(h) 50 ml *0·2M Sodium ascorbate

(i) 100 ml 0·3M Potasium phosphate pH 8·0

(j) 100 ml 0·3M NH$_4$ Cl

(k) 100 ml 0·3M Na arsenate pH 8·0

(l) 20 ml 0·02M ADP (Sigma)

(m) 10 ml 10^{-4}M dichlorophenyl dimethylurea (DCMU; 'Diuron' from Du Pont) in ethanol

(n) 10 ml 10^{-2} and 10^{-4}M carbonyl cyanide M−chlorophenylhydrozone (C.C.C.P.) in ethanol (Calbiochem or Boehringer)

(o) 10 ml 0·05M ATP (BDH, Sigma)

(p) Chloroplasts equivalent to 2–3 mg chlorophyll per group.

All solutions should be stored in a refrigerator. Nos. (f), (g), (h), (l) and (n) should be prepared on the day they are used.

Those marked * are required only if methyl viologen is used as electron acceptor.

Type of group
Pairs.

Instructions to the student
Set up the oxygen electrode (see Section 10.2) to measure oxygen evolution if you are using potassium ferricyanide as electron acceptor, or oxygen uptake if you are using methyl viologen. For each experiment you will make a basic reaction mixture in the electrode to which you will add the reagents being studied with a micro-syringe during the course of the reaction. The basic reaction mixture should contain

0·2 ml 1M Sorbitol

0·1 ml 1M Tris-HCl pH 8·0

0·1 ml 0·1M MgCl

0·1 ml 0·05M K ferricyanide

Chloroplasts = 100 μg chlorophyll

H$_2$O to 2·0ml.

If methyl viologen is used as electron acceptor, omit the ferricyanide and add

0·1 ml 0·01M NaN$_3$

0·1 ml 0·002M methyl viologen.

The sodium azide is added to inhibit any catalase present in the chloroplasts.

Measure the rate of oxygen evolution in the basic system and investigate the requirements for light and an electron acceptor. You may need to vary the amount of chloroplasts you add to obtain suitable rates.

To demonstrate photosynthetic control make the following addition to the basic reaction mixture while following the progress of electron transport:

1 Nothing, i.e., only basic reaction mixture

2 20 μl ADP 0·02M

3 20 μl potassium phosphate 0·3M

Investigate the effect of the following compounds on electron transport under phosphorylating and non-phosphorylating conditions:

1 50 μl dichloro phenyl dimethyl urea (DCMU) 10^{-4}M

2 20 μl sodium arsenate 0·3M

3 20 μl ammonium chloride 0·3M

4 10 μl CCCP 10^{-5}M

5 10 μl CCCP 10^{-2}M

6 10 μl ATP 5 x 10^{-2}M

If you are using methyl viologen as acceptor measure the effect of adding ascorbate (0·2M, 20 μl) and dichlorophenol indophenol (0·001M, 10 μl) to DCMU-inhibited chloroplasts. Investigate whether you can demonstrate photosynthetic control in this system. Try any other combinations of reagents which you think may produce interesting results.

Recording results and Discussion
Keep the original charts which you obtain from the oxygen electrode as a record of the experiment. From these charts measure the rates of electron flow under the different conditions used and calculate the photosynthetic control ratio of the chloroplasts you used. Think about the effect of the different inhibitors and co-factors you have used, relate the effects you have observed to the models of the chloroplast electron transport system which you have been given. Do your results agree with these models? Can you suggest any experiments which would be a better test of these models?

Extensions and applications
The type of experiment described here is widely used in photosynthesis research on the mechanism of electron transport and ATP synthesis. It is a convenient tool for preliminary screening of possible electron transport inhibitors for use as herbicides and for studies of the mode of action of antibiotics and other compounds which affect membrane structure.

Further reading
San Pietro, A. (ed)
1972 'Methods in Enzymology', Vol. XXIII, Academic Press, London

Rabinowitch, E. and Govindjee,
1969 'Photosynthesis', Wiley, London

'Annual Reviews of Biochemistry' and 'Plant Physiol-
    ogy'
The experiment is based on a paper by
Telfer, A. and Evans, M.C.W.
    1971 FEBS Letters, 14, 241

Hints for the teacher
The experiment is designed using spinach chloroplasts
but peas could be used equally well. Any of the
chloroplast preparations described in Section 9.7
can be used but preparation 3 is impractical for a
large class because of the small yield of chloroplasts.
Whole chloroplasts from preparation 2 gives a good
photosynthetic control ratios which decrease with
time. Broken, washed chloroplasts give lower ratios
but show little change over a 3 hr period. If the
students try to do the experiment using chloroplasts
which they make themselves, it is advisable to supply
some chloroplasts, as some groups may make inactive
ones.

M.C.W. EVANS
University College
LONDON

Section 10

Physiological Chemistry

1 Respiration and fermentation in yeast

Time required

4 hr. The development of respiratory activity in anaerobically grown cells is best seen if the experiment is extended over 5 hr by taking an hour off for lunch in the middle of the experiment.

Assumed knowledge

A-level chemistry and the first term of a first-year course on biology of cells. The subjects dealt with are cell structure, properties of macromolecules, respiration, photosynthesis and membrane phenomena.

Theoretical background

Respiration and fermentation in yeast. Yeast oxidises glucose to pyruvic acid. In respiring cells this pyruvic acid is oxidised completely to CO_2 via the tricarboxylic acid cycle and the respiratory chain.

$$glucose + 6\,O_2 \rightarrow 6\,CO_2 + 6\,H_2O.$$

If the yeast does not contain fully developed mitochondria, or if the activities of the mitochondrial enzymes are limited either by lack of oxygen or by the addition of an inhibitor, then the pyruvic acid is fermented to CO_2 and ethanol:

$$glucose \rightarrow 2\ ethanol + 2\,CO_2.$$

This class involves investigations of the relationship between respiration and fermentation in yeast.

Necessary equipment

Each pair of students will need:

10 ml anaerobically grown yeast
10 ml aerobically grown yeast
4 manometers and flasks together with 4 spaces in a water bath (Gallenkamp)
1 tube of a mixture (1:1) of vaseline and lanolin (BDH)
4 papers for centre wells
10 ml 10% KOH for centre well
2 Pasteur pipettes with rubber teats
Graduated pipettes, 2 x 5·0 ml and 2 x 1·0 ml

blowing ball for filling pipettes (Aspirette from Jencons)

ml 1·33M glucose in 0·067M KH_2PO_4 (pH 4·5)

ml 1·6mM sodium azide (BDH) in 0·067M KH_2PO_4 (pH 4·5)

plastic wash bottle with distilled water

pe of group

ne experiment should be carried out by pairs of udents.

structions to the student

espiration of glucose results in an uptake of oxygen d a production of CO_2 in roughly equal amounts. lcoholic fermentation of glucose is characterised by e production of CO_2 in the absence of any oxygen otake. These exchanges of gas may be measured by anometric techniques.

Manometric techniques are described separately at e end of this experiment.

You are given two cultures of the Yeast, *Torupsis utilis* (National Type Culture Collection). One lture was grown in air and the other anaerobically. oth cultures are harvested on the morning of the xperiment, when the cells are washed free of growth edium and re-suspended in 0·067M KH_2PO_4 (pH 5). After harvesting the cultures are kept at 2°C.

Start the following experiment with anaerobically grown yeast *at once*.

(a) Neighbouring pairs of students must share a thermobarometer. Set up a thermobarometer, containing 3·5 ml water, at once.

(b) In this, and in the subsequent experiment, it is important to prepare identical samples of yeast suspension. This can be done by careful pipetting from a homogeneous suspension. Ensure that the suspension is homogeneous by *shaking* it *thoroughly* before *sampling*.

(c) Pipette 2·5 ml of anaerobically grown yeast suspension and 0·5 ml of 1·33M glucose into each of two reaction flasks. Put one of the divided filter papers and a few drops of KOH (0·1 to 0·2 ml.) in the centre well of one of the two flasks. The alkali must be added with *extreme care* and *none* must be allowed to reach the main compartment of the vessel. Add the alkali with a Pasteur pipette, after removing any excess alkali from the tip of the pipette. Mount the flasks on their manometers and put them in the bath. The flasks should be shaken at about 120 oscillations per minute. This rate should *not be*

exceeded as fast shaking may lead to contamination of the main compartment with alkali from the centre well. Allow 15 min for equilibration and then determine the initial rates of CO_2 production and oxygen uptake. Remember that you are measuring rates of gas exchange and that your apparatus must remain airtight. Plot your results. Take enough readings to give a smooth curve. Allow at least 10 min between readings. Once you have determined the initial rates of gas exchange, open the taps and leave the flasks shaking in the bath until you come back from lunch. Then re-determine the rates and see whether they have altered during the experiment.

2 Pipette 2·5 ml of aerobically grown yeast into a reaction flask. Pipette 0·5 ml of 1·33M glucose into the side arm and add filter paper and alkali to the centre well. Mount the flask on its manometer, put in the bath and allow to equilibrate. Determine the rate of oxygen uptake. When you have done this, tip the glucose from the side arm into the main compartment and determine the effect of glucose on oxygen uptake. When tipping, take great care not to mix the contents of the flask with those of the centre well. If you have time, remove the flask from the manometer and carefully add 0·2 ml of 1·6mM sodium azide to the cell suspension.

*Azide is very poisonous—
do not pipette by mouth.*

Remount the flask and after equilibration determine the effect of azide upon the rate of oxygen uptake. Azide will inhibit the enzyme cytochrome oxidase. This enzyme catalyses the final step in the transfer of hydrogen from $NADH_2$ to oxygen.

3 The above arrangements will leave one spare flask and manometer for each two pairs of students. One of you should use this to determine the rate of gas exchange in the absence of KOH under the conditions described under (2). This will enable you to calculate CO_2 production as well as oxygen uptake.

The arrangement of manometers by neighbouring pairs should be:

Pair 1   *Manometer*   *Contents*
         1   Anaerobically grown yeast with KOH

155

	2	Anaerobically grown yeast *without* KOH
	3	Aerobically grown yeast with KOH
	4	Thermobarometer
Pair 2	5	Anaerobically grown yeast with KOH
	6	Anaerobically grown yeast *without* KOH
	7	Aerobically grown yeast with KOH
	8	Aerobically grown yeast *without* KOH

Recording results

Present your experimental data concisely and write a brief explanation of your results. This explanation should cover

1 Any differences in the behaviour of the two cultures
2 Any effect of incubating anaerobically grown cells in air for five hours
3 Any effect of azide

Record your results on the sheets provided. Correct the readings for changes in the thermobarometer and determine the differences in pressure per interval during the time between readings. This will enable you to follow the rates of gas exchange during the experiment and will avoid the processing of an overwhelming amount of data at the end of the experiment.

Discussion

Pay particular attention to the following points.

1 Have you obtained accurate measurements of the rates of gas exchange for all your samples?
2 What may you say about the pathways of respiration in aerobically grown yeast and in anaerobically grown yeast? Explain the differences in the rates of gas exchange of the two types of yeast culture.
3 Explain the effects of transfer to air on the gas exchange of the anaerobically grown yeast.
4 Compare the behaviour of aerobically grown yeast in the presence of azide with that of the freshly harvested anaerobically grown culture.

Extension and applications

1 The study of the changes that occur when anaerobically grown yeast is transferred to air is a powerful method for investigating the development of mitochondria.
2 The results of studies with inhibitors may be used as an indication that a particular sequence is operating *in vivo*.
3 Manometry can be used for following the course of a wide range of reactions involving gas exchange, e.g. respiration, fermentation, photosynthesis, and the activities of certain enzymes *in vitro*.

Hints for the teacher

The experiment

1 Typical results are:

| | Gas exchange ($\mu l/hr/2\cdot 5$ ml yeast) | |
	Oxygen uptake	CO_2 production
Anaerobically grown yeast		
at 11·30 hr	99	222
at 16·00 hr	191	227
Aerobically grown yeast		
without glucose	50	56
with glucose	223	247
with glucose and azide	25	75

156

The data for each pair of students will be based upon measurements of single samples. The reliability of the data can be established by comparing results from the class as a whole. This is possible if the yeast is properly sampled. This requires care in dispensing the yeast to each pair of students and in the student's sampling of the yeast. Yeast sediments and thus great care must be taken to shake the cultures before sampling. The major mistakes are those commonly found in manometry:

(a) Failure to place the correct solutions in the appropriate flasks or compartments.
(b) Leaks.
(c) Contamination of the yeast with KOH.

The extent to which respiratory activity develops in the anaerobically grown yeast varies. Careful attention must be given to the methods of growing and harvesting the cultures.

Growth of yeast. *Torulopsis utilis* is grown on a medium that contains, per 1 litre: 10 g Difco Bacto yeast extract, 54 g glucose, 1·44 g $MgSO_4.7H_2O$, 1·0 g KH_2PO_4, 1·2 g $(NH_4)_2SO_4$, 0·5 g NaCl, 0·87 g $CaCl_2.6H_2O$ and 0·5 ml 1% $FeCl_3$. The medium is sterilised by autoclaving (15 lb/in^2 for 15 min). The yeast is grown in 200 ml of medium in a 2 litre Erlenmeyer flask. An inoculum of 8·0 ml of yeast suspension is added to the 200 ml of medium. The inoculum is prepared by transferring a small number of yeast cells from the slope to 8·0 ml of the above medium in a test tube. This is incubated at 30°C for 24 hr and then used immediately. After the 200 ml of medium in the 2 litre flasks have been inoculated the cultures are gassed and incubated on a rotary shaker at 25°C. The aerobic cultures are gassed with a continuous stream of sterile air throughout the growing period. The anaerobic cultures are immediately flushed with sterile, oxygen-free nitrogen, and are gassed in this way throughout growth.

The aerobic cultures are harvested after 48 hr growth and the anaerobic cultures are harvested after 7 days growth. The aerobic cultures are harvested by centrifugation and washed thrice with 20–30 volumes of 0·067M KH_2PO_4 (at pH 4·5). The same basic procedure is used for the anaerobic cultures except that the harvesting is carried out at 2°C with pre-cooled buffer and that after centrifuging the yeast is re-suspended under a stream of nitrogen. It is essential to reduce the time between harvesting of the anaerobic cultures and the beginning of the experiment to a bare minimum. Even when this is done a certain amount of respiratory activity develops in the anaerobically grown yeast. This can be reduced by keeping the cultures at 2°C until they are added to the manometer flasks. Both anaerobically and aerobically grown yeast should be re-suspended so that a 10-fold dilution of the suspension given to the students gives a reading of 55 in an EEL 'Unigalvo' nephelometer without a filter— alternatively use a spectrophotometer at O.D. 540. In general, a class of 100 students needs one aerobic culture and 12 anaerobic cultures.

Appendix
Manometry

General

Manometry may be used to measure the rates of reactions in which gases are evolved or absorbed. The technique is widely used and may be applied to the measurement of gas exchange by tissue slices, suspensions of cells, isolated enzymes, and chemical reactions. The principle of the technique is that a reaction is made to take place in a closed apparatus that is attached to a pressure gauge or manometer. The temperature and volume of the apparatus are kept constant and any gas output or uptake can be *measured* as changes in *pressure*. The volumes of gas exchanged may be calculated from the changes in pressure.

The apparatus consists of a manometer and a reaction flask. The manometer is a U-tube filled with a liquid that has a specific gravity of 1·034. Thus 10 000 mm of the liquid are equivalent to 760 mm/Hg. The level of the liquid in the manometer can be adjusted by a screw clip on the reservoir. One end of the manometer tube is open to the atmosphere and the other end is attached to a reaction flask. The reaction flask and the limb of the manometer to which the flask is attached form a closed system the volume of which can be kept constant by adjusting the level of the manometer fluid. Under these conditions changes in pressure caused by a reaction in the flask can be measured as changes in the height of the fluid in the open limb of the manometer. There are three compartments in the reaction flasks that you will use. The main compartment is used for the biological material and the medium in which the material is suspended. The centre well is used for substances that you wish to keep permanently separated from the biological material and the side arm is

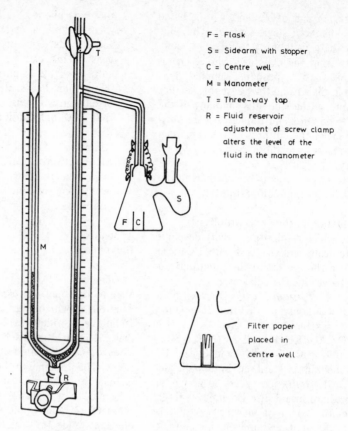

F = Flask

S = Sidearm with stopper

C = Centre well

M = Manometer

T = Three-way tap

R = Fluid reservoir
    adjustment of screw clamp
    alters the level of the
    fluid in the manometer

Filter paper
placed in
centre well

Figure 1 The Warburg constant volume respirometer

used for substances that you wish to tip into the main compartment during the course of an experiment. The flask must be shaken in a water bath at constant temperature. This is to ensure that the gas phase in the flask is kept in equilibrium with the liquid phase and that the temperature of the flask contents is kept constant.

Thermobarometer

The level of the manometer fluid will be affected by changes in atmospheric pressure and any changes in the temperature of the contents of the closed limb of the manometer and the flask. The extent of these changes must be determined and the value obtained must then be subtracted from the total change in pressure. The difference will represent the pressure change caused by the activity of the material in the flask. This correction is made by using a thermobarometer. A thermobarometer consists of a mano-

meter and a flask that contains only water. The thermobarometer is treated, in all other respects, exactly the same way as the manometer and flask that contains the biological material. Thus the thermobarometer will record changes in pressure that are common to all flasks but are not due to the activity of the biological material.

Procedure

1 Take great care of the manometers and the flasks. Keep the manometer upright. Do not knock either the manometer or the flask on the side of the bath. Do not put any strain on the manometer. If you turn the tap make sure that you turn the tap with one hand and hold the tap housing with the other. When putting the flask on to the manometer and when removing the flask, hold the neck of the manometer firmly in your spare hand.

2 Prepare the flasks by adding the solutions and the material. Take particular care when putting alkali in the centre well. Put the filter paper into the centre well first and then add the alkali with a fine Pasteur pipette. Wipe any excess alkali off the pipette with a piece of tissue. No alkali must be allowed to drop into the main compartment or to spray onto the sides of the flask.

3 Put the stoppers in the side arms of the reaction flasks. Mount the flasks on the manometers. Make sure that each flask is attached to its own manometer and is closed by its own stopper. This is important because the volume of the space in the flask and the manometer has been measured and is used to convert changes in pressure to changes in volume. The join between the flask and stopper and that between the flask and manometer must be airtight. Grease the stopper and the cone of the manometer *before* these joins are made. Your demonstrator will show you how to make these joins. Attach the springs that will hold the flasks in position.

4 Put each manometer separately onto the shaker. Make sure that the tap on the manometer is open. This tap is a 'tail-tap' and the only position in which it is shut is when the mark on the handle is uppermost.

5 Start the shaker and allow the contents of the flask to equilibrate to the temperature of the bath. After 10–15 min for equilibration test each manometer for leaks. You can detect any major leak by closing the manometer tap when the levels of the fluid are high and then lowering the level in the open limb of the manometer. If the system is airtight this procedure will create a negative pressure inside the manometer. If there is a leak the levels of the fluid in the limbs will equalise.
Leaks and unclosed taps are the commonest mistakes that are made by beginners.

6 After you have checked for leaks, close the taps and begin measurements. When you close the taps adjust the fluid in the open limbs of the manometers so as to give:
(a) A positive pressure in the thermobarometer, in flasks in which you expect gas consumption, and in flasks in which you expect little change in pressure
(b) A negative pressure in flasks in which you expect an output of gas.
The readings are taken as follows. Stop the shaker and adjust the level of the fluid in the closed limb of the manometer to the centre mark (150 on our manometers). Twist the adjusting screw back and forth through a small angle a few times before making the final adjustment. This will ensure that the fluid is running freely in the tube. Next read the level of the fluid in the open limb of the manometer. The readings should be taken at the level of the bottom of the meniscus. Avoid parallax errors. Re-start the shaker as soon as the reading has been made. After a measured interval repeat the above procedure. Every time you take a reading you must remember to read the thermobarometer. You should appreciate that by bringing the level of the fluid to the same mark in the closed limb of the manometer you ensure that your readings are made at a constant volume.

7 Take sufficient readings to obtain a smooth curve from which to calculate the rate of gas exchange. The intervals between readings should be 10–15 min.

8 When you have finished your readings *open* the taps of all the manometers and leave them on the bath.

Conversion of changes in pressure to changes in volume
The readings from the manometers represent changes in pressure. These values may be converted to changes in volume (measured in μ litres) by multiplying the pressure changes (after correction for any changes in the thermobarometer) by the appropriate *Flask Factor*. This factor K is given by

$$K = \frac{V_g \dfrac{273}{T} + V_f \alpha}{P_0}$$

where V_g = Volume (ml) of gas phase in flask and manometer to the 150 mm mark.
V_f = Volume (ml) of fluid in the flask.
α = Solubility (ml/gas/ml of liquid at standard pressure at temperature T) of gas involved in the reaction liquid.
P_0 = Standard pressure (10 000 mm of manometer fluid).
T = Absolute temperature of bath.

The derivation of the above equation is given in 3rd edn. Dixon (1951) 'Manometric Methods' pp.11–13 Cambridge University Press. You are provided with factors for the flasks that you use.
N.B. You must appreciate that the flask factor will vary with the volume of the flask, the volume of

READINGS

Time (min)		0	15	30	45	60	75				
F	12	190	174	162	147	134	120				
L	TB	207	206	209	210	212	214				
A	21	212	210	212	213	214	215				
S											
K											
N											
O											
Thermobarometer readings – Thermobarometer readings at zero time											
		0	−1	+2	+3	+5	+7				

CORRECTED READINGS

Time (min)		0	15	30	45	60	75				
F	12	190	175	160	144	129	113				
L											
A	21	212	211	210	210	209	208				
S											
K											
N											
O											

CHANGES IN PRESSURE

Time (min)		0-15	15-30	30-45	45-60	60-75					
F	12	−15	−15	−16	−15	−16					
L											
A	21	−1	−1	0	−1	−1					
S											
K											
N											
O											

he liquid in the flask, the gas that is measured, and the temperature.

Measurement of oxygen uptake

Oxygen uptake may be measured by determining the net change in pressure in a reaction flask that has alkali in the centre well. The CO_2 that is produced is absorbed by the alkali and the fall in the level of the manometer fluid will represent the oxygen uptake of the sample. A small piece of filter paper is generally put in the centre well in order to give a greater area for CO_2 absorption. If h represents the change in level of the manometer fluid, then the oxygen uptake (μ litres) = $h \times kO_2$, where kO_2 is the flask factor for oxygen.

Measurement of carbon dioxide production

Duplicate samples are put in separate flasks. The first sample is used to measure the rate of oxygen uptake as described above. The second sample is incubated without any alkali in the centre well and the manometer records the net change in pressure. This is the resultant of increase in pressure due to CO_2 production and decrease in pressure due to oxygen uptake. As the two samples of material have the same rate of oxygen uptake, the decrease in pressure caused by oxygen uptake in the flask without alkali can be calculated from the rate of oxygen uptake determined in the first flask. Once the change in pressure due to oxygen uptake and the net change in pressure in the second flask are known it is possible to calculate the change in pressure due to CO_2 production. This change in pressure is finally converted to a change in volume by multiplying by the flask factor for CO_2.

Treatment of data

1 Record your readings on the printed sheets provided.
2 Correct the measurements for any changes in the thermobarometer.
3 Determine the rates of changes in pressure.
4 Calculate the rates of changes in volume.

Specimen experiment

Duplicate samples of 20 discs of carrot storage tissue were set up. Alkali was added to flask 12 but not to flask 21. Flask factors for oxygen: flask 12, 2·31; flask 21, 2·13: for CO_2 for flask 21, 2·32.
The readings obtained are shown in the Table. (p.160)

Oxygen uptake in Flask 12
From the experimental data, rate of change of

pressure is −62 mm (−61 + [−1]) manometer fluid/hr. Therefore rate of change of volume is 62 x 2·31 = 143 μl oxygen/hr.

CO_2 production in Flask 21
(a) From flask 12 we know that the sample in flask 21 is absorbing oxygen at the rate of 143 μl/hr. Therefore change in pressure in flask 21 that would be caused by oxygen uptake is given by

$$143 = hO_2 \text{ x Flask 21 factor for oxygen } (2 \cdot 13)$$

$$hO_2 = -67 \text{ mm/hr}$$

(b) Experimental data for flask 21 shows net change in pressure of −3 mm/hr. Thus change in pressure due to CO_2 production is given by

$$-3 = -67 + hCO_2$$

$$hCO_2 = +64$$

(c) If change in pressure due to CO_2 production in flask 21 is +64 then volume of CO_2 production is given by multiplication by flask factor for CO_2, i.e.

$$64 \text{ x } 2 \cdot 32 = 148 \ \mu l \ CO_2/hr.$$

T. ap REES
Cambridge

2 Kinetics of oxygen utilisation by yeast and *E. coli* using an oxygen electrode

Time required
3–4 hr.

Assumed knowledge
General aspects of respiration and substrate uptake by cells.

Theoretical background
The aim of the experiment is to familiarize the student with some properties of cell respiration and substrate uptake which can be easily demonstrated using yeast and *E. coli*. In addition, the use of the Rank oxygen electrode will show the principals of a continuous monitoring system for assaying the oxygen concentration in the reaction mixture.

The substrates selected for use will show the selective uptake and utilisation characteristics of, for example, glucose and succinate, and of the D– and L–forms of glutamate.

The Rank oxygen electrode
This electrode is designed for following the uptake or production of oxygen by cell suspensions, subcellular particles or enzyme systems. The principle of operation is that first described by Clark (1956).

Oxygen diffuses through a thin (0·0005in) teflon membrane and is reduced at a platinum (half-cell) surface immediately in contact with the membrane

$$O_2 + 2e^- + 2H^+ \rightarrow H_2O_2$$

$$H_2O_2 + 2e^- + 2H^+ \rightarrow 2H_2O$$

The other half-cell is also incorporated in the base of the incubation vessel and is composed of a Ag–AgCl electrode.

Electrical connections and recording
The platinum electrode is polarised at 0·6 V with respect to the Ag–AgCl. The current which flows under these conditions is about 1 μ A in stirred air-saturated water at 30°C.

The current flowing is proportional to the activity (partial pressure) of oxygen in solution over a wide range. A suitable circuit is shown in Figs. 2 and 3.

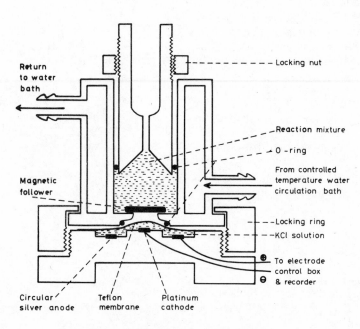

Figure 1 The Rank oxygen electrode

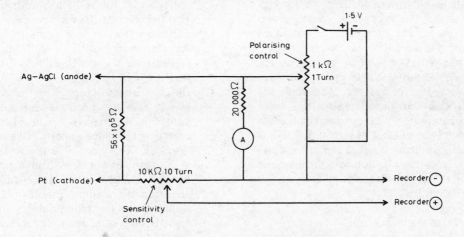

Figure 2 Circuit for normal box without back off

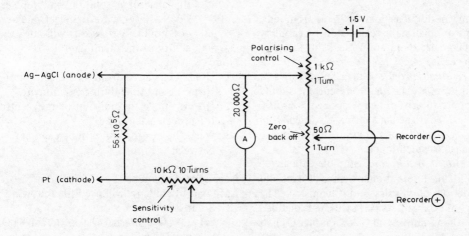

Figure 3 Circuit for box with zero back off for use in
photosynthesis.

The choice of value for the sensitivity control resistor will depend on the span of the recorder. A suitable value is given by

$$R = 2 \times \text{span of recorder (in mV)} \times 10^3 \text{ ohms.}$$

Ideally R should not exceed about 2kΩ, since this will alter the polarising voltage.

Suitable recorders
Any potentiometric recorder with a sensitivity better than 20 mV full scale and which will accept a source impedence of 2kΩ/mV span is suitable. For the best results a response time of 1–2 s is advisable.

Method of use
Connect the Ag–AgCl electrode to the *positive* side of the potential divider and the platinum to the *negative*. Add air-saturated medium to the incubation vessel; place the perspex cap in position and, after ensuring that no air bubbles are trapped, switch on the magnetic stirrer. Adjust the sensitivity control to give a suitable deflection on the recorder. When steady state has been achieved proceed with the

163

experiment. All additions should be made through the small hole in the perspex cap. Be careful not to introduce air bubbles. The experiment is concluded when the suspension becomes anaerobic (if respiration is being followed); that is when the current falls to zero, or very nearly so. To start the next experiment remove the perspex cap and wash the vessel out by sucking out the contents with a water pump. Add distilled water, suck out again and repeat as necessary. A drawn-out piece of glass tubing with a short piece of rubber tubing attached is a useful water remover. There is no need to turn off the magnetic stirrer between experiments once a suitable stirring rate has been established.

1 *Do not*
   (a) Overtighten the locking nut
   (b) Switch off the electrode between experiments
   The electrode may be left on all day without harm.
2 Suitable magnetic 'fleas' may be made from pins, paper clips or small nails sealed in glass capillary tubing. Make sure there are no leaks in the glass seals at the ends.
3 *Test for residual current.* Add a few crystals of sodium dithionite. The electrode current should fall within 5 s to zero or nearly so; if it does not, change the membrane and check electrical connections. A 'leaky' membrane often leads to noise. Change the membrane.
4 *Oxygen content of solutions.* Suitable data are to be found in the International Critical Tables and the Handbook of Physics and Chemistry. The solubility of oxygen in air-saturated distilled water is $0 \cdot 257 \, \mu\text{moles/ml}$ ($257 \, \mu\text{molar O}_2$) at $25°\text{C}$.
   Standardisation and salt effect:
   Dixon and Kleppe, Biochem. Biophys. Acta, 96, 357, 1965.
5 *Operation of the electrode*
   (a) Set the dial on the recorder to the zero position and switch it on to the calibrated mode of operation (rather than variable) and set the pen to zero using the coarse and fine zero controls of the recorder.
   (b) Then make sure there is 4 ml water in the electrode vessel and switch on the polarising voltage at the control box and increase it until $0 \cdot 65 \, \text{V}$ ($35 \, \mu\text{A}$) are registered on the meter.
   (c) Turn the control knob fully clockwise and sensitivity knob fully anticlockwise.
   (d) Turn on the magnetic stirrer, set the dial on

the recorder to the 10 mV position and see that the recorder still reads zero.
   (e) Now turn the sensitivity control until the recorder reads full scale deflection. Your recorder should now be calibrated ready for use.
   (f) Check the zero current by removing the water from the electrode and replacing it with $0 \cdot 01\text{M}$ borax (4 ml) in which a few crystals of sodium sulphite have just been dissolved. The recorder will now read near zero; if it does not, tell a demonstrator. The borax sodium sulphite mixture will produce an anaerobic solution. Instead of borax and sulphite, sodium hydrosulphite (dithionite) crystals may be added to the 4 ml water to make the solution anaerobic. Wash well afterwards.
   (g) Your electrode is now ready for use, remove the anaerobic solution and replace it with water. Never leave the electrode vessel without liquid in it as this will alter the permeability of the membrane.

Note
1 Use a constant stirring rate and do not turn off stirrer at any time during experiment.
2 Do not move the position of the electrode once it has been established.
3 Wash the electrode reactions chamber 3–4 times between each run.

Necessary equipment
Each pair of students will need the following.
1 Solutions
   (a) 1% (w/v) fresh yeast in $0 \cdot 02\text{M}$ KPO_4 buffer pH $7 \cdot 0 \rightarrow 20$ ml
   (b) 1% (w/v) yeast (in phosphate buffer) aerated vigorously for 24 hours $\rightarrow$ about 20 ml
   (c) *E. coli* suspension in $0 \cdot 02\text{M}$ tris buffer, pH $7 \cdot 5 \rightarrow$ about 20 ml. (Cells previously grown anaerobically in Penassay broth (antibiotic medium 3, Difco) for 14 hr at $37°\text{C}$).
   (d) $0 \cdot 02\text{M}$ phosphate buffer pH $7 \cdot 0$ and $0 \cdot 02\text{M}$ tris buffer pH $7 \cdot 5$
2 Substrates
   (a) 1 M glucose about 2 ml
   (b) 1 M succinate about 2 ml
   (c) 1 M L-glutamate about 2 ml
   (d) 1 M D-glutamate about 2 ml
3 Oxygen electrode and control box (Rank Bros.)
4 Small magnetic flea
5 Teflon membrane (Rank Bros.) (2cm^2) and Lens tissue (Gallenkamp)

6 1 recorder
7 1 water aspirator
8 1 magnetic stirrer (Rank Bros, Gallenkamp)
9 Assorted microlitre syringes (Hamilton, S.G.E., Gallenkamp)
10 Ice buckets and ice
11 Saturated KCl solution
12 0·01M Borax solution
13 Crystals of sodium sulphide
14 Sodium hydrosulphite (dithionite) (BDH)

Type of group
Students work singly or pairs.

Instructions to the student
Setting up the electrode.
1 Detach the base of the incubation vessel by unscrewing the perspex locking nut.
2 Add sufficient 1M KCl or saturated KCl to wet the Ag and platinum electrodes.
3 Cut a 1 cm square of lens tissue and make about a 1 mm diameter hole in it with scissors. Place the tissue over the platinum electrode so that the hole is over the electrode.
4 Cut a 1 cm square piece of teflon and place over the lens tissue and lock this in place over the platinum electrode by putting the incubation vessel in place and gently screwing down the locking nut. Take care that no air bubbles are trapped, and that the membrane is not twisted. The oxygen electrode is now ready for operation and is placed on the magnetic stirrer.

Recording Results
Calculate the rates of O_2 uptake in μmoles O_2 respired/minute/ml cell suspension. Tabulate. Add 2 or 3 representative portions of well documented chart into your report.

Discussion
The student should think carefully as to what the data means. The respiration rates before and after addition of substrate to starved and fresh yeast tell a significant amount about respiration. The D– and L–forms of substrates may be differently utilised by yeasts and bacteria, as may the three different substrates.

Extensions and applications
Limiting amounts of substrates may be added to the reaction mixture and then the stoichiometry of oxygen uptake for each substrate calculated. This should be compared with the theoretical values.

Inhibitors of respiration, e.g. azide, HCN, may be tested. The oxygen electrode is used increasingly in biochemical, medical and biological research for continuous and rapid monitoring of oxygen concentrations in solution. The Rank electrode is convenient and relatively inexpensive. Another popular type is the Clark electrode made by YSI, Yellow Springs, Ohio—obtainable from V.A. Howe in England.

Further reading
Packer,
   1967 'Experiments in Cell Physiology', p. 154, Academic Press, New York
Lessler, M.A. and Brierly, G.P.
   1970 'Oxygen Electrode Measurements in Biochemical Analysis', in 'Methods of Biochemical Analysis', 17, 1
Delieu, T. and Walker, D.A.
   1972 New Phytologist, 71, 201.

Hints for the teacher
Ordinary baker's yeast may be used and the E. coli can either be bought as the frozen paste or easily cultured. The oxygen electrodes and recorders should be carefully checked before the start of the experiment if they are to be used immediately. However, the assembly of the electrode by the students themselves does teach the principal of the O_2 electrode very well.

D.O. HALL
King's College
London

3 A simple polarographic method for investigating the effects of temperature and pCO_2 on the oxygen-haemoglobin dissociation curve

Time required
This experiment can be easily carried out in 3 hours in the laboratory with the required apparatus, when 100 ml fully oxygenated fresh heparinised blood (per student pair) and 100 ml totally deoxygenated under nitrogen are previously prepared.

Assumed knowledge
Although it is not essential that an oxygen-haemoglobin dissociation curve has been determined using the Van Slyke gas analysis apparatus, students are advised to familiarise themselves with the sigmoid curves obtained under varying conditions, and understand the basic concepts of oxygen carriage by the erythrocytes and plasma of the blood.

Theoretical background
Whereas oxygen-haemoglobin dissociation curves have been obtained using the laborious gas analysis technique, which measures the volume of total oxygen content released from a sample of blood which had been previously equilibrated at a prese' oxygen tension, this method simply measures polaro graphically the oxygen tension of blood at a pre determined level of saturation.

The method works simply because the oxyger carried by the haemoglobin inside the red blood cells must be in equilibrium with that carried in the dissolved form in physical solution:

$$Hb + \text{dissolved } O_2 \rightleftharpoons HbO_2$$

The quantity of gas dissolved depends upon the partial pressure of the oxygen in the gas phase and also on the solubility coefficient of the oxygen in the solution in question:

$$\text{Vol.} = \text{pressure·x solubility coefficient.}$$

Necessary equipment
A Clark-type oxygen electrode can measure the oxygen tension in physical solution, thereby deter mining the tension or pressure in the gas phase with which it is in equilibrium. The senior elements of the YSl 5331 oxygen probe (Shandon Scientific Co. Ltd) are separated from their environment by a thir membrane which is permeable to oxygen which car enter the interior of the sensor. When a suitable polarising voltage (0·8 V) is applied across the cell oxygen will react at the cathode, causing a current to flow through the cell. The sensor therefore measures the oxygen pressure because the force causing oxyger to diffuse through the membrane is proportional to the pressure of oxygen outside the membrane. Any alteration in current due to a change in oxyger

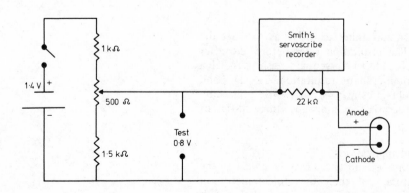

Figure 1

sion can be monitored with a potentiometric
order. For this purpose the circuit shown in Fig. 1
ictions adequately. The probe is fitted with a jack
ig.

Therefore when blood, with variable percentage
uration, is isolated from any gaseous phase, it is
ssible to measure the corresponding O_2 tensions.

tructions to the student

e electrode is set up as in the diagram (Fig. 2) in
ml fresh blood e.g. Rabbit saturated with oxygen,
lated from the air with a film of liquid paraffin
DH) and vigorously stirred without disturbance of
paraffin layer. A hot plate magnetic stirrer
illenkamp) is used for this purpose. Blood exhaust-
ly reduced by bubbling a slow stream of N_2
ough a stock volume is contained in a vessel with
outlet at the bottom. To the outlet is attached a
rt length of Portex tubing (Hoslab) for outflow
trolled by a spring clip. A small amount of blood
run out to waste and then 100 ml are run into a
ker below a paraffin layer. This exhaustively
uced blood is drawn into a burette containing
ut 3 ml liquid paraffin in the tip, by applying
ative pressure at the other end. With the burette
below the tap filled with reduced blood, it is
ced in position below the level of the liquid
affin in the 100 ml beaker as in Fig. 2.

Having set the desired temperature of the stirred
od, and oxygen tension is recording uniformly in a
table position on the recorder, the system is ready
titration. Small amounts of reduced blood (1 or
l at a time) are accurately delivered and the
centage of the total change of oxygen tension of
blood is measured on the recorder. The total
gen change is measured in arbitrary units on the
order paper when the electrode is taken from
0% air-saturated blood and placed in the exhaustiv-
reduced blood again under liquid paraffin. The
sitivity of the recording apparatus is set so that
deflection is nearly full scale. This total deflec-
n must be measured either before the beginning or
the end of the titration and all other deflections
asured as a percentage of this. When the electrode
noved from one sample to another care must be
en to ensure that no air bubbles remain in the
on of the sensor tip.

ording results

nple readings when total arbitrary units change
ween saturated and unsaturated blood = 96.

Starting Vol. Sat. blood	Vol. Unsat. Added	New Total Vol.	Calculated % Sat.	% of Total Units Change
20 ml	-	-	-	-
20 ml	1 ml	21 ml	95·0	27
21 ml	1 ml	22 ml	91·3	46
22 ml	2 ml	24 ml	83·3	59
24 ml	4 ml	28 ml	etc.	etc.

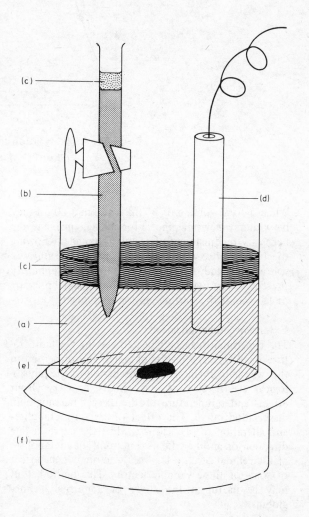

Figure 2 (a) Stirred blood; (b) reduced blood; (c) liquid paraffin; (d) O_2 sensor; (e) stirring slug; (f) heater-stirrer.

167

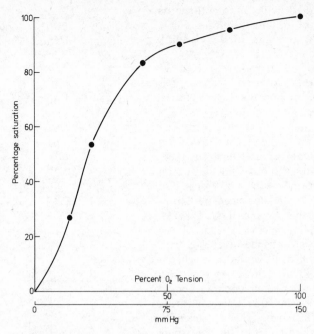

Figure 3 The relationship between percentage saturation of blood and oxygen tension.

When a total volume of 40 ml is reached 20 ml may be withdrawn with a pipette for convenience, and the necessary readjustment of calculation made. Knowing the tension of oxygen with which the 100% saturated was equilibrated, the results may then be graphically presented as in Fig. 3, if for example a partial pressure of 150 mm is the starting tension of oxygen.

Extensions and applications

The experiment can be repeated varying the temperature or pCO_2. The results obtained and presented in this way lend themselves to discussion of the physiological significance of the effects of CO_2 (Bohr effect) and temperature on the oxygen-haemoglobin dissociation curve. The effect of certain metabolites and alteration of pH also is to shift the curve in one direction or another. Bearing in mind the metabolism at the cellular level, it is possible to predict the effect of some of these variable factors. The method itself may be useful in the study of abnormal haemoglobins.

Further reading

Ganong, H.F.
1969 'Review of Medical Physiology', 4th ed., Blackwell, London, Chap. 35

Edwards, M.J. and Martin, R.J.
1966 J. Applied Physiol., 21, 1898

Hints to the teacher

In preparation for this experiment a stock of blo[od] should be made as fully reduced as possible bubbling a stream of nitrogen, containing the amou[nt] of CO_2 required, through the flask for approximat[ely] 3 hr before the experiment is due to start. A stock saturated blood may also be prepared by bubbl[ing] oxygen, with the correct CO_2 content, through similar flask. If equilibration with room air conditi[ons] is required, blood can be saturated by swirling sm[all] volumes at a time over the inside surface of a 2 li[tre] conical flask.

A kit with instructions for assembly of membrane of the oxygen probe can be obtained fr[om] Shandon Scientific Co. Ltd.

Any further details required of the practical circ[uit] can be obtained from W. Barry.

V.A. KNIG[HT]
W. BAR[RY]
Card[iff]

Measurement of the amount of ascorbic acid in cabbage

Time required
2 hr.

Assumed knowledge
A-level chemistry and the first term of a first year course on biology of cells are assumed. The subjects dealt with during the first term of this course are cell structure, properties of macromolecules, respiration, photosynthesis, and membrane phenomena.

Theoretical background
Ascorbic acid (vitamin C) is found in varying amounts in a wide variety of plant tissues. The role of ascorbic acid in plant metabolism is not yet understood. The determination of the ascorbate content of cabbage illustrates the general problems involved in the measurement of cell contents. Firstly, the ascorbic acid must be extracted from the cells. This can be done by breaking the tissue in a medium suitable for extraction. Most tissues of higher plants can be homogenised by grinding in a pestle and mortar. The addition of a little clean sand will make this process easier. The extraction of the ascorbic acid must be complete and must not be accompanied by loss of ascorbic acid.

Many plant tissues contain the enzyme ascorbic acid oxidase. This enzyme catalyses the oxidation of ascorbic acid to dehydroascorbic acid. When cells are disrupted by grinding, the activity of ascorbic acid oxidase may be sufficient to catalyse the oxidation of all the ascorbic acid originally present in the tissue. In order to avoid such a loss of ascorbic acid you are advised to extract the tissue with 5% metaphosphoric acid. This treatment will inactivate the oxidase. Ensure that the tissue is well immersed in the metaphosphoric acid before you begin grinding. The homogenate may be filtered by passing through a wad of glass wool.

Ascorbic acid has the following formula:

$$O=C-C=C-C-C-CH_2OH$$

with $-O-$ bridging the first and fourth carbons, H on the end carbon, and OH, OH, H, OH below the respective carbons.

The hydrogen atoms of the two enol groups are readily oxidised and thus ascorbic acid is a strong reducing agent. Ascorbic acid can be measured by means of its reducing property. The dye, 2,6-dichlorophenol-indophenol is blue in alkali and pink in acid. This dye is reduced by ascorbic acid to the colourless leuco form according to:

$$O = \underset{Cl}{\overset{Cl}{\diamond}} = N - \diamond - OH + 2H$$

oxidised form (blue or pink)

$$\rightarrow HO - \underset{Cl}{\overset{Cl}{\diamond}} - \underset{N}{\overset{H}{|}} - \diamond - OH$$

reduced or leuco form (colourless)

You may assume that ascorbic acid is the only substance present in extracts of cabbage that will reduce the dye at a pH below pH 4·0. Therefore you can measure the amount of ascorbic acid in an extract by titration against a solution of dye. You have been given a standardised solution of dye. The amount of ascorbic acid equivalent to 1·0 ml of the dye is written on the blackboard. A pink colour that persists for 15 s after the addition of one drop of dye should be taken as the end point. The reported values for the ascorbic acid content of cabbages vary over the range 20 to 60 mg/100 g fresh weight.

Necessary equipment
Each student will need:
1 cabbage—avoid red cabbages and prefer those with firm white hearts
1 pestle and mortar (13 cm diameter)
1 x 100 ml measuring cylinder
3 x 150 ml conical flasks
1 x 2in funnel
1 x 4in funnel
2 x 50 ml volumetric flasks
2 stirring rods
2 x 10 ml graduated pipettes
1 x 50 ml burette
1 burette stand, clamp and bosshead
1 each of the following sizes of beaker: 600 ml, 250 ml, 100 ml, 50 ml
1 flat white tile
1 plastic wash bottle of distilled water
1 blowing ball for filling pipettes (e.g. Aspirette from Jencons)
200 ml 5% metaphosphoric acid (BDH)
100 ml 0·025% 2,6-dichlorophenol-indophenol (BDH)
20 g acid-washed sand (BDH)

2 g glass wool
1 wax pencil
1 carving knife and cutting board
1 hot plate
Access to balance for weighing 1–60 g

Type of group
Each student should work alone under the guidance of a demonstrator. Each demonstrator has charge of 12–15 students.

Instructions to the student
Measurement of the amount of ascorbic acid in cabbage

You are asked to answer the following questions. Firstly, what is the ascorbic acid content, in mg/100 g fresh weight, of the cabbage that has been provided for you? Secondly, does boiling of the cabbage in water cause destruction of the ascorbic acid initially present in the tissue?

Do this experiment on your own. If you need advice speak to a demonstrator and not to your neighbour.

Our present attempts to understand the mechanisms that control cell metabolism depend to a significant extent upon our ability to measure the amounts of enzyme substrates present in a cell at a particular time. This experiment introduces you to the difficulties involved in making, and in interpreting, such measurements. *Do not be deceived by the apparent simplicity of this experiment.* The experiment involves problems of sampling, replication, and avoidance of artifacts caused by preparation of cell extracts. Please bear in mind that at the end of the day each of you should have obtained sufficient data to provide trustworthy answers to the questions asked. If you find yourself pressed for time, complete the first part of the practical properly and leave the second question. When you weigh out the cabbage, put it on a filter paper, *do not* weigh it directly on the balance pan.

Hints for the teacher
1 The experiment is designed to make students think and work independently. Thus the students are required to work alone and are not given the schedule until they enter the laboratory.
2 The major mistakes that are made are:
   (a) Failure to sample the cabbage correctly
   (b) Failure to take replicate samples and failure to consider variation between replicate samples
   (c) Incomplete extraction of ascorbate from the tissue

(d) Loss of ascorbate through failure to inactivate ascorbic acid oxidase. The sample must be thoroughly immersed in phosphoric acid before homogenisation.
(e) The addition of cabbage to boiling water followed by boiling for 5 min or so does not destroy appreciable amounts of ascorbate. Some ascorbate may be lost if the cabbage is added to cold water and the temperature raised gradually to 100°. Both methods of boiling release ascorbate into the water. The unwary student often forgets to assay the water and concludes that boiling destroys ascorbate. It is important to realise that the cabbage should be boiled in the absence of phosphoric acid but that acid must be added before the ascorbate is titrated.

3 The method used to assay ascorbate is described by Roe, J.M. in 'Methods of Biochemical Analysis' ed. Glick, D., Vol. 1, p. 115 Interscience Publishers, New York, 1954.
4 The dichlorophenol-indophenol is standardised by titration against a freshly prepared solution of ascorbic acid (200 mg/litre) in 5% HPO_3.

T. ap REES
Cambridge

5 Uptake of phosphate and glucose by carrot cells

Time required

A good experiment can be completed in 4 hr. It is best if the incubation in the solutes is prolonged by persuading students to set up the experiment in 2 hr before lunch, to take 1 hr for lunch, and then to return for 2 hr in the afternoon.

Assumed knowledge

A-level chemistry, and the first term of a first-year course on biology of cells are assumed. The subjects dealt with during the first term of this course are cell structure, properties of macromolecules, respiration, photosynthesis and membrane phenomena.

Theoretical background

It is vital for the students to have been taught about active transport and the use of the Nernst equation.

The uptake of phosphate and glucose by carrot tissue

Diffusion of uncharged molecules depends upon differences in chemical activity. The diffusion of ions, because of their charge, depends upon differences in chemical activity and differences in electrical potential. Thus ions diffuse along a gradient of electrochemical potential that is the resultant of two driving forces, that of chemical activity and that of electrical potential. Cells are able to move some ions against a gradient of electrochemical potential. In addition there is evidence that some cells can move uncharged molecules against a gradient of chemical potential. The movement of ions against a gradient of electrochemical potential or of uncharged molecules against a gradient of chemical potential is called active transport. This process is dependent upon a decrease in free energy of some metabolic process. Both diffusion and active transport contribute to solute uptake by cells. In non-photosynthetic cells the energy for active transport is provided by cellular oxidations. The way in which oxidation and active transport are linked is not known. Two major hypotheses have been put forward. One is that the energy of cellular oxidations is made available to transport systems in a general way in the form of ATP or some closely related compound. The other

hypothesis is that the active transport of anions is specifically linked to the process of hydrogen transport in the respiratory chain.

In this experiment you are asked to:

1 Determine the rates of uptake of glucose and phosphate by disks of carrot storage tissue.
2 Consider whether any uptake is an active process.
3 Investigate whether measurement of the effects of respiratory inhibitors on the uptake of glucose and phosphate can distinguish between the hypotheses put forward above in explanation of the linkage between respiration and active uptake. The two inhibitors are azide and 2,4-dinitrophenol. Their action is discussed below.

Azide inhibits cytochrome oxidase. Azide at 2×10^{-3} M inhibits the oxygen uptake of carrot tissue by 50% of the control value. DNP is said to uncouple hydrogen transport and ATP formation in the respiratory chain. The transfer of hydrogen from $NADH_2$ to oxygen is linked to the formation of ATP from ADP and inorganic phosphate. This linkage is obligatory in the sense that hydrogen transfer is dependent upon a supply of ADP and phosphate. DNP disrupts this linkage so that in the presence of DNP no ATP is formed and hydrogen transfer is not impeded by lack of ADP and phosphate. Thus treatment with DNP will cause a sharp decrease in the production of ATP during respiration. In cells where respiration is limited by the availability of ADP and inorganic phosphate, DNP may speed up hydrogen transfer to oxygen. This happens in carrot where DNP at 5×10^{-5} M stimulates oxygen uptake by 140% of the control value.

Necessary equipment

The equipment required will depend upon the type of experiment that an individual student designs. The major variable is the number of samples. Thus there should be ample reserves of boiling tubes, places for incubating the samples, and test tubes.

The following equipment per student will cover most experiments:

Access to EEL portable colorimeter with filters No. 608 and 626 or spectrophotometer—one between two students
12 colorimeter tubes (8 x 1·4 cm)
14 boiling tubes (15 x 2·4 cm) in racks
18 test tubes (15 x 1·4 cm) in racks
2 thin glass stirring rods (7in long)
12 glass condenser bulbs to fit the test tubes
2 x 10 ml measuring cylinders
1 plastic wash bottle with distilled water
1 blowing ball for filling pipettes (Aspirette from Jencons)

1 Petri dish (diameter 9 cm)
20 filter papers (diameter 12·5 cm)
1 magic marker or other means of labelling tubes
Graduated pipettes 2 x 1·0 ml, 2 x 2·0 ml, 4 x 5·0 ml, and 2 x 10·0 ml
300 discs of carrot
100 ml solution A—80μg/ml glucose, 1μmole/ml phosphate
50 ml solution B—80μg/ml glucose, 1μmole/ml phosphate + 2 x 10^{-3} M sodium azide (BDH)
50 ml solution C—80μg/ml glucose, 1μmole/ml phosphate + 5 x 10^{-5} M DNP (BDH)
50 ml acid molybdate reagent (BDH)
50 ml 12% trichloroacetic acid (BDH)
20 ml Somogyi's copper reagent (Dissolve 4g $CuSO_4$. $5H_2O$, 24 g anhydrous Na_2CO_3, 16 g NaKtartrate and 180 g anhydrous Na_2SO_4 in 1 litre water.)
    or Folin and Wu's alkaline copper solution (BDH)
20 ml arseno-molybdate reagent (BDH)
10 ml of the following standard solutions:
    0·25 μmole KH_2PO_4/ml + 20 μg glucose/ml
    0·50 μmole KH_2PO_4/ml + 40 μg glucose/ml
    0·75 μmole KH_2PO_4/ml + 60 μg glucose/ml
100 ml of a solution that contains 1 μmole KH_2PO_4/ml and 80 μg glucose/ml
20 ml 5 x 10^{-5} M dinitrophenol (DNP) for use as blank in colorimetric assay
Access to boiling water bath and space therein to incubate 12 test tubes
Access to water bath and shaker at 30°C.

Type of group
The experiment can be done individually or by groups of students.

Instructions to the student
You are provided with:
1 Slices of the storage tissue of carrots. These contain phosphate at 2·4 millimolar and 15 mg glucose per g fresh weight. The potential across the cell membrane is −80 mV.
2 Three solutions, all of which contain glucose at 80 μg/ml and phosphate at 1 μmole/ml. Solution A contains nothing other than glucose and phosphate. Solution B contains sodium azide at 2 x 10^{-3} M.
    Solution C contains 2,4-dinitrophenol (DNP) at 5 x 10^{-5} M.
3 Instructions and apparatus for the determination of glucose and phosphate.
The assays for phosphate and glucose are sensitive enough to detect uptake of both solutes when

30 discs are incubated in 5 ml of solution for 60 min. Take 0·5 ml of the solution for glucose measurements and 2·0 ml for the measurements of phosphate.

Azide and DNP are powerful poisons handle both compounds with great care

Assay of inorganic phosphate
This is assayed by the method described in Section 10.6. The assays should be carried out in colorimeter tubes. Add to each tube:
    2·0 ml of phosphate solution
    *4·0 ml of trichloroacetic acid
    †3·0 ml of acid molybdate reagent
    1·0 ml of water

Stir well, allow the colour to develop for about 10 minutes and measure in the EEL colorimeter with filter 608 or at 660nm. Prepare a standard curve from the phosphate solutions provided.

Azide and glucose do not interfere with this assay. The colour produced by phosphate in the presence of DNP should be measured against a blank that contains the same amount of DNP.

Assay of glucose
You can assay glucose by virtue of its ability to reduce copper. The amount of copper that is reduced will be determined by measuring the colour formed by the reaction between cuprous copper and arseno-molybdate. To measure the amount of glucose in the medium in which the discs have been suspended you should put 0·5 ml of the medium in one of the test tubes. Add 0·5 ml water and then add 1·0 ml of Somogyi's alkaline copper reagent. Cover the tube with a glass stopper and leave it in a boiling water bath for 15 min. Cool and then add 1·0 ml of the arsenomolybdate reagent. Shake the tube well, when the effervescence has stopped, make the total volume up to 10 ml and measure the colour in the EEL colorimeter with filter 626 or at 680nm. You are provided with glucose solutions that contain 80 μg/ml, 60 μg/ml, and 20 μg/ml.

Do not pipette the Somogyi reagent or the arsenomolybdate reagent by mouth

Discussion
Pay particular attention to the following points:

---

*Trichloroacetic acid is a dangerous and corrosive poison.
†The acid molybdate reagent is made up in 10N H_2SO_4 and contains the ferrous sulphate

1 Did the cells take up glucose and phosphate?
2 Have you obtained a reliable estimate of the rates of accumulation?
3 Can you decide whether any uptake was an active process?
4 Can you draw any firm conclusions from the effects of the inhibitors?
5 What are the limitations of this type of experiment?
6 Can you design a more incisive experiment?

Extensions and applications
Carefully carried out, this type of experiment yields useful and accurate information about solute accumulation but not about rates of influx.

Hints for the teacher
The experiment
1 If 30 discs are incubated in 5·0 ml of solution at 30°C for 3 hr, rates of uptake for phosphate and glucose of 0·5–1·0 μmole and 100–200 μg per sample per hour, respectively, are obtained. The greatest complication is anaerobiosis. This will cause leakage of solutes from the cells. Anaerobiosis occurs readily if the sample size is increased, if the ratio of carrot discs to solution is increased, and often if the incubation is prolonged beyond 3 hr.
2 The effects of the inhibitors are complex and vary with the batch of carrots and the length of incubation. Azide almost always causes marked leakage of glucose and appreciable leakage of phosphate. This leakage becomes more marked with longer incubations. Generally DNP inhibits the accumulation of both phosphate and glucose. In longer incubations DNP may cause leakage. These effects are useful in that they cause the students to think about the validity of their experimental data, the relationships between fluxes and accumulation, and the relationship between metabolism and solute movement.
3 The form of the experiment is left to the student. The most common errors are failure to take replicate samples, failure to measure accumulation at different times, and over-ambition that results in the use of more samples than can be analysed in the time available.

Preparation of carrot discs
Cork borers (No. 6, diameter 1 cm, Gallenkamp) are used to remove, vertically, cylinders of storage tissue from carrot roots. The outer 1 cm of the root should not be used but all the rest is satisfactory. The bruised ends of the cylinders of tissue are cut off and the remainder of the cylinders are sliced mechanically into discs that are about 1 cm x 1 mm. The discs are rinsed in distilled water and then placed in a Florence flask (round, flat-bottomed flasks, Gallenkamp) of distilled water. The discs are kept circulating in this Florence flask by a continuous stream of air that is directed against the bottom part of the flask wall. The flask must be full of water before the discs will circulate. The discs are starved in this way at 25° for 5 days. The water in the flask is changed three times on the first day and twice a day thereafter. It is essential that the discs do not become anaerobic during this starvation and great care must be taken to ensure constant aeration and circulation. Avoid excessive overcrowding. It is particularly important not to maltreat the carrot discs. They should not be crushed or allowed to dry out or left in a heap on the bench.

Incubation of samples
The essential need is for a system that will shake large numbers of boiling tubes at 30°C.

Assay of phosphate
See Section 10.6.

Assay of glucose
It is important to ensure that the reaction mixtures are placed in *boiling* water and that it is kept boiling for the whole of the incubation. The assay is described in detail by Hestrin, Feingold and Schramm in Methods in Enzymology, Vol. 1, p. 231 (edited by S.P. Colowick and N.O. Kaplan, Academic Press, 1955).

T. ap REES
Cambridge.

6 Changes in the activity of fructose-1, 6 diphosphate during the differentiation of photosynthetic cells

Time required
4 hr.

Assumed knowledge
A-level chemistry, and the first term of a first-year course on the biology of cells.

Theoretical background
The formation of a photosynthetic cell is one of the most important and obvious forms of cell differentiation. The most striking aspect of this differentiation is the synthesis of photosynthetic pigments. At the same time the cell synthesises enzymes needed for the photosynthetic reduction of CO_2. The extent of such syntheses can be judged by comparing enzyme activities in photosynthetic and non-photosynthetic cells. The enzyme fructose-1,6-diphosphatase catalyses the reaction

fructose-1,6-diphosphate + H_2O → fructose-6-phosphate + H_3PO_4.

This reaction is an important step in the photosynthetic fixation of CO_2.

Necessary equipment
Each student will need:
Sufficient 7-day-old seedlings of mustard to give about 2 g cotyledons
6 sheets of filter paper (15 cm diameter)
2 Petri dishes (6 cm diameter)
1 single-edged razor blade
*2 Small pestles and mortars (5 cm diameter)
*2 x 25 ml beakers
*8 x 15 ml plastic centrifuge tubes
1 x 5 ml graduated pipette
2 x 10 ml measuring cylinders
4 x 1·0 ml graduated pipettes
3 x 5·0 ml graduated pipettes
1 blowing ball for filling pipettes (Aspirette from Jencons)

8 colorimeter tubes (8 x 1·4 cm)
1 EEL Portable Colorimeter with filter No. 608 or spectrophotometer
1 plastic wash bottle with distilled water
Access to balance for weighing 500–3000 mg
Access to bench centrifuge—MSE Minor
Place to shake four plastic centrifuge tubes at 25°C
2·5 ml 25 mM Fructose-1,6-diphosphate (BDH)
70 ml extraction medium (see 'Instructions to the student') (Keep Cold)
50 ml acid molybdate reagent
10 ml each of 0·5 mM, 1·0 mM, 1·5 mM and 2·0 mM KH_2PO_4.
60 ml 12% trichloroacetic acid (BDH)

1 The equipment required depends upon the number of assays undertaken. The above list is sufficient for one extract of cotyledons and one extract of hypocotyls.

2 The temperature of the extracts is kept low by extracting in pre-cooled medium with pre-cooled equipment and by storing the extracts in ice. All equipment marked * is placed in ice on an ice bucket and kept in the cold room until just before the student requires it.

3 The enzyme assays are carried out at 25°C. A convenient means of shaking large numbers of centrifuge tubes at 25°C is needed. One possibility is to adapt a rectangular water bath, such as is sometimes used for manometry, so that the tubes are held in spring clips attached to a metal bar shaken horizontally just above the water.

Type of group
This experiment can be done individually or in pairs.

Instructions to the student
You are given a number of 7-day-old seedlings of mustard (*Brassica nigra*). Design and complete an experiment that compares the activity of fructose-1, 6-diphosphatase in the photosynthetic and non-photosynthetic parts of the seedlings. You are advised to use the lower part of the hypocotyl as non-photosynthetic tissue. Do not begin this experiment until you have discussed your plan with your demonstrator. Take note of the following:

1 Fructose-1, 6-diphosphatase may be extracted in a solution that contains 100 μmoles tris buffer at pH 8·8 (2-Amino-2-(hydroxy-methyl)-propane-1:3 diol) (BDH), 5 μmoles $MgCl_2$ and 600 μg EDTA (sodium salt of diaminoethanetetra-acetic acid) (BDH) per ml. Cell debris can be

174

removed from the extract by centrifuging at speed 10 for 3 min.

2 The enzyme may be assayed by measuring the release of phosphate from fructose-1,6-diphosphate. The assay mixture should contain 0·5 ml enzyme extract (corresponding to 100—200 mg fresh weight of tissue)

0·2 ml of fructose-1,6-diphosphate at 25 μmoles/ml

0·3 ml of extraction medium.

Incubate in a plastic centrifuge tube at 25°C for 10 min.

3 The amount of inorganic phosphate in the fructose-1,6-diphosphate solution will be told to you.

4 Appropriate control measurements must be devised and taken into account in the calculation of the results.

Assay of inorganic phosphate

This assay of phosphate depends upon the formation of a blue complex between phosphate and molybdic acid in the presence of a reducing agent. Ferrous sulphate is used as a reducing agent. The blue molybdate complex is determined colorimetrically. At the end of the incubation period add to the mixture:

4 ml of 12% trichloroacetic acid
3 ml of acid molybdate reagent
2 ml of distilled water.

Stir with a glass rod, leave for 10 min, centrifuge at speed 9 for 10 min, and measure the colour in an EEL colorimeter with filter 608.

You are supplied with phosphate solutions containing 2 μmole/ml, 1·5 μmole/ml, 1·0 μmole/ml and 0·5 μmole/ml. Prepare a standard curve by taking 1·0 ml of each of these solutions and use it to determine the amount of phosphate in the incubation mixtures.

N.B.

1 Trichloroeactic acid is a dangerous and corrosive poison.

2 The acid molybdate reagent is made up in 10N H_2SO_4, and contains the ferrous sulphate.

Discussion

Pay attention to the following:

1 Are your measurements reliable estimates of the fructose-1,6-diphosphatase activity of the two types of tissue?

2 Are there significant differences between the activities found in the two tissues?

3 What is the most likely explanation of your results?

4 What experiments would you do to test your hypothesis?

Extensions and applications

The careful measurement of enzyme activities in tissues at different stages of differentiation is a valuable means of estimating the biochemical changes that cause and accompany differentiation.

Hints for the teacher

The experiment

1 Students should have had experience of extracting enzymes from plant cells before attempting the quantitative comparison of two tissues as required in this experiment.

2 Representative values for cotyledons and hypocotyls are 0·24 and 0·06 μmoles phosphate per 0·10 g fresh weight per 10 min.

Significant differences from these figures could be due to:

(a) Loss of enzyme during extraction. The extracts must be prepared quickly, kept cold, and assayed quickly.

(b) Incomplete extraction of the enzyme due to failure to homogenize the samples.

(c) Method of sampling. There is almost always some chloroplast development in the hypocotyl and this is more marked at the top than the bottom. Relatively high values for the enzyme are obtained if only the upper parts of the hypocotyl are taken. The activity in the cotyledons, when expressed on a fresh weight basis, is lowered if portions of the hypoctyls are included in the sample. Excision of the cotyledons should be done carefully.

3 The most common fault in the design of the experiment is failure to take into account the inorganic phosphate present in the extracts. Great care is necessary to ensure that appropriate controls are measured.

Growth of mustard

'Micafil' (vermiculite) is neutralised with dilute nitric acid, soaked in Knop's solution (see Section 2.6), autoclaved, and placed in plastic trays (12in x 8in) to a depth of 2in. In each tray, 30 g of seed are sown. The seed is only just covered. The trays are then placed at 18—20°C and are illuminated for 16 hr each day. The seedlings are watered on days 3, 5 and 7.

Assay of phosphate
This assay is essentially that of Taussky and Shott (Journal of Biological Chemistry, 202, 675, 1953).

Fructose-1, 6-diphosphate
It will be necessary to determine the amount of phosphate present in the fructose-1, 6-diphosphate solutions. This value can be given to the class. Alternatively the determination can be left to the class as part of the experiment.

T. ap REES
Cambridge.

7 Dye reduction by isolated chloroplasts

Time required
About 3 hr.

Assumed knowledge
Ability to isolate chloroplasts (see Section 9.7) and use of simple spectrophotometer.

Theoretical background
Carefully isolated chloroplasts are capable of light-dependent reduction of the blue dye, 2,6-dichloro-phenol-indophenol (DCPIP) which can be measured colorimetrically—concommitant O_2 evolution from photosystem II occurs but it is not measured in this experiment.

The rate of dye reduction is measured under various conditions, so plan the experiment carefully and do not begin until you are sure how you are going to complete the experiment.

Necessary equipment
Each pair of students will require the following
Phosphate buffer 0·1M pH 7·3 → 50 ml
0·35M NaCl in 0·01M tris buffer pH 7·3 → 50 ml
0·035M NaCl → 20 ml
0·3mM DCPIP (BDH) → 20 ml
$(NH_4)_2SO_4$ 10^{-4}M → 10 ml
10^{-5}M DCMU → 5 ml
80% acetone in water → 25 ml
Fresh spinach or peas
Centrifuge with 15 ml slots in head (MSE)
Phase microscopes + centering telescopes, oil immersion
Spectrophotometer

Glassware and others
2 beakers, 250 ml
5 centrifuge tubes, 15 ml
5 centrifuge tubes, 100 ml
1 x 50 ml cylinder
2 x 10 ml graduated pipettes
2 x 1 ml graduated pipettes
Glass rods and absorbent cotton wool
2 slides, 6 cover slips
Muslin 4 layers
Parafilm (Gallenkamp)

6 spectrophotometer tubes
12 test tubes
1 x 150 W lamp
Timer
Aluminium foil to weigh fresh spinach and use in dark control
Filter paper and filter funnel
Chilled mortar and pestle (Gallenkamp) and fine sand (BDH)
Ice bucket and ice.

Type of group
Students work singly or in pairs.

Instructions to the student
Isolation of the chloroplasts
Care is needed in the isolation procedure as one is trying to break the cells without damaging the complex multi-enzyme system contained in the chloroplast. Mechanical damage to the chloroplasts can be reduced by careful homogenisation, so avoid frothing and stop when most of the tissue has been homogenised. Osmotic damage to the chloroplasts can be reduced by preparing them in an isotonic medium (0·35M NaCl) which is buffered. Since denaturation of enzymes is a major problem in experiments with cell free systems, work quickly and assay the enzyme(s) as soon as possible after isolation. Work at 0–5°C, keeping all extracts, solutions and materials at this temperature—use ice liberally!

Weight out 10 g of healthy looking spinach leaf tissue, excluding stems and larger veins. Break up and homogenise in a chilled mortar with 20 ml of cold 0·35M NaCl + 0·01M tris buffer pH 7·3—add a small amount of sand (teaspoon) to facilitate grinding. Keep samples of the homogenate and subsequent four fractions and examine under the phase contrast. Filter and squeeze the homogenate through four layers of muslin into a beaker. Divide the filtrate between two 15 ml plastic centrifuge tubes and centrifuge at 900 rev/min (approx. 200 g) for 2 min. This will sediment unbroken cells, cell walls, etc. Decant the supernatant fluid into another 2 centrifuge tubes and centrifuge at 2 500 rev/min (approx 1 000g) for 5 min. Carefully decant the supernatant and resuspend the sedimented chloroplasts in a total volume of 10 ml of 0·035M NaCl as follows:

Use a small knob of cotton wool on the end of a glass rod or a small fine paint brush; initially add 0·5 ml of 0·035M NaCl to the tube, gently resuspend to an even consistency, and then add more solution up to 10 ml. The hypotonic 0·035M NaCl disrupts the chloroplasts, giving a suspension of broken chloroplasts, which should have a very active electron transport activity.

Measurement of dye reduction
The reactions should be run in colorimeter tubes and followed in the spectrophotometer. (Choose the best wavelength to follow the reduction of the dye by measuring a quick absorption spectrum of the dye.)
To each tube add:
1 0·1M phosphate buffer pH 7·3
2 Chloroplasts, approx. 0·1–0·2 ml
3 0·3mM DCPIP, 0·5 ml (Can be varied if necessary. This is added last)
Final volume = 10 ml

Gently mix by inverting over parafilm. Measure the optical density immediately. Illuminate by placing in a test tube rack about 60 cm from a 150 W lamp. Follow the change in optical density with rapid readings every minute; the rate should be linear for about 4–5 min. Find the optimal measuring conditions for your chloroplast preparation and then standardise on this. Then vary reaction parameters individually.

Recording results
Include 2–3 representative graphs showing how the reactions proceed. Tabulate all the results for easy comparisons. Remember to use only the *initial* rates of reaction.

Discussion
How does light intensity affect the rate of the photosynthetic reactions measured? How do the uncoupler (ammonia) and the inhibitor of oxygen evolution (DCMU) function? Are there any other herbicides or insectides which inhibit chloroplast reactions?

Extensions and applications
The reduction of the dye DCPIP is used in following the Photosystem II activity of chloroplasts. The reduction may be monitored continuously in a recording spectrophotometer with a source of illumination. In this way the action of various uncouplers, inhibitors, etc., may be investigated and the rates of various partial reactions easily followed. This is used extensively for screening prospective herbicides and assaying the phototoxicity of new insectides, drugs, etc.

Further reading
Hall, D.O. and Rao, K.K.
   1972 'Photosynthesis', Studies in Biology Series, No. 37, Edward Arnold, London
Gregory, R.P.F.
   1972 'The Biochemistry of Photosynthesis', John Wiley, New York

Hints to the teacher
This experiment can be shortened or lengthened. The students learn to design and perform an experiment by taking into account many different aspects of the assay procedure and properties of chloroplasts.

D.O. HALL
King's College
London.

8 Oxidative phosphorylation in rat liver mitochondria

1 Measurement of ADP:O ratios
2 Mode of action of uncoupling agents and energy-transfer inhibitors

Time required
If the isolated mitochondria are supplied and no calibrations are attempted, both sections can be completed in 2 hr. Calibration of oxygen concentration and ADP concentration will each take a further hour.

Assumed knowledge
This is a more advanced practical than that of Section 9.4 and requires a greater background knowledge. It is assumed that the student has already completed the previous experiment on electron transport and is familiar with the reactions involved in oxidative phosphorylation as described by:

Lehninger, A.L.,
   1971 'Biochemistry', Worth, New York
Griffiths, D.E.
   1965 'Essays in Biochemistry', Vol. 1, p.97, Eds. Campbell, P.N. and Greville, G.D . Academic Press. London.

Theoretical background
In this experiment the student will observe the extent to which the phosphorylating system controls the rate of electron transport. This is done by adding a small amount of ADP to respiring mitochondria; this will allow phosphorylation, which is accompanied by rapid electron transport (*State 3* respiration). Once the small amount of ADP has been converted to ATP the rate of electron transport is decreased (*State 4* respiration); this phenomenon is known as respiratory control and represents a control mechanism to prevent excessive oxidation of substrates when there is no demand for energy. The effectiveness of this control mechanism is measured by the respiratory control ratio which is obtained by dividing the rate of electron transport in State 3 by that in State 4. Values for the respiratory control ratio in excess of 4

are considered quite good. By going through the State 3 and State 4 cycle it is also possible to calculate the stoichiometry between moles of ADP phosphorylated and atoms of oxygen reduced (the ADP:O ratio). To do this draw a straight line through the most rapid rate of oxygen uptake which occurs just after the ADP is added (State 3) and another straight line through the lower rate that occurs when the ADP is exhausted (State 4) and measure the vertical height between the point at which these two lines intersect and the position on the chart at which ADP was added. Assuming the amount of oxygen dissolved in the assay medium to be 0·24 μmoles per ml it is possible to calculate the amount of oxygen represented by the height of the vertical line just constructed; this is then the oxygen used while the ADP was being phosphorylated. Since you know how much ADP was added you can calculate the ADP:O ratio and hence the number of energy-conserving sites associated with the oxidation of the substrate used. This data can then be combined with data from the previous experiment in order to build up a complete picture of the processes involved in electron transport and oxidative phosphorylation. The way in which dinitrophenol and oligomycin interact with the process of oxidative phosphorylation will also be observed.

Necessary equipment

A supply of fresh rat liver mitochondria, prepared as described in Section 9.4. It is essential that these be of good quality; this is more important in this experiment than in the earlier experiment.

An oxygen electrode, recorder and magnetic stirrer (Rank Bros.)

3 x 25 μl syringes (SGE)

3 x 100 μl syringes (SGE)

Several 1 and 5 ml pipettes.

Solutions required

Reaction mixture—the same as in Section 9.4, except the ADP is omitted.

5 ml succinate 1·0 M Na salt pH 7·0

5 ml β-hydroxybutyrate 1·0 M Na salt pH 7·0

2 ml ascorbate 0·5 M Na salt pH 7·0

2 ml tetramethylphenylenediamine (TMPD) (BDH) 0·02 M

1 ml dinitrophenol (BDH) 0·001 M pH 6·5

0·1 ml oligomycin (Sigma) 1 mg/ml (ethanol)

1 ml ADP (Sigma) 0·025 M pH 7·5.

Type of group

This experiment is best carried out by pairs of students to each electrode.

Instructions to the student

Set up the electrode as already desc
9.4 and 10.2, add 20 μl of succina
30—60 s. When the trace is linear add 10 μl of ADP
and wait; the slope of the line should increase for a time and then decrease. Once it has decreased wait for about 60 s and add another 10 μl aliquot of ADP, the cycle should be repeated. Carry out this experiment with both β-hydroxybutyrate and ascorbate + TMPD as electron donors. In the latter case the stimulation of the rate of oxidation by ADP is not so marked but should be apparent especially if you look along the line of the trace. Because of the lower ADP : O ratio with ascorbate + TMPD you will only be able to make one ADP addition.

Next set up an experiment with succinate as the electron donor and bring the mitochondria into State 4 (this means the ADP is exhausted) by adding 5 μl of ADP and waiting for the rate to slow up; then add 20μl of dinitrophenol; the rate of oxygen utilisation should now be permanently stimulated. Repeat the experiment but after initiating the rate with succinate add 100 μl of ADP (this is sufficient to sustain the rapid State 3 rate to the bottom of the chart). As soon as a linear rate has been established add 5 μl of oligomycin; when the rate is again linear add 10 μl of ADP and after 40—60 s add 20 μl of dinitrophenol.

Finally set up the electrode so that the mitochondria are oxidising succinate in the absence of ADP; then add 20 μl of dinitrophenol and after 60 s add 5 μl of oligomycin.

Recording results

All the rates should be calculated as nmoles of oxygen consumed per mg of protein per minute, and entered into the table supplied.

Treatment		Rate of oxygen reduced n moles/ min/mg protein
1 Succinate	State 2	
	State 3	
	State 4	
	ADP : O ratio	
2 β-Hydroxybutyrate		
	State 2	
	State 3	
	State 4	
	ADP : O ratio	

Treatment		Rate of oxygen reduced n moles/ min/mg protein
3	Ascorbate/TMPD	
	State 2	
	State 3	
	State 4	
	ADP : O ratio	
4	Succinate State 4	
	State 4 + DNP	
5	Succinate State 3	
	State 3 + oligomycin	
	State 3 + olig. + DNP	
6	Succinate State 2	
	State 2 + DNP	
	State 2 + DNP + olig.	

ADP and AMP concentration which is necessary if the preparation contains any adenylate kinase. The concentration of dissolved oxygen is best determined by liberating oxygen from hydrogen peroxide by catalase as described by Dixon, M. and Kleppe, K. (1965) Biochim. Biophys. Acta, 96, 357. If plant mitochondria are used or submitochondrial particles are available that will oxidise exogenous NADH, then the complete oxidation of a small single addition of NADH can be used to calibrate the oxygen solubility: see Chappell, J.B. (1964), Biochem. J., 90, 225.

J.M. PALMER
Imperial College
London

Discussion

As in the previous experiment (Section 9.4) it is useful exercise to draw up a scheme which relates the number of sites of phosphorylation with the position at which each substrate donates electrons to the electron transport chain and then compare your scheme with the normally accepted scheme. If the previous experiment has already been completed the data from both should be discussed together.

In the calculation of the ADP:O ratio any non-phosphorylating rate existing before ADP was added has been ignored. This is always done, however in your opinion is it a valid method of calculating the ADP:O ratio?

Further reading

Same as in the previous experiment. In addition the student should consult
Chance, B. and Williams, G.R.
1956 Advances in Enzymology, 17, 65–134.

Hints for the teacher

This is an easier experiment than Section 9.4 as far as the student is concerned, however, it requires higher-quality mitochondria in order to obtain a sufficiently high level of respiratory control, and it is advisable to have a technician prepare the mitochondria and have them tested before the class starts. If reasonable respiratory control is not attained, addition of Bovine Serum Albumen to 1% (w/v) in the mitochondrial suspension may help considerably. The complexity of the experiment can be increased by calibrating the ADP concentration using Boehringer Test Kit 15980 TAAB; this will give both the

9 Separation of proteins

	Sephadex	DEAE
Catalase suspension (crystalline from beef liver; Sigma)	0·05 ml	0·08 ml
Cytochrome c 5mg/ml (Sigma)	0·5 ml	0·5 ml
Ferritin CdSO$_4$, 1/10th dilution of Sigma material	0·5 ml	0·08 ml

Time required
5–7 hr, depending on how the experiment is organised.

Assumed knowledge
For second-year students familiar with spectrophotometers, basic biochemical techniques and chromatography.

Theoretical background
Proteins are charged molecules of high molecular weight. Many separation techniques utilise one, or both, of these properties.

DEAE-cellulose is a substituted cellulose (paper) and contains diethylaminoethane [$-CH_2N(C_2H_5)_2$] groups attached at random to the glucose units of the cellulose. The resulting DEAE-cellulose is basic:

$$\text{cellulose} - CH_2N(C_2H_5)_2 + HCl \rightleftharpoons$$
$$\text{cellulose} - Ch_2N^+H(C_2H_5)_2 + Cl^-$$

and will act as an anion exchanger, in the same fashion as will Dowex I (Cl). However, whereas Dowex I (Cl) will denature all but the smallest polypeptides, even very large proteins can be passed with safety through a DEAE-cellulose column.

Columns of Sephadex beads act on rather different principles. Sephadex beads consist of cross-linked dextran molecules and a Sephadex column contains liquid which is inside the beads and liquid which is outside the beads. A molecule in solution can enter the liquid inside the bead if it is small enough to pass through the interstices of the dextran network which define the bead. The degree of cross-linking of the dextran molecules can be controlled, thus providing a range of beads which will exclude molecules of various defined sizes. Thus, whereas a DEAE-column separates molecules on the basis of their charge, a Sephadex column separates molecules on the basis of their size and shape (which is, for globular proteins, a function of molecular weight).

Necessary equipment
Required per pair of students
Protein solution. Best results have been obtained if slightly different proportions of standard proteins are mixed and applied to the two columns.

Gradient-making apparatus and stirrer (MSE)

2 glass columns, approximately 23 cm x 1·5 cm (Gallenkamp)

1 fraction collector with tubes (Gallenkamp)

Suspend 4 g DEAE-cellulose (BDH) in 10mM tris buffer pH 8 and adjust pH to 8·0 with acid or alkali using a pH meter. Allow the suspension to settle, pour off the fines (supernatant) and resuspend the settled material in fresh 10mM tris buffer pH 8, check pH and use this to pour the column.

3 g Sephadex G-200 (Pharmacia) swollen in 10mM tris buffer pH 8 plus 0·2M KCl for 48 hr.

Required per class
20 vol. H$_2$O$_2$

10mM tris pH 8–2 litres

10mM tris pH 8 plus 0·5M KCl–2 litres

10mM tris pH 8 plus 0·2M KCl–5 litres

Polyacrylamide gel apparatus and power pack (Shandon)

7% Polyacrylamide gel in suitable tubes to fit apparatus

Sodium dithionite

U/V spectrophotometer with quartz cells (Pye Unicam)

Type of group
We find that three fraction collectors are as many as one demonstrator can control and so the class is divided into thirds.

Instructions to the student
You are provided with a mixture of ferritin, cytochrome c and catalase. The first two of these proteins are red, the last yellow. Catalase is an enzyme catalysing the reaction:

$$2 H_2O_2 \rightarrow 2 H_2O + O_2$$

and may be detected by adding 20 vol. H$_2$O$_2$ and observing whether gas is given off instantaneously or not. Cytochrome c may be detected by adding dithionite and observing the characteristic colour change. Any protein may be detected by OD$_{280nm}$ measurements.

181

Examine the behaviour of the protein mixture on columns of DEAE-cellulose and Sephadex.

Sephadex G-200 column
Sephadex beads have to be swollen for 48 hr in buffer before use. Pour a 10 cm column as demonstrated. Take great care not to disturb the top of the column when loading with the protein mixture and wash in the protein with portions of tris/KCl buffer. Collect 2 ml fractions from the moment of loading the column and elute with 10 mM tris pH 8 containing 0·2 M KCl. Collect about 15 fractions and measure the OD_{280nm} of each.

DEAE-cellulose column
The column is poured in the same manner as the Sephadex column. Load the protein solution on with great care. Do not disturb the top layer of DEAE-cellulose and wash the protein in with small portions of buffer (3 x 1·0 ml). Start collecting 5·0 ml fractions from the moment you start to load the column. Fill the column up with 10 mM tris pH 8 and connect to a reservoir. After about 8 fractions (40 ml) have been collected you will need to start applying a KCl gradient.

Use 200 ml (or if you only have a little time 150 ml; this usually gives adequate results) of 0·5 M KCl in 10 mM tris pH 8 and 200 ml of 10 mM tris pH 8 in the apparatus provided—whose use will be demonstrated to you. Collect about 50 fractions in all and measure the OD_{280nm} of each.

Polyacrylamide Electrophoresis
The voltages and currents used in this apparatus can be lethal! Do not touch or use.

The demonstrator will subject the suspension to this method of analysis at a running pH of 8–9. You will be expected to make notes of this technique and the results obtained.

Recording Results
Plot the OD_{280nm} of each fraction as a function of fraction number (which is equivalent to elution volume). You should get a numer of peaks. Identify each.

N.B. The proteins used have the following molecular weights:

Cytochrome c	11 000;
Ferritin	~ 500 000;
Catalase	250 000;

Discussion
Are the results of these three methods self consistent? What happened to the cytochrome c in the poly-acrylamide gel apparatus?

Extensions and applications
All these methods of separating proteins are routinely used in research.

Further reading
'Methods of Enzymology' Academic Press, various volumes, contains further details of all these techniques.

Hints for the teacher
It is best if the students start with the Sephadex column (Booklet available from Pharmacia) and some of them begin analysing fractions from these whilst others set up the DEAE-cellulose column. The polyacrylamide gel analysis is best done as a demonstration whilst the DEAE-cellulose column separation is attempted. The original recipe of Ornstein and Davis, (1964) Ann. N.Y. Acad. Sci., 121, 305, using 7% gels and alkaline running buffers works well.

J.M. ASHWORTH
Essex.

10 The partial purification and assay of ferredoxin from plants and bacteria

Time required
4 hr.

Assumed knowledge
Use of oxygen electrode or spectrophotometer. Theoretical understanding of the basis of protein purification, particularly the use of selective precipitation by organic solvents, and of ion-exchange chromatography.

Theoretical knowledge
Proteins with active centres containing equivalent amounts of non-haem iron and 'inorganic' sulphur are apparently universally distributed in the electron transport systems of all organisms. There are several different groups of these so-called iron-sulphur proteins; those with characteristic electron paramagnetic resonance spectra in the reduced form with g perpendicular around $g = 1.94$ are called ferredoxins. The ferredoxin in aerobic electron transport systems such as mitochondria are tightly bound to the membranes and are extremely difficult to isolate. One of the ferredoxins involved in the chloroplast electron transport chain, which is a red protein containing 2 iron and 2 sulphur atoms per molecule, is readily soluble, as is one of the ferredoxins found in strictly anaerobic bacteria such as *Clostridium pasteurianum* or the anaerobic photosynthetic bacteria such as *Chromatium*. In contrast to the 2Fe-containing plant-type ferredoxins, these bacterial ferredoxins are black proteins with eight iron and eight sulphur atoms per molecule. Ferredoxins can easily be partially purified by organic solvent precipitation and chromatography on DEAE-cellulose. Their biological activity can be measured by their ability to catalyse NADP reduction by chloroplasts.

Necessary equipment
1. 1 oxygen electrode for each group, or access to one good ultraviolet spectrophotometer for each 2 or 3 groups. The spectrophotometer must be capable of giving readings at 340 nm with a turbid sample. The Unicam SP800 or equivalent will work, an SP600 will not.
2. Light source. Slide projector or reflector flood lamp (100 W) with heat filter (e.g. 1 litre Roux bottle (Gallenkamp) filled with water). 1 for each electrode, or spectrophotometer.
3. Atomix (MSE) or equivalent blender for every three groups.
4. High-speed refrigerated centrifuge (required only for bacterial preparation) (e.g., MSE 18).
5. MSE super minor or equivalent bench centrifuge for the plant preparation.
6. Recording spectrophotometer. This is desirable but not essential.
7. Cell-breaking apparatus for the preparation of a cell-free bacterial extract. Sonication (Dawe Soniprobe), French press (Aminco), or any other standard procedure for the preparation of bacterial extracts will be suitable.
8. For each group:
   (a) 2 x 100 ml centrifuge tubes
   (b) Test tube rack; 12 test tubes
   (c) Pipettes 0.5 ml, 1.0 ml, 5 ml, 10 ml
   (d) Microlitre syringe 25 μl (Hamilton, SGE)
   (e) Nylon bag or butter muslin. See Section 9.7.
   (f) 3 spectrophotometer cuvettes if using spectrophotometric assay
   (g) 3 x 250 ml beakers
   (h) 2 x 100 ml measuring cylinders
   (i) 1 x 25 ml measuring cylinder
   (j) Glass wool (Gallenkamp)
   (k) 1 chromatography column (Gallenkamp). These should be about 2.5 cm diameter glass columns. If commercial columns are not available, simple columns may be made by drawing the end of a 25 cm length of tubing to about 5 mm diameter and using a short piece of flexible tubing with a screw clip to close off the column. A plug of glass wool can be used as a sinter.
   Solutions: for class of 20–25
   (a) 0.02M potassium phosphate pH 7.5, 10 litres
   (b) 0.2M NaCl in 0.02M potassium phosphate pH 7.5, 6 litres
   (c) 0.8M NaCl in 0.2M potassium phosphate pH 7.5, 4 litres
   (d) 1M tris-HCl pH 8.0, 100 ml
   (e) 0.5M $MgCl_2$, 100 ml
   (f) 0.1M NH_4Cl, 100 ml
   (g) 0.01M NADP (Sigma, BDH) 10 ml
   (h) Ferredoxin standard solution, if available. (1 mg sufficient for 10 groups; Sigma). Not essential.
   (i) DEAE-cellulose (Reeve-Angel; BDH);

183

Whatman DE 23 pre-treated according to the manufacturers instructions and suspended in 0·02M potassium phosphate pH 7·5 containing 0·3M NaCl.

(j) Spinach or other suitable plant materials

(k) *Clostridium pasturianum* or other suitable anaerobic bacterial cell paste

(l) Acetone at -20°C

(m) Spinach chloroplasts 1 mg/ml, 20-30 ml.

Type of group

Students work in pairs.

Instructions to the student

Setting up the DEAE-cellulose column

Put a plug of glass wool in the bottom of the chromatography column. Half fill the column with 0·2M NaCl in 0·02M phosphate buffer. Stir the DEAE-cellulose suspension supplied and pour into the column. Allow the buffer to run out slowly; the DEAE-cellulose will settle on the glass wool to form a column, which should be 4–5 cm high. If the DEAE-cellulose you put in is not enough, add more suspension before the buffer level is half way down the glass column. When you have sufficient cellulose, allow the buffer to run down until it just reaches the surface of the cellulose; *do not* let it go completely dry. Carefully pipette 0·02M phosphate buffer onto the column allowing it to run down the sides so that it does not disturb the cellulose; wash this buffer through the column until all the salt has gone, about 100 ml should be enough. The column is now equilibrated and ready for use.

Preparation of cell-free extracts

1 *Plant ferredoxin.* Cut 40 g of leaves into small pieces and blend with 80 ml of 0·02M phosphate buffer for 1 min in the blender. Filter the suspension through the nylon bag or cheese cloth to remove large debris. Add 40 ml of cold acetone to the solution slowly, stirring all the time—this gives a final acetone concentration of about 33%. Allow the solution to stand on ice for 5 min and then centrifuge at 6000 rev/min in the MSE super minor for 10 min. The supernatant contains ferredoxin.

2 *Bacterial ferredoxin.* Suspend 20 g of cell paste in 40 ml of 0·02M phosphate buffer. Disrupt the cells by the procedure you have been told to use. Add 20 ml of cold acetone slowly to the broken cell suspension, (final acetone concentration about 33%) stirring continuously and allow to stand for 5 min, on ice. Centrifuge for 10 min at 18 000 rev/min (40 000g) in MSE 18 or equivalent refrigerated centrifuge. The supernatant contains ferredoxin.

3 *Column chromatography.* Carefully pipette the ferredoxin-containing supernatant on to the column without disturbing the DEAE-cellulose and allow it to run through the column. Wash the column with 0·2M NaCl in 0·02M phosphate buffer until the pass solution is essentially colourless. Very carefully pipette 0·8M NaCl in 0·02M phosphate buffer onto the top of the column and allow it to run slowly through the column eluting the ferredoxin. When the extract is put on the column a dark band should collect at the top of the column. This band is the ferredoxin and other acidic compounds. When you add the 0·8M NaCl solution a band should move from the top of the column containing the ferredoxin. When this band gets near the bottom of the column collect 2–3 ml fractions. The dark fractions will contain the ferredoxin. Assay these fractions for ferredoxin activity in NADP reduction by chloroplasts. Take an absorption spectrum in the visible region of the sample containing the most ferredoxin and repeat this after reducing the ferredoxin by adding a few crystals of sodium dithionite to the curvette. A decrease in optical density between 400 and 500nm is an indication of the presence of ferredoxin in the sample. It is unlikely to be pure enough after this brief purification to give a good spectrum.

4 *Ferredoxin assay procedures.* The rate of NADP reduction by spinach chloroplasts is dependent on ferredoxin concentration. The amount of ferredoxin in a preparation may be estimated on the basis of its effect on the rate of electron transport in spinach chloroplasts with NADP as electron acceptor. The chloroplasts used are broken and washed to remove any endogenous ferredoxin. The assay may be done either by following the light-dependent reduction of NADP spectrophotometrically or O_2 evolution with an oxygen electrode.

5 *Reaction mixture for the oxygen electrode.*

1M Tris-HCl buffer pH 8·0, 0·2 ml

0·05M $MgCl_2$, 0·1 ml

0·1M NH_4Cl, 0·1 ml

0·01M NADP, 0·1 ml

Chloroplasts = 100 μg chlorophyll

Ferredoxin preparation = 0 to 50 μg

Water to 3·0 ml

Measure the rate of oxygen evolution on illumination at different ferredoxin concentrations.

6 *Reaction mixture for the spectrophotometric assay*

1M Tris—HCl buffer pH 8·0, 0·2 ml

0·05M $MgCl_2$, 0·1 ml

0·1M NH_4 Cl, 0·1 ml

0·25M Sodium ascorbate, 0·1 ml
0·002M dichlorophenol indophenol (BDH) 0·1 ml
0·01M NADP (Sigma; BDH) 0·1 ml
Chloroplasts (heated for 5min at 50°C) giving a
    final amount of 40 μg of chlorophyll
Ferredoxin preparation = 0 to 50 μg
Water to 3·0 ml

Place the reaction mixture in a 1 cm cuvette and read
the OD_{340} against a blank minus ferredoxin. Illuminate for 3 min and again read the OD_{340}, repeat for a
second 3 min period. The rate of NADP reduction is
proportional to the ferredoxin concentration. Under
the conditions described an optical density change of
2 is equivalent to the reduction of 1 μmole of NADP.

Recording results

Describe what you see during the preparation. Calculate how much ferredoxin you have extracted either
in arbitrary units or if a standard is available as mg of
ferredoxin.

Discussion and extensions

The experiment is an introduction to protein purification techniques commonly used in enzyme isolation.
The assay procedure is an example of the complex
assays which may have to be carried out in the
isolation of biologically active materials. The procedures described are the initial stages normally used
in the complete purification of ferredoxin.

Further reading

San Pietro, A (ed.)
    1972 'Methods in Enzymology', Vol. XXIII,
    Academic Press, London
Reviews of ferredoxin function in Annual Review of
Biochemistry, Ann. Review of Microbiology and in
Advances in Microbial Physiology.

Hints for the teacher

Broken, washed spinach chloroplasts, prepared by
method 1 in Section 9.7 must be supplied.
Preparation time 45 min. The best plant material is
spinach—spinach beet will not do. Others such as
lettuce may be used but often contain polyphenol
which interferes with the column chromatography.
The best bacterial material is *Clostridium pasteurianum* (Carnahan and Castle, J. Bact., 75, 121, 1958);
most other clostridia can be used, e.g. *C. welchii*
(Sigma), as can some other strict anaerobes and
photosynthetic bacteria (Cell cultures from National
Collection of Industrial Bacteria, Aberdeen.) Strepto-
cocci or anaerobically grown *Escherichia coli* will not
work.

M.C.W. EVANS
University College
London.

11 Purification of proteins by column chromatography. Separation of catalase and glucose oxidase from a fungal extract

Time required
1½–2 hr: Further experiments 1–2 hr.

Assumed knowledge
Students are expected to know what an enzyme is.

Theoretical background
Diethylaminoethyl (DEAE)-cellulose is a medium suitable for chromatography of many types of proteins, and consists of cellulose chemically substituted with $-OCH_2CH_2N^+H(C_2H_5)_2$ groups. Proteins which contain negatively charged $-COO^-$ groups tend to bind to the DEAE-cellulose by electrostatic attraction, and their affinity can be decreased by (1) lowering the pH, which decreases the net negative charge on the proteins, or (2) by increasing the concentration of salts, which compete with the protein for charges on the cellulose. In this experiment, a mixture of proteins is applied to a column of DEAE-cellulose in a dilute buffer so that they are tightly adsorbed to it. The proteins are then selectively removed with buffers of increasing power of elution.

Glucose oxidase is an enzyme from moulds and bacteria which catalyses the reaction

$$D\text{-Glucose} + O_2 = D\text{-Gluconolactone} + H_2O_2.$$

Catalase is an enzyme found in almost all aerobic cells which catalyses the dismutation of hydrogen peroxide:

$$2H_2O_2 = 2H_2O + O_2.$$

Under certain circumstances it can catalyse the oxidation by hydrogen peroxide of other substrates such as ethanol:

$$CH_3CH_2OH + H_2O_2 = CH_3CHO + 2H_2O.$$

Necessary equipment
Glass column, about 1 cm diameter, at least 10 cm long, with a glass sinter or plug of glass wool at the bottom to retain the DEAE-cellulose (Gallenkamp). Close off at the bottom with a length of plastic tube and a pinch-clip.
A rack with about 15 test tubes
Fluted filter paper and funnel
Glass rod
4 × 100 ml beakers,
Pasteur pipettes
DEAE-cellulose, such as Whatman DE 23 precycled according to the manufacturer's instructions and equilibrated with buffer B1. A crude glucose oxidase preparation from *Aspergillus*, such as product No. 39016 from BDH (this contains catalase as a contaminant).
Buffers:
   B1 : 20mM sodium acetate, 8mM acetic acid
   B2: 40mM sodium acetate, 40mM acetic acid
   B3: 100mM sodium acetate, 100mM acetic acid
   10 vol. hydrogen peroxide
   Solid glucose

Type of group
Individual students in a large class can do the main experiment. Ancillary experiments may be limited by the amount of equipment available.

Instructions to the student
Pour a slurry of DEAE-cellulose into the column and allow to settle. The final bed height should be 3–4 cm. Wash the column with about 10 ml of B1.

Do not allow the column to run dry, as this may introduce bubbles. The top of the DEAE-cellulose must be kept flat. If it is not, stir up the top few millimetres in buffer with a glass rod and allow to settle. Subsequently, to avoid disturbing the DEAE-cellulose, always add buffers carefully with a pipette down the walls of the column.

Dissolve 0·5 g of *Aspergillus* extract in 25 ml water with stirring, then filter. Pass the clear extract through the column, and note that the coloured material forms a band at the top of the column. When the solution has run through, wash the column with a total of 10 ml of B1. Begin to collect the effluent solution in a series of test tubes. When the buffer B1 has run through, wash with 10 ml of B2. A band of greenish material should run down the column—try and collect this band in a minimum volume in one test tube. Next wash with 10 ml of B3, when a yellow

186

band should run down the column—again collect this concentrated band in one test tube.

Test for catalase: to a sample of the solution in each test tube add 1 drop of hydrogen peroxide. Catalase causes a vigorous effervescence due to release of O_2 gas.

Test for glucose oxidase: Separate each yellow-coloured fraction into two test tubes, and add about 50 mg/ml of solid glucose to one of them, shaking until it dissolves. Compare the colour of the two tubes. Glucose oxidase contains flavin-adenine dinucleotide (FAD), which is bleached on reduction by glucose.

Recording results
Record the results of the tests on the effluent fractions in the form of a table, using symbols such as —, not detected; +, detectable; ++, present in high concentration, for the two enzymes.

Discussion
Given that the pK of acetic acid is 4·7, calculate the pH of buffers B1, B2 and B3, and show that these buffers would be expected to have increasing eluting power.

How does the oxidation of glucose by glucose oxidase differ from the normal pathways of degradation of glucose in aerobic cells?

Extensions and applications
1 By using an oxygen electrode, the reactions catalysed by these enzymes may be further investigated.

Set up the oxygen electrode as described in Section 10.2 and place in it 1·8 ml of buffer B1, 0·1 ml 1M D-glucose and then $50 \mu l$ of the glucose oxidase solution. Record the rate of oxygen uptake. Add 0·1 ml of catalase; if the glucose oxidase fraction is free of catalase, there should be an increase in oxygen concentration due to decomposition of H_2O_2, followed by a rate of oxygen uptake which is one-half of the previous rate. Add $10 \mu l$ of ethanol; the rate should increase. Explain your observations in terms of the reactions of the two enzymes.

2 With the use of a spectrophotometer it is possible to investigate the prosthetic groups of the two proteins. Glucose oxidase contains FAD, and has absorption maxima at 450 and 380 nm; observe the effect of glucose on the spectrum. Catalase contains a haem group (a complex of porphyrin and ferric iron) and has an absorption maximum at 406 nm.

Further reading
The method is adapted from
Pazur, J.H. and Kleppe, K.
   1964 Biochemistry, 3, 578

For further reading:
Knight, C.S.
   1967 'Column chromatography', Adv. Chromatography, 4, 61
Bentley, R.
   1963 Glucose oxidase in 'The Enzymes', Vol. 1 ed. by Boyer, P.D., Lardy, H. and Myrbäck, K., Academic Press, New York, 567
Keilin D., and Hartree, E.F.
   1945 'Catalase', Biochem J., 39, 293
Nicholls, P. and Schonbaum, G.R.
   1971 In 'The Enzymes', Vol. 8, ed. by Boyer, P.D., Lardy, H. and Myrbäck, K., Academic Press, New York, p.147

Hints for the teacher
The precycling and equilibration of DEAE-cellulose takes 2–3 hr; this time can be reduced by using the pre-swollen microgranular form, Whatman DE52, which however gives slower flow rates.

R. CAMMACK
King's College
London.

187

12 Adsorption and thin-layer chromatography

Time required
3–5 hr.

Assumed knowledge
For second-year students familiar with spectrophoto-meters.

Theoretical background
Adsorption chromatography consists of separating substances by passing a solution containing a mixture of them through a column of finely powdered adsorbent material contained in a glass tube and then washing ('developing') the column with solvent. The adsorption of a substance is a function of its chemical structure, the nature of the adsorbent and of the solvent. Thus the adsorption affinity will be related to the possession and number of certain types of bond and groups in the molecule. The same substance may be more strongly adsorbed on one adsorbent than on another and the adsorption also depends upon the solvent used for developing. The more polar the solvent the less strongly will the substance be adsorbed and this is the basis of the elution of substances from a column. By gradually increasing the polarity of the solvent different substances can be eluted separately.

Recently it has proved possible to obtain very thin uniform layers of adsorbents on glass supporting plates. Such 'thin layers' of adsorbents act as does the column of adsorbent and the thin-layer plate is used in the same way as the filter paper in paper chromatography. However, whereas paper chromatography is most useful for hydrophilic molecules, thin layer chromatography (TLC) is best suited for mixtures of hydrophobic molecules.

Necessary equipment
Required per pair of students
1 glass column (Jencons, Gallenkamp)
2 x 250 ml beakers
2 small TLC Plates (Gallenkamp) made from 30g Kieselgel (BDH) + 60 ml water, 0·025in thick

Required per class
150 g Spence-type 'H' alumina (dry) (Koch-Light)

Acetone
Light petroleum (40–60°C boiling point) (BDH)
Ether
Methanol
Isopropyl ether (Koch-Light)
Hot-water baths at approximately 60°
Microlitre syringe or pipette (Gallenkamp & S.G.E.)
Compressed-air blowers, retort stands, clamps and bosses
Waring or Atomix blenders (MSE)
Vegetables, fruits and flowers in season. (Dock leaves [*Rumex sp*] give particularly good results.)

Type of group
Pairs of students usually do this experiment with us and any number can be catered for.

Instructions to the student
Weigh out 15 g of Spence-type 'H' alumina in a 250 ml beaker. Add 0·3 ml water and stir well with a glass rod. Add about 50 ml of light petroleum (40–60°C boiling point) and stir well. Insert a glass wool plug in the chromatography tube and clamp in a vertical position. Place a glass funnel in the top and pour in the alumina slurry. Tap the tube gently until the top of the alumina column is level and keep it covered with solvent at all times.

Extract the given vegetable, flower or fruit in a little acetone e.g. 1/1 (w/v) in a blender. Place the acetone extract in a beaker in a water bath in a fume hood and blow off the acetone with a stream of air. Dissolve the extract in about 10 ml of light petroleum. Separate off the aqueous layer, blow off the light petroleum and re-dissolve the extract in about 5 ml of light petroleum. Allow the column to flow until there is about 1 cm of solvent above the adsorbent bed and then pipette the lipid extract very gently onto the column. When almost all the extract has become adsorbed on to the column, wash in with 1 ml quantities of solvent and top up the tube with solvent. There will be some separation of the components of the extract even with a non-polar solvent like light petroleum, but after passing about 50 ml of light petroleum through the column it is best to change to solvents of increasing polarity. Start with 2% ether in the light petroleum and after you have judged that optimal development and elution has occurred with this mixture, increase the polarity to 4% ether. Continue in this way through 8, 10, 20, 50 and 100% ether and separate the individual bands or zones of the chromatogram. In the case of a leaf lipid, the chlorophylls will probably remain adsorbed; these can be removed with a 10% methanol–

petroleum mixture.

Collect the bands as they come off, and concentrate 1–2 ml portions by evaporating off the solvent in a fume hood. These concentrates may be redissolved in light petroleum for analysis on TLC plates. Spot on a TLC plate with a microlitre syringe samples of the fractions you obtain and the original extract. Develop the TLC with 50% ether: 50% petroleum.

Recording results
Record the appearance of the adsorption column at various stages and make a sketch drawing of the TLC plate. If you have the time or facilities run spectra of the various fractions in the visible (400–700 nm) range.

Discussion
Did you keep your extract in strong sunlight whilst doing this experiment? What would happen if you irradiated the components likely to be present in leaves and fruits?

Extensions and applications
Adsorption chromatography and TLC are widely used as quantitative and qualitative methods of separating hydrophobic compounds in organic chemistry and the pharmaceutical industry.

Further reading
Text books of practical organic chemistry describe these procedures in detail. e.g.
E. Merck, A.G.
   1970 'Chromatography', 2nd ed.; Darmstadt
Stock, R. and Rice, C.B.F.
   1967 'Chromatographic methods', Chapman and
   Hall, London

Hints for the teacher
The separation scheme above is a general one and will need modifying slightly depending on the vegetable, flower or fruit. It is best to find out by preliminary experiment what mixture of solvents works best for the material available. The description above should apply to most leaf lipid extracts. The solvents used are highly volatile and care must be taken to avoid fires. No smoking must be allowed in the laboratory.

J.M. ASHWORTH
Essex.

Section 11

Growth

1 Bacterial growth and the effects of chloramphenicol and penicillin

Time required
3–4 hr on first day, ½ hr on second or subsequent day.

Assumed knowledge
Some knowledge of bacterial growth and metabolism is of value but not absolutely essential. Practically, a moderate efficiency with a pipette and an understanding of sterile techniques are the main requirements.

Theoretical background
When bacteria are inoculated into a liquid medium, a period of time may elapse during which no growth takes place. This is the *lag phase* and may be due to the need to synthesise new enzymes before growth can occur or to resynthesise enzymes which have decayed as the cells have aged in their previous environment.

The lag phase is gradually superseded by a period of exponential growth in which the equation $m = m_0 e^{kt}$ is followed. Here m is the mass of cells/ml at time t, m_0 the mass/ml at zero time, and k is a constant which may therefore be used to characterise the rate of growth. A more convenient way of expressing the growth rate is by the *mean generation time*, which is the time taken for the mass of cells in a culture to double. If T is the mean generation time, then $2m_0 = m_0 e^{kt}$, whence $T = \dfrac{(\log_e 2)}{k}$. Because growth is exponential, the $\log_{10}$ of the mass of cells/ml plotted (on linear scale) against time gives a straight line and the mean generation time is most simply obtained as the time taken for $\log_{10}$ to increase by $\log_{10} 2$, i.e. 0.3.

In general, the optical density (OD) of a culture (or more strictly the turbidity measured with a nephelometer) is proportional to the mass of cells/ml; other measurements which can be used to determine the mean generation time are 'numbers of cells' or any one characteristic cell component (such as total nitrogen content of protein or nucleic acids).

If OD is plotted directly onto a log scale (on semi-log graph paper) then the mean generation time is simply the time taken for the value of the ordinate to double. This period of growth is therefore called either the *exponential* or *log phase*.

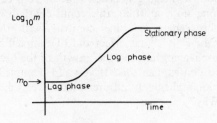

Figure 1

The exponential phase of growth does not extend indefinitely. Eventually, due either to the complete utilisation of an essential raw material in the medium or to the building up in the culture of substances toxic to the cells, the rate of growth falls, eventually to zero, and the cultures enter the *stationary phase*.

Many antibiotics act by inhibiting the formation of one or other of the main types of macromolecule of the bacterial cell. Chloramphenicol stops protein synthesis; penicillin inhibits the formation of mucopeptide which is the constituent of the cell wall that gives the organism its characteristic shape and prevents the fragile cell membrane from bursting as a result of the high osmotic pressure produced by the solutes in the cytoplasm. Growing cultures treated with penicillin will therefore lyse so that both the turbidity and the number of viable cells will decrease.

Chloramphenicol, on the other hand, will stop growth, but moderate concentrations will not kill the cells so that viability remains constant.

Necessary equipment
Each pair of students will need the following:
3 flasks (or large 'T' tubes) for 50 ml broth
30 x 1 ml pipettes
5 x 10 ml pipettes
Several sterile Pasteur pipettes
24 sterile 1 oz universal bottles (Gallenkamp) (or test tubes)
24 sterile 1 oz universal bottles (or sterile test tubes containing 4·5 ml of sterile distilled water, for serial dilutions).
4 nutrient agar plates (Difco, Oxoid)
Wax pencil or similar

Lysol jar (disinfectant)
Bunsen
40% formaldehyde in dropping bottle
Shaking water baths for all flasks or 'T' tubes
Hilger spectrophotometers (or similar to read turbidities), 1 per pair or 2 pairs of students

Type of group
Probably best to work in pairs. We accommodate 15 pairs but availability of instruments to measure OD, shaking water baths, etc, only limiting factor.

Instructions to the student
In this experiment, the growth of *E. coli* is followed by OD measurements and the effects of chloramphenicol and ampicillin (a penicillin that is active against Gram-negative bacteria such as *E. coli*) on growth and viability are investigated.

Take three flasks, each containing 50 ml of nutrient broth and label them 'control', 'pen' and 'CM'. Warm to 37°C in a water bath and then (zero time) inoculate each flask with 1·0 ml of the culture of *E. coli*. Mix and transfer a 6 ml sample from each flask into a 1 oz bottle. Then put the flask back in the shaker. Add 2 drops of 40% formaldehyde to each sample to kill the cells and read the OD on an Hilger spectrophotometer (52 filter, 520 nm.)

N.B. Two practical points:
1 Use 4 pipettes to take the turbidity samples; label these and retain, each standing in a separate bottle, for further sampling from the appropriate flasks.
2 You are working with rather dense bacterial suspensions which need good aeration if they are to grow properly; try to prevent your neighbours from turning the shaker off for long periods when samples are being taken!

At 30 min take further samples for turbidity measurements.

At 60 min take another set of samples. Then add 1·0 ml of sterile water to 'control'.

1·0 ml of ampicillin solution 1·0 mg/ml (Penbritin, Beecham) to 'pen'.

1·0 ml of chloramphenicol (Sigma) solution (1·5 mg/ml) to 'CM' and also take out 0·5 ml from the control flask for serial dilution (*don't* add formaldehyde to this!). Add to the first bottle (10^{-1}) of the serial dilution series, and proceed as indicated in the section headed 'serial dilutions'.

At 90, 120 and 150 min take further samples for OD measurements from each flask and at 150 min also remove 0·5 ml from the 'control', 'pen' and 'CM' flasks for serial dilutions. If time is available add 0·5 ml of ampicillin to the CM tube, return to the water bath and make further OD readings at 180 and 210 min.

Serial dilutions for viable counts

The cultures that you are using contain of the order of $10^7 - 10^8$ cells/ml. To count the individual cells (by their ability to form colonies) considerable dilution is necessary. This is achieved by a series of 10-fold dilutions, each of which involves transferring 0·5 ml by sterile pipette into 4·5 ml of sterile water. After swirling the mix, a *fresh* sterile pipette is used for the next dilution.

Thus, the 0·5 ml sample of the control culture taken at 60 min is diluted into 4·5 ml of water to give a dilution of 10^{-1} and this process repeated up to a dilution of 10^{-6} (label bottles).

With a glass marking pencil draw lines to divide a nutrient agar plate into quarters (mark the base and not the lid), label the plate with your initials and '60 min control'. Use a sterile Pasteur pipette to suck up some of the 10^{-6} dilution. With the Pasteur pipette held vertically and with the tip about 2 in above one of the marked quarters of the nutrient agar plate, allow a single drop to fall on to the plate; rock the plate gently to spread a little, move the pipette to a different quarter and allow another single drop to fall. Each drop is 0·025 ml. With a fresh Pasteur pipette suck up some of the 10^{-5} dilution and allow a single drop to fall on each of the two remaining quarters. *Do not move the plate.*

Similarly place two standard drops from each of the 10^{-6} and 10^{-5} dilution tubes for the '150 min control' and for the '150 min CM' on previously labelled nutrient agar plates. For the '150 min Pen' sample place one drop from each of the 10^{-4}, 10^{-3} and 10^{-2} dilution tubes and one drop from the undiluted culture in the quarters of a further nutrient agar plate.

Leave the four plates undisturbed on your bench the right way up for at least half an hour until the liquid has soaked in and the plates are dry. Then put them *inverted* into the baskets provided. They will be incubated at 37°C overnight, then stored in the cold and given back to you next time you attend.

Recording results

It is good practice to record your OD readings directly onto a table as they are taken. At the end of the experiment these readings should be plotted onto a graph with OD as abscissa (preferably on a log scale) against time. The mean generation time can be read directly from the straight part of the graph (if OD is plotted on a log scale). The viable counts should enable you to correlate optical density with viable count.

Discussion

What was the mean generation time of your culture? How long do you think this could continue? How does this generation time compare with that of eukaryotic cells? What effect did (1) ampicillin, (2) chloramphenicol have on the growth of the bacteria? How do you explain this? If you added ampicillin to the chloramphenicol culture what effect did this have? Why?

Extensions and application and further reading

This subject can be followed up in any standard textbook. See, for example,
Davis, W., Dulbecco, R., Eisen, F., Ginsberg, H. and Wood, A.
 1970 'Microbiology', Harper and Row, New York

Hints for the teacher

This class has been found to be surprisingly successful. We use *E. coli* MRE 600 (Whatman Biochemicals, MRE or Sigma). From a freeze-dried culture we resuscitate on plate and slope, take 1 loop of overnight growth, inoculate into 50 ml broth (Difco, Oxoid) in a 250 ml flask, incubate in shaking waterbath at 37°C overnight and then dilute 1 in 2 to 1 in 3 so that 1 ml of this added to 50 ml of broth gives an OD reading (i.e. zero time reading for class) of about 0·02.

It may be better to incubate the nutrient agar plates for the viable counts say at 30°C or even room temperature, depending on timing of the 'second day'. Incubation at 37°C can quickly lead to overgrowth of the colonies within the spread drops.

It is a good idea to have at least one complete run through of the experiment, so that allowances can subsequently be made for any variations of local equipment or conditions.

E. SIDEBOTTOM
K.G.H. DYKE
D.G. WILD
Oxford.

2 An examination of some effects of antimicrobial substances on cultured human cells

Time required
A minimum of two periods of 3 hr each is required—one to set up the experiment for minimum antibacterial concentration, and one a week later for the cell experiments. This means all apparatus and medium being given to the student sterile and ready to use, as well as a 24 hr culture of bacteria and 12 flasks of cells. It means too the examination of only morphological change as an index of toxicity. In addition to the two days mentioned for setting up the experiments, short periods will be required for observing the effects of the drugs on the cells. A maximum of 40 hr has been available for students treating such topics as honours year 'projects'. In this time they can set up a statistically satisfactory number of replicate experiments, and examine indices of toxicity other than morphology—e.g., effects of drug on cell numbers and culture protein.

Thus any time from 6 to 40 hr can be spent on the exercise.

Assumed knowledge
Technique of aseptic handling; techniques of sterilisation of media and glassware; techniques of finding minimum inhibitory concentration of a substance for common bacteria or moulds; ability to use a phase-contrast microscope; knowledge of the histology of living, unstained, human cells; ability to use a Coulter counter, or failing this, a haemocytometer; ability to carry out protein estimations, including the use of a spectrophotometer.

Theoretical background
Many substances which kill bacteria and moulds are unsuitable for use as preservatives of pharmaceutical preparations, or for use clinically, either topically or systemically, because of their toxicity. It is of interest to compare the toxicity of antimicrobial substances to micro-organisms and to cultured, human cells.

Necessary equipment
A laboratory and protective clothing suitable for the handling of micro-organisms (this part of the work must be carried out in a laboratory apart from that in which cell cultures are handled)

Cultures of common micro-organisms such as *Staphylococcus aureus, Streptococcus faecalis, Proteus vulgaris, Pseudomonas aeruginosa, Bacillus subtilis,* etc., obtainable from the National Collection of Type Cultures (for student use choose non-pathogenic strains and supply as 24 hr cultures)

Media for growing micro-organisms, e.g. nutrient broth (Oxoid)

Antimicrobial substances such as phenol, chlorocresol (BDH), cetrimide (BDH), benzyl alcohol (BDH), ethyl alcohol, acriflavine (BDH), penicillin, streptomycin (Glaxo), sulpha drugs (BDH), etc., given as concentrated solutions to save laboratory time in dissolving them

20 x 1oz. screw-capped sterile bottles each, for carrying out the antimicrobial tests

Sterile Pasteur pipettes, 1 ml and 10 ml—several each, to allow for any accidentally contaminated on unwrapping.

Incubators at temperatures appropriate to the micro-organisms used—generally 37°C, but some moulds etc. prefer 25°C.

Bunsen burners.

Cultures of human cells—H.Ep.2 or WI-38 are useful, obtainable from Flow, Wellcome, Biocult

Media for growing the cells—for the H.Ep.2 cells this is Medium 199 + 10% horse serum, and for the WI-38 cells this is EMB medium + 10% calf serum, both obtainable from the same sources as the cells, or from BDH, and bought as 10 times concentrates without antibiotics or bicarbonate (stock medium contains antibiotics and test medium of course does not); trypsin — 1:250 (Difco) and used as Paul, J., Cell and Tissue Culture, 4th Edition, Livingstone, Edinburgh, 1970, p. 216

Falcon flasks, 30 ml, obtainable from Falcon Plastics, or 2 oz. flat, screw-capped bottles with cap liners made from Dow-Corning Silastic rubber sheeting, 502-5, non-reinforced extra firm grade, 0·120in thick, obtainable from Messrs. Down Brothers, Mayer and Phelps

Stoppered cylinders for making up the medium

Membrane filters for sterilising the medium, obtainable from Oxoid, grade 0·45

Membrane filter holders, e.g. Sterifix, Baird & Tatlock

Hot-air oven to sterilise glassware mentioned above

Phase-contrast, inverted microscopes, one with a 35mm camera attached

Ilford PAN F film and ID 11 Developer

Acetic acid stop bath
Amfix fixer (or Johnson's reversal development kit)
Developing tank
Washer (omit these five items if not doing film developing in lab.)
Coulter counter (a haemocytometer chamber may be used but is not recommended as it is much more time consuming)
Cell-counting diluent as Paul, above, p. 352
SP 600 spectrophotometer
Reagents for protein determination, e.g. by the method of Oyama, V.I. and Eagle, H. (1956), Proc. Soc. Exp. Biol. Med., 91, 305–7.

Type of group
Exercise can be carried out by individual students, preferably not more than 6 per member of staff.

Instructions to the student
Find the minimum inhibitory concentration of the given antimicrobial substance for the given culture of micro-organisms, carrying out the experiment in 10 bottles of nutrient bacterial medium and 10 bottles of tissue culture medium. Many micro-organisms will be found to be more resistant to the antimicrobial substance in the tissue culture medium.

On the second day, use the answer obtained above to decide a suitable range of concentrations to apply to the human cells. Generally three concentrations, in geometric progression, can be used. Make up the tissue culture medium from the concentrate to include these three concentrations of drug.

If the cultures have been given ready, check for normality by microscope, and then replace the medium with test medium. Incubate. Use three flasks or bottles for each of the three concentrations, and a further three as controls with medium not containing drug. A suitable volume per flask is 5 ml and per bottle 10 ml. Make sure the caps are tightly screwed on to preserve the pH.

If the cultures have not been given ready, they will need to be trypsinised from stock cultures. To do this, prepare a solution of trypsin according to Paul (1970), above, and filter to sterilise. Add a few ml to the culture from which the medium has been poured off, rock gently, pour off, replace bottle in incubator for a few minutes, and when the cells come off the glass, add appropriate medium and distribute the suspension into the three flasks as above. Incubate.

Examine the cultures periodically, and draw or photograph. If a more extensive exercise is being undertaken, more bottles of cells will be required, e.g. three per concentration per day for a week, for cell counts or for culture protein estimations.

Counting: drain, rinse, trypsinise, suspend in diluent of known background count, and then count.

Protein estimation: drain, rinse, and estimate protein for example by method of Oyama and Eagle, above.

Recording results
Write a brief account of your work; this should be adequate for allowing repetition of the work. Include drawings, photographic prints or reversal-developed transparencies of the cells, and labelled diagrams of any apparatus you have used for the first time. Show numerical results as tables. If the results seem more readily appreciated in graphical form, include graphs also.

Draw conclusions about the relative toxicity of the compound studied to the micro-organisms and to the human cells used.

Extensions and applications
The exercise can be adapted for other drugs—e.g. anti-cancer drugs on normal and malignant cells, or can be carried out by more advanced students, using other drugs for other effects, e.g. enzyme induction, porphyrin production, effects on chromosomes, effects on cell respiration, development of drug resistance, etc. According to the extent and depth of the investigation, and lack of previous information on the topic selected, the same basic theme can be an M.Sc. or Ph.D. topic.

Further reading
Dawson, M.
1972 'Cellular Pharmacology', C.C. Thomas, Springfield, Illinois

Hints for the teacher
Materials must be ordered in good time if not in stock. Any supplier of cells may occasionally be unable to supply a particular cell type. If the cells come in early they can be kept growing, but preferably for not more than about two months or cell division will be less frequent.

Students are liable to come dropping in at odd times between other classes to see how their cells are getting on and ask about anything unusual. This can be time consuming, though not to be discouraged. The preparation time is very variable, depending on whether only a short amount of class time is available, necessitating everything being given ready, or whether the students themselves can do some of the preparatory work.

MARY DAWSON
Strathclyde.

3 Growth experiments with *E. coli* and its mutants

Time required
4 hr.

Assumed knowledge
For second year students familiar with the basic techniques of microbiology and spectrophotometers.

Theoretical background
Bacteria have a high metabolic rate and can grow readily on a wide range of simple organic compounds as their sole source of carbon. They have thus been very useful for biochemical studies. The discovery that many bacterial strains have a parasexual system for genetic recombination has meant that genetic studies can supplement biochemical investigations of, particularly, the nature and control of metabolic processes.

In these experiments you will practise some of the simpler manipulations used by biochemists when handling micro-organisms and will carry out simple tests designed to illustrate the usefulness of one species of bacteria *Escherichia coli K12* in particular.

Necessary equipment
Each pair of students will require the following.
1 Sterile solutions of:
10 ml each of methionine (2mg/ml), tryptophan (2mg/ml) threonine + leucine (2mg of each/ml) thymine (2mg/ml), histidine (2mg/ml , glucose (2mg/ml), arginine (2mg/ml) (all BDH)
10 ml glucose-6-phosphate (2mg/ml) ⎫ Sterilise by
10 ml pyruvate (2mg/ml) ⎬ filtration
⎭ (Millipore)
20 Sterile 50 ml Erlenmeyer flasks needed per practical
2 x 10 ml sterile pipettes
5 x 1 ml pipettes
2 Non-sterile solutions of: (all BDH)
10 ml 0·1M glucose
10 ml 0·1M glucose-6-phosphate
10 ml 0·1M pyruvate (sodium)
10 ml 0·1M aspartate (aspartic acid DL)
10 ml 0·1M glutamate (L-glutamic acid)
10 ml 0·1M glyoxalate (sodium)
10 ml 0·1M succinate

3 Tubes containing growth medium
4 x 20mM acetate growth medium (white cotton wool plug)
4 x 10mM glucose growth medium (yellow cotton wool plug)
2 x 20mM citrate growth medium (blue cotton wool plug)
4 Agar plates
2 x 50mM acetate
2 x 50mM glucose containing bromothymol blue (2 drops), 10mM phosphate buffer pH 7·3, 20mM NH_4Cl, and trace metals, in 1·5% agar
4 x 10mM glucose growth medium containing appropriate supplements
5 Growth medium is made up from:
(a) *Basal media*
KH_2PO_4 19 g
$NaHPO_4$ 51·1 g (anhydrous) ⎫ Up to
NH_4Cl 27 g ⎪ 2 litres
Vit. B_1 1 mg (BDH) ⎬ with
Ca Cl_2 50 ml (8g/litre hydrated ⎪ water
Salts solution 100 ml ⎭
(b) *Salts solution*
0·4 g Mn Cl_2 $4H_2O$ ⎫
0·4 g Fe SO_4 $7H_2O$ ⎬1 litre
8·0 g Mg SO_4 $7H_2O$ ⎭
Dilute $\frac{1}{5}$ with carbon source and water and add 1·5% Ionagar (Difco) to solidify when necessary. Supplements are added (3ml/ 100 ml medium) as appropriate. This medium is used in all cases except the bromothymol blue plates where a low buffer is needed.
6 Solid pyrogallol (BDH)
7 100 ml 50% K_2CO_3 in three dropper bottles
8 Spectrophotometer (Unicam SP 600)
9 Glass spectrophotometer cells
10 Shaking water baths at 37°C to hold 50 ml flasks
11 Bunsen burners and platinum loops (Gallenkamp)
12 Suspension of *E. coli* and mutants and *Klebsiella aerogenes* grown overnight on nutrient broth
13 Plates of *E. coli* mutants (K2, K1.1.2.5^c)

Type of group
4 pairs of students. The number of pairs is limited only by the size of the shaking water baths.

Instructions to the student
WARNING
None of the organisms you will be handling is pathogenic but all bacteria should be handled with care. Always take sterile precautions when opening

195

tubes, flasks etc., containing bacterial suspensions; pipette properly and hence avoid the risk of accidentally swallowing bacterial suspensions; **always wash your hands before leaving the laboratory.**

Growth studies

1 *Carbon requirements.* You are provided with plugged test tubes containing sterile growth medium with either glucose, citrate or acetate as sole carbon source and suspensions of *E. coli* and *K. aerogenes*.

Inoculate the following tubes with a loopful of *E. coli* suspensions:

    2 acetate tubes (white plug)
    2 glucose tubes (yellow plug)
    1 citrate tube (blue plug)

and inoculate *one* of each of the 3 types of tube with a loopful of the *K. aerogenes* suspension.

Take one of the *acetate* tubes inoculated with *E. coli* and cut off the top of the cotton wool plug. Push the plug into the tube until it is about 1½ in above the surface of the liquid. Place a spatula tip-full of solid pyrogallol onto the top of the plug and drop onto it 2–3 drops of 50% K_2CO_3 solution. Place a small cotton wool plug in the top of the tube and finally stopper the tube tightly with a rubber bung. The tube is now anaerobic. Treat one of the *glucose* tubes inoculated with *E. coli* in a similar fashion.

Label all tubes with your name and place the anaerobic tubes in the test tube rack provided and the other tubes in the slope basket: these will be incubated aerobically at 37°C for you.

You are provided with two plates containing respectively 50mM acetate and 50mM glucose-agar medium that is low in buffering capacity [10mM Na/KPO_4], and containing the pH indicator bromothymol blue. Streak out some of the *E. coli* suspension on both these plates. Label the plates with your name and place them on one side.

2 *Growth tests on mutant strains of E. coli.* It is possible to expose bacteria to mutagenic agents and select for mutant organisms having an altered metabolism. Analyses of such mutants have led to insights into both the nature and the control of metabolic pathways.

(a) You will be given 20mM glucose agar plates containing three of the following four supplements: tryptophan, histidine, arginine, (threonine + leucine), which have been thinly seeded with a suspension of a mutant of *E. coli* K12, called K2. Drops of a solution of the fourth supplement can now be placed on the surface of the agar and their ability to stimulate (or inhibit) growth determined. Look carefully at the plates

after they have been incubated at 37°C: how do you interpret your results?

(b) Take an agar plate, containing 30mM acetate medium supplemented with methionine and thymine, which has been 'seeded' with a mutant of *E. coli* K12 (K1.1.2.5^c) grown up in nutrient broth. Place a drop of (i) glucose, (ii) glucose-6-phosphate, (iii) pyruvate at three marked points. 0·5M solutions work best; try three separate loopfuls and wait for each to soak in before the next application. *Remember to flame your loop after each application.* Examine the plates after incubation at 37°C. What is the effect of these 'natural' food materials on the growth of the mutant?

Use of growth experiments to investigate metabolic controls

You are provided with a flask containing *E. coli*. K1.1.2.5^c (a derivative of K12) growing on 30mM acetate medium supplemented with methionine and thymine. Put 10 ml of the culture into each of four flasks provided using sterile techniques. Label them with your name and place them into a shaking water-bath at 37°C. After about 5 min take a reading at 680 nm from each flask: this is done by pouring approximately 3 ml into a cuvette, reading at 680 nm against a water blank, and *pouring the suspension back into the correct flask.* Repeat the process 30 min after the first reading: if all cultures are growing (as indicated by an increase in the extinction at 680 nm), record the reading and add to

Flask 1	:	0·5 ml of water
Flask 2	:	0·5 ml of 0·1M-glucose
Flask 3	:	0·5 ml of 0·1M-glucose-6-phosphate
Flask 4	:	0·5 ml of 0·1M-pyruvate

Continue to take readings at 30–45 min intervals. (It does not matter when you take readings, as long as you know and record the experimental time). If growth slows markedly or stops, one of a variety of compounds should now be added. It would be best if *one* group of students were to add *one* compound to all its flasks, other groups can add others. Thus,

Group 1 might add 0·5 ml of 0·1M aspartate to each flask

Group 2 might add 0·5 ml of 0·1M glutamate to each flask

Group 3 might add 0·5 ml of 0·1M glyoxylate to each flask

Group 4 might add 0·5 ml of 0·1M succinate to each flask

(and other compounds can be suggested by the

demonstrators).

Continue to take readings at 30—45 min intervals.

Recording results

Examine the growth tubes and determine the amount of growth according to an arbitrary scale from 0 = no growth to ++++ = good growth. Note the colour of the bromothymol blue plates. Examine your mutant plates and note whether growth or inhibition has occurred. Plot the OD 680 readings from your flasks on semi-log graph paper against time of incubation.

Discussion

Are the results of your growth-tube experiments reasonable? What light do they throw on the colour changes you observed in the grown family blue plates? How do you think it is possible for metabolites such as glucose-6-phosphate, glucose and pyruvate to cause growth inhibition?

Extensions and applications

Experiments of this kind are widely used to help define problems in the field of metabolic regulation. Examples of the application of this kind of experimental approach can be found in the review by H.L. Kornberg (Biochem. J., 99, 1, 1966) and these kinds of experiment were first described by Kornberg and Smith (Nature, 224, 1261, 1969). This paper describes how such observations may be used to obtain further mutants.

Further reading

Articles on the growth and cultivation of bacteria and a variety of different types of mutant are contained in

Norris, J.R., Robbins, D.W. (eds.)
    1969 'Methods in Microbiology', Vol. 1, Academic Press, London

Droop, M.R. and Wood, E.J.F.
    1968 'Advances in Microbiology', Academic Press, London

Hints for the teacher

The students should start this experiment at the section headed 'Use of growth experiments to investigate metabolic controls'. If possible the growth experiment should be followed for 6—8 hr to obtain reliable growth curves. The rest of this experiment can be done in 2—3 hr whilst growth is followed.

E. coli. K12 strain K1.1.2.5^c is a derivative of the H fr H strain first described by Hayes, and lacks phospho*enol*pyruvate carboxylase and phospho*enol*pyruvate synthase activity. It is constitutive for isocitrate lyase activity and the section 'Use of growth experiments . . .' can be interpreted in terms of the fine control of the glyoxylate cycle.

E. coli and mutants can be obtained from National Collection of Industrial Bacteria.

It is not necessary to use *E. coli* K2 (Section 2(a)); any multiple auxotroph will do provided the growth plates used contain the appropriate supplements.

J.M. ASHWORTH
Essex.

4 Bacterial conjugation; isolation of recombinants

Time required
¼ hr on first day, ¼ hr on third day and 5 min on fourth day to observe results.

Assumed knowledge
None, except facility with a bacteriological wire loop.

Theoretical background
When cultures of certain strains of bacteria are mixed they interact and produce hybrid cells with some of the characteristics of each of the original strains. It has been shown that the roles of the two strains are different: if one of the cultures is exposed to a lethal dose of ultraviolet light before mixing it is still able to contribute genetic characteristics to viable hybrid progeny, but if the other is so treated there are no viable hybrids. The dispensable type is known as the genetic donor or male strain, and the other is the recipient or female strain.

During conjugation between Hfr or F^+ (male) and F^- (female) cells of *E. coli*, all or part of the chromosome of the male passes into the female cell and is immediately expressed in the form of proteins. Furthermore the new genes may be incorporated into the female chromosome, replacing the original genes and resulting in a hybrid or recombinant chromosome bearing some of the genes of each parent.

Each Hfr strain has a characteristic point on its chromosome at which transfer always starts, and proceeds in the same direction, so that the genes received by the F^- cell depend on the speed and time of transfer. The average time taken to transfer a particular gene can be determined by interrupting the conjugation by vigorous stirring (which separates the pairs, breaking the DNA passing between them) at various times and looking for F^- cells that now contain this gene. In this way the relative positions of genes on the bacterial chromosome have been mapped.

The contribution of the donor to the recombinant will be limited by the extent of transfer and by the frequency of recombinational events, i.e. how much of the male DNA is actually incorporated into the female chromosome, replacing the homologous sequences of female DNA. Therefore we can get an indication of which of the parents is the recipient by examining the recombinant bacteria to see which parent they most resemble.

Necessary equipment
Each student will require the following.
First day
About 1 ml of each of two fresh cultures of *E. coli*
(a) Hfr, labelled 'X' or 'Y' and (b) F^-, labelled 'Y' or 'X'
1 wire loop
1 Petri dish or minimal agar (Oxoid)

Third day
1 ml of 'X' and 1 ml of 'Y'
Wire loop
1 small sterile tube, small bottle of sterile broth (Oxoid)
1 sterile pipette (1 ml or Pasteur)
2 Petri dishes of nutrient agar (Oxoid)
1 ml of streptomycin solution (Glaxo)
1 ml of phage suspension or access to a central supply

Type of group
Individual students in large classes.

Instructions to the student
First day
You are given two cultures of *E. coli*, labelled 'X' and 'Y'. One is an Hfr and the other an F^- strain. Neither will grow on minimal agar (containing only salts and glucose). The Hfr are met^-, that is they have a mutation in a gene which is concerned with the synthesis of methionine, and the F^- cells are leu^- (they need leucine for growth). Both have other requirements, which are included in the plates.

The effect of conjugation between the two strains can be observed by cross-streaking on minimal agar. Make a broad streak of culture X with a loop across a minimal agar plate and allow a few minutes to dry. Across this, at right angles to it, streak another loopful of X and one of Y, each with a single sweep of the loop (flame and cool the loop before each operation, of course). Mark the positions and direction of the streaks. Incubate at $37°C$ for two days.

3rd day
One of the parental strains, X and Y, has a mutation which renders it resistant to the antibiotic streptomycin, and one lacks a receptor site needed for the adsorption of a particular bacteriophage. Therefore test the parents and the recombinant progeny as follows to see which parent the recombinant resembles.
1 With a sterile loop make a band of streptomycin

198

along a diameter of a nutrient agar plate and allow it a few minutes to dry. Pick an isolated recombinant colony from your recombination plate (after 2 days' incubation) with a sterile loop and disperse it in about 0·5 ml of nutrient broth in a sterile tube. Then streak a loopful across the streptomycin band. Keep the suspension for the phage-sensitivity test. The original cultures X and Y should also be streaked across the streptomycin at different places on the same plate, each labelled clearly. Incubate the plates overnight at 37°C.

2 Prepare another plate by streaking the phage suspension in a band across it and allowing it to dry. The three strains of bacteria are then cross-streaked over it, each with a single stroke of the loop (in order to display a sharp cut-off in the case of a sensitive strain).

Fourth day
Examine the incubated plates to determine streptomycin and phage-sensitivity.

Recording results
Make diagrams of the plates you have prepared, labelling the materials streaked and directions of streaking, if relevant.

Discussion
Which of the cultures X and Y is probably the genetic donor and which the recipient? Why do the results of these tests not provide conclusive evidence of identification? A more convincing demonstration may be obtained by mixing suitable male and female cultures on a minimal agar plate containing streptomycin. A male strain will conjugate and transfer its DNA even if it is sensitive to streptomycin, i.e. unable to grow in the presence of streptomycin. Thus a sensitive Hfr and a resistant F⁻ will produce recombinant colonies, but a resistant Hfr with a sensitive F⁻ will not.

A simple way to distinguish male from female bacteria is to test their sensitivity to a male specific phage. This you have in fact already done since the phage used was the RNA phage R17. The interpretation of your results did not depend upon this knowledge, but will, hopefully, be strengthened by it.

Extensions and applications
The experiment demonstrates the power of a selective medium (minimal agar) to reveal the presence of a small number of cells of a particular character amongst a much larger population not possessing that character.

It also demonstrates the simple technique of cross-streaking which has many applications in the detection of sensitivity to inhibitors, dependence on growth factors and other interactions. The left-hand end of a streak of bacteria formed from left to right across the inhibitor provides a control area of growth and also, if needed, a source of cells uncontaminated by inhibitor.

Further reading
Hayes, W.H.
1968 'The Genetics of Bacteria and their Viruses', Blackwell, London

Hints for the teacher
The strains of *E. coli* are chosen so that the F⁻ is defective in a gene that is transferred early during conjugation. The F⁻ used was EMG28 which is resistant to streptomycin and auxotrophic for threonine, leucine, proline and thiamine. The Hfr strain was Hfr_1 (K12) which is sensitive to streptomycin, requires methionine and biotin, and transfers the threonine and leucine genes early. Information concerning suitable strains of bacteria and their sources may be found in Clowes, W. and Hayes, W.H. (eds.) 'Experiments in Microbial Genetics', Blackwell London, 1968.

Maintain stocks on nutrient agar slopes at 4°C, and subculture about twice a year.

The cultures are grown in nutrient broth to an optical density at 660 nm of about 0·4 (approximately 4×10^8 cells/ml). Shortly before the class they are diluted in broth to an optical density of 0·04 and dispensed in 1 ml aliquots into sterile plugged tubes.

Streaking of bacteria suspended in nutrient broth on minimal agar yields a faint background of microcolonies. This may be reduced by diluting the growing culture with water instead of with broth, immediately before streaking.

Minimal agar plates contained tris-buffered mineral salts medium, sodium pyruvate (0·8 g/litre, probably unnecessary) and glucose (5 g/litre), supplemented with proline (10 μg/ml), thiamine (1·0 μg/ml) and biotin (0·1 μg/ml, probably unnecessary).

Phage R17 was grown in *E. coli* Hfr (see Section 6.1), freed of bacteria by centrifugation and used for cross-streaking at a concentration of about 10^{10} p.f.u. (plaque forming units/ml; probably considerable excess). When Hfr cells are streaked across R17 a substantial number of resistant colonies grow up. Maximum contrast is obtained after 8—10 hr incubation at 37°C.

Streptomycin was dissolved in sterile water at 2 mg/ml.

For nutrient broth, agar and minimal agar, see Meynell, A. and Meynell, A., 'Theory and Practice in Experimental Bacteriology', Cambridge University Press, 1965. We use Oxoid Nutrient Broth No.2 and Oxoid Agar No.3.

M.L. FENWICK
Oxford.

5 The cell cycle and induction of synchrony

Time required
Two successive days and several hours on a later day.

Assumed knowledge
Aseptic technique and the problems involved in the use of radio-isotopes.

Theoretical background
Under given conditions cell lines show a fixed generation time (and doubling time); that is, they will divide at fixed time intervals. However, growth conditions vary continuously and a vessel of fixed size can accommodate only a narrow range of cell concentrations. Experiments to study growth parameters are therefore not as easy to perform with cultured cells as with micro-organisms.

Animal cells pass through a cycle of events between divisions which, in culture, takes about a day to complete. Of this time only 5-8 hr is spent making DNA (S-phase). Specific inhibition of DNA synthesis has no effect on those cells not making DNA and they proceed round the cycle until they reach the beginning of S-phase where they accumulate, i.e. they are chemically synchronised and, when the inhibition is reversed, they will all start together to make DNA.

N.B. Take care to distinguish between *rate* and *extent*.

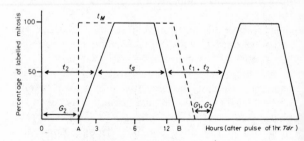

Figure 1 (See Section 4.4)

Necessary equipment
Hot (37°C) room containing an incubator and a bench
Bottle of L929 cells
Cell counter (Coulter or haemocytometer) (Gallenkamp)

Pipettes   0·1 ml sterile   (2)
          1·0 ml sterile   (7) per student
          5 ml sterile   (5)
          10 ml sterile   (8)

30 disposable dishes (5 cm diam.) and trays to support them

Eppendorf automatic pipettes, 100 μl, and sterile tips (V.A. Howe)

4 oz bottles (sterile)

Trypsin/versene * (5 ml)

EC_{10}† (100 ml) (Biocult, Flow)

BSS (Earles, pH 7·4) (300 ml) (Biocult, Flow)

Deoxycytidine (10^{-3}M) (1 ml) (Sigma)

Glycine (10^{-2}M) (1 ml) (BDH)

Adenosine (7mM) (1 ml) (BDH)

Aminopterin (10^{-5}M) (2ml) (Serva)

^{3}H-thymidine 30 μCi/μmole/ml (2 ml) (Radiochemical Centre, Amersham)

Thymidine (10^{-3}M) (1 ml) (BDH)

5% Trichloracetic acid (TCA)

Ethanol and ether

Hair drier

Toluene scintillator (5 g PPO/litre Analar toluene) (Koch-Light)

Hyamine hydroxide solution (Koch-Light)

Giemsa reagent (Gurr)

Scintillation counter

Type of group

The experiment is best carried out by pairs of students with no more than 10 pairs being involved at any one time. One demonstrator per 6 students is recommended.

Instructions to the student

You are provided with a Roux bottle of L929 cells in early stationary phase. Aseptically trypsinise the cells in the morning as follows:

Carefully remove the medium from the bottle and replace with 2–3 ml trypsin/versene solution. Incu-

---

* The trypsin/versene contains per 200 ml

   0·1 g trypsin (1:250) (BDH)   0·18 g Na_2HPO_4
   0·12 g sodium citrate   0·03 g KH_2PO_4
   1·52 g sodium chloride   0·03 g versene
              0·03 g potassium chloride
adjusted to pH 7·8 and sterilised by filtration.

† EC_{10} is Eagles minimum essential medium containing 10% calf serum and penicillin and streptomycin. It is buffered with Hepes buffer. This is important as students will be constantly opening the incubator door and it is impossible to maintain the correct p_{CO_2} under these conditions. It is for a similar reason that the incubator should be sited in a hot room, i.e. to maintain a constant temperature.

---

bate (37°C) for 3–5 min until the cells are loosened from the glass. Transfer the cells to a sterile 4 oz bottle and make up to 25 ml with EC_{10}. Pipette to obtain a single cell suspension and count the cells.

Set up 30 x 5 cm dishes each containing 6 x 10^5 cells in 3 ml supplemented growth medium as follows:

EC_{10}	100 ml
deoxycytidine (10^{-3}M)	1 ml
glycine (10^{-2}M)	1 ml
adenosine (7mM)	1 ml

Incubate the dishes at 37°C in a humidified incubator.

At 5 p.m. add 0·05 ml 10^{-5}M aminopterin/dish.

Reverse the inhibition 16 hr later:

To 10 dishes add 0·1 ml 10^{-3}M thymidine.

To 20 dishes add 0·1 ml ^{3}H-thymidine (30μCi/μmole/ml). *Do not allow the cells to cool.* To do this use an Eppendorf micropipette with a sterile tip.

Harvest 2 radioactive and 1 non-radioactive dish throughout the day, i.e. after 0, ½, 1, 2, ... , 8 hr as follows

Radioactive dishes

Suck off the radioactive medium (put in a suitable waste container) and wash four times with 3 ml cold BSS. Throughout these procedures take care not to wash the cells from the dish. Wash (on ice) with 5 ml portions of 5% TCA (4 x 5 min) and then twice with 5 ml absolute ethanol. Air dry. Dissolve the cells in 1 ml 0·3N NaOH (37°C). Take 0·5 ml samples to dryness, e.g. blowing air stream, in scintillation vials. Add 0·5 ml hyamine hydroxide, and cap. Heat to 60°C for 10 min and allow to cool. Add 5 ml toluene scintillator. Estimate radioactivity in the liquid scintillation spectrometer.

Non-radioactive dishes

Remove the medium and wash twice with BSS, twice with 5% TCA and twice with water. Stain with Giemsa reagent for 10 min. Rinse with water and air dry.

Examine under the microscope and assess the number of cells in mitosis.

Variations

The cells may be grown on glass cover slips within the dishes. On harvesting, the cover slips would be dipped successively into the various wash solutions and finally the cells would be solubilised directly in scintillation vials by addition of hyamine hydroxide.

Cover slips, fixed to slides, can be stained or used in autoradiographic studies.

Recording results and Discussion

The results should be presented graphically and from them it is possible to calculate the amount of DNA made per cell. How does this correlate with the expected value?

How is aminopterin working, and what is the role of the glycine, adenosine and deoxycytidine?

To what extent do you think this statement is true: 'specific inhibition of DNA synthesis has no effect on those cells not making DNA'? (see 'Theoretical background')

Extensions and applications

The composition of the cell is changing in an ordered way throughout the cell cycle. A study of the control of these periodic changes is therefore a study of gene regulation which is fundamental to cell biology. This experiment forms the basis to much of the work using synchronised cell cultures.

Further reading

Mitchison, J.M.
  1971 'The Biology of the Cell Cycle', Cambridge University Press

Hints for the teacher

What is required are basically the cells (from Flow or Biocult) and medium. A confluent Roux bottle of L929 cells provides more than enough cells, which respond well to the synchronisation procedure. The medium should not have been changed for 48 hr. Care should be taken in supervising the use of the tritiated thymidine and in the disposal of all contaminated materials.

R.L.P. ADAMS
Glasgow.

6 Synchronous division in a *Tetrahymena* culture using heat shocks

Time required

Providing a number of heat shocks have been given, the actual minimum time required in a class experiment is 2½ hr. A longer period of time enables one to examine behaviour of the culture after the first peak of synchronous division.

Assumed knowledge

The student should be able to use a haemocytometer and make measurements using a graduated eyepiece. The practical can be used to teach sterile sampling techniques, but sampling can be done without any precautions.

Theoretical background

When cultures of protozoa or animal cells are examined at random, only a small percentage of cells are in the process of cell division. Thus the mitotic index, the ratio of the number of cells dividing to the number of cells in the sample, is always low. However, during many studies of cell synthesis or behaviour, it is an advantage to have all cells at the same stage in their cell cycle. One means of achieving cell synchrony is to halt cells in some way at a sensitive stage in the cycle, for example deprive them of an essential component of replication or halt cells at a temperature-sensitive stage. In the ciliate *Tetrahymena*, considerable degree of synchrony can be achieved by successive heat shocks over a period of time.[1] The cells are grown at 33°C for ½ hr, then placed at 27°C for ½hr, back to 33°C, then 27°C, etc. for a period of 6–7 hr, finally remaining at 27°C. After approximately 80 min at this temperature, a considerable percentage of the cells in the culture will divide, to be followed by a second, smaller peak of synchrony another 60 min later. When *Tetrahymena* are about to divide, an indentation appears in the mid-line (ciliates divide transversely) which gradually forms a definite 'waist' before daughter cells separate.

Necessary equipment

Each student will require the following.

1 x 250 ml Erlenmeyer flask containing 100 ml of a culture of *Tetrahymena pyriformis*, which has

already been subjected to a succession of heat shocks

A supply of sterilised Pasteur pipettes or glass tubing cut to 25 cm, plugged at both ends with cotton wool and sterilised. These can then be pulled and ends sealed in flame

Teats (2)

Bunsen burner

Diamond pencil (Gallenkamp)

Slides and cover slips

Solid watchglasses and lids (6)

Microscope (standard optics or phase contrast)

Graduated eyepiece (Gallenkamp)

Stage micrometer slide (optional, 2–3 per class)

Haemocytometer (Gallenkamp)

10 ml 5% formalin with pipette and teat

Stopclock

Water bath at 27°C ⎫
Water bath at 33°C ⎭ One between 4 students

Racks for Erlenmeyer flasks, or means of holding down flasks in water bath

Non-absorbent cotton wool

Arrange to have all *Tetrahymena* cultures at 27°C

Type of group
The experiment is designed for individual students;

Instructions to the student
Your culture of *Tetrahymena* has been subjected to six 30 min heat shocks at 33°C, interspersed with six 30 min periods at 27°C, and is at 27°C now.

1 Take a sterile Pasteur pipette, break off sealed end using diamond pencil, and withdraw a sample from the culture using sterile techniques as follows: with bunsen burner alight, take flask and withdraw bung holding flask between thumb and forefinger of left hand and bung between 3rd and 4th fingers. Flame lip of flask, shake and withdraw sample with pipette in right hand, flame lip of flask and replace bung. Get a demonstrator to show you. Points to watch are (a) do not put bung down on bench, (b) make sure bung is not near flame and (c) it may sound obvious, but do not flame bung! Place sample in a solid watchglass and flask into water bath at 33°C and note time (check time for this, i.e. put back at 27°C if still within 27°C period).

2 Look at the cells in the watchglass and add 1 drop of 5% formalin (BDH). Keep this pipette *separate*. Take a sample and place on a slide. Cover and make measurements of length and breadth using graduated eyepiece. Place another sample on haemocytometer and count number of cells in sample, and the number of dividing cells (i.e. mitotic index).

3 Exactly 30 min later, take *Tetrahymena* culture from water bath at 33°C, shake and sample using sterile techniques and another Pasteur pipette. Place *Tetrahymena* culture in waterbath at 27°C and sample of cells into another watchglass. Repeat size measurements and estimate mitotic index.

4 30 min later, sample *Tetrahymena* culture *but* leave at 27°C. Make size measurements and mitotic index estimation.

5 30 min later, repeat sampling procedure and *leave* culture at 27°C.

6 The *Tetrahymena* culture has now been 60 min at 27°C after last shock treatment. From this time you may expect the peak of synchrony to arrive, so sample every 10 min from now, and every 5 min should dividing cells start to become frequent. It is important to shake the culture gently before sampling since dividing cells tend to settle on the bottom of the flask. You will know if the peak of synchrony has passed because your total cell count will increase and cell size will decrease. Providing the samples are taken, killed in formalin and labelled, measurements, etc., can be made later.

7 If sufficient time is available, sample 30 min and 60 min after the first peak of synchrony.

Recording results
A graph should be drawn of mitotic index versus time, indicating duration of heat shock treatment. The cell-size measurements are best presented as histograms, using arbitrary units, or calibrating the graduated eyepiece using a stage micrometer slide (generally too few cells will have been measured to warrant statistical treatment).

Discussion and extensions
Consider which of the many events of cell cycle might be temperature sensitive and thus how the cells are held back at this point. Compare your % synchrony with other members of the class—are they all constant within experimental error or widely different? If latter, what errors might arise during the experiment and how many cells should one count to obtain a significant result? Can you consider the cells normal after the induction of synchronous division in this way, and how could you test this experimentally?

This method of inducing synchronous division is not in general use in tissue cultures, where aminopterin, or excess thymidine treatments are favoured. Nevertheless, it has been widely used with *Tetrahymena* and a considerable volume of biochemical information has been collected. The method is suitable for *Tetrahymena* because the cell cycle takes about 3 hr at 27°C. Other protozoa have been less easy, e.g. amoeba cell cycle takes 2½ days at 17°C and period of time required at higher temperature to induce synchrony has not been determined successfully.

Further reading
Scherbaum, O. and Zeuthen, E.
    1955 'Temperature-induced Synchronous Divisions in the Ciliate Protozoan *Tetrahymena pyriformis* growing in synthetic and Proteose-peptone Media', Exp. Cell Res., Supple. 3, 312
Scherbaum, O.
    1960 'Synchronous Division of Microorganisms', Ann. Rev. Microbiol., 14, 283

Hints for the teacher
The success of this experiment depends on getting in a sufficient number of temperature shocks. Seven periods at 33°C will usually give 60% cells in synchronous division, 5 periods at 33°C usually shows a reduction to 25%, which is disappointing to the student. A culture of *Tetrahymena pyriformis* (we use W strain) can be obtained from Cambridge Culture Collection of Algae and Protozoa (36, Storey's Way, Cambridge) and should be grown in 1% proteose-peptone (1% proteose-peptone [Difco]; 0·25% yeast extract [Difco]) sterilised at 15 lb/in² for 15 min. Sterile techniques should be used to inoculate the initial culture and throughout growth. A 500 ml Erlenmeyer flask containing 250 ml proteose-peptone may be grown at 27°C for 1–2 days, or 17°C for 2–3 days before sub-culturing. This flask will provide enough material to inoculate 25 experimental 250 ml flasks containing 100 ml proteose-peptone. The experimental flasks should be inoculated the afternoon previous to the practical and placed in waterbath overnight at 27°C. They should then be transferred simultaneously to 33°C at the start of the temperature shock treatment, and thereafter changed over at 30 min intervals. It is usually better to make one person responsible for this part of the preparation.

The students do not need to sample using sterile techniques, since bacterial contamination does not interfere with the result; however, the practical serves as a useful way of introducing these sampling methods. Sterile pipettes can be made from glass provided by the students, if time. Have plenty of spare non-absorbent cotton wool in case of accidents!

SHIRLEY E. HAWKINS
King's College, London

7 The growth of asynchronous and synchronous cultures of a fission yeast *Schizosaccharomyces pombe*

Time required
The experiments can be completed in two 3 hr sessions, the first working with the exponentially growing culture and learning the methods, and the second using the synchronous culture.

A method of staining and preserving the cells is described but its application would require further practical time.

Assumed knowledge
Use of haemocytometer and cell size measurement with an eyepiece graticule.

Theoretical background
The growth of a culture of *Schizosaccharomyces pombe* is similar to that of a bacterial culture in having a lag phase, exponential growth phase and a stationary phase (see section 11.1). The individual cells of the yeast are round-ended cylinders and they grow simply by increasing in length while maintaining a constant diameter. The length increases linearly during the first 75% of the cell cycle and then remains constant for the remainder of the cycle. During this terminal period the nucleus divides and a septum called the cell plate is formed between the two daughter cells.

The growth of a culture may be followed by measurement of cell number, optical density, cell-plate index and cell length. Cell number gives a direct measure of the increase in population. Optical density is due to the amount of light scattered by the culture and, although this is not directly related to cell number, it gives a measure of the increase in total cell mass. The cell-plate index indicates the proportion of cells near division; in an exponentially growing culture it should remain constant. As *S. pombe* divides by transverse fission, daughter cells will be short and mature cells long; in an exponentially growing culture the distribution of cell sizes should emain constant. These four parameters are used first to study the growth of an exponentially growing culture and then one where cell division has been synchronised. This synchrony is achieved by selecting young (short) cells from an exponential culture by centrifugation.

Necessary equipment

Materials
Schizosaccharomyces pombe culture—Strain 132 (National Collection of Yeast Cultures)
Malt Extract Broth (Oxoid)

Apparatus
Each group of students will need the following.
6 Pasteur pipettes
1 haemocytometer
2 tally counters
Microscope + eyepiece graticule for each student
The following will need to be available in the laboratory
Waterbaths at 32°C—preferably with shakers
Calibration slides for eyepiece graticules
Spectrophotometers.

Type of group
Pairs or groups of three.

Instructions to the student

The exponential culture
1 *Sampling.* Samples are removed from the cultures in the water baths at 15 min intervals. The measurements described below must be made as soon as possible and must in any case be completed before the next sample is removed. As yeast cells tend to clump it is essential to shake the culture vigorously before removing each sample (about 0·5 ml) with a Pasteur pipette. Do not remove the culture from the water bath for longer than necessary. The culture's growth must be followed for 2—2·5 hr.

2 *Cell number* The number of cells in the culture is counted in a haemocytometer with all the standard precautions associated with the use of that instrument. Counting is best carried out using a x40 objective. The daughter cells are separated physiologically by the cell plate before they become detached from each other, but the precise moment at which one cell becomes two is not easy to determine. We shall adopt the convention of counting bodies with cell plates as one cell, and counting two cells when the two daughter cells have clearly rounded off their ends although they may be still attached to each other.

3 *Cell length.* The progression of a cell through the cell cycle is indicated by its length. Using an eyepiece graticule and an oil-immersion objective, preferably with phase optics, measure the length of 50 cells chosen at random throughout the culture, and construct a histogram to show the distribution of lengths in the culture. Also calculate the mean cell length.

4 *Cell-plate index.* The frequency of division in the culture may be estimated from the number of cells containing a cell plate. The cell-plate index is merely the percentage of cells in this stage and can be determined by scoring the number of cell plates in 200 cells. Alternatively, the less experienced operator might find it an advantage to fix some cells from each sample so that they may be stained and studied later (see below). The cell plate index of an exponentially growing culture is around 7%. If a significantly higher value is obtained it is probably due to the erroneous scoring of rugosities (wrinkles, lumps) on the cell surface as cell plates.

Staining of cells for cell-plate counting; a few drops of the culture are dried on a slide on a hot plate (90—100°C) and then washed to remove medium. A smear is drawn over the cells using 10% Indian ink (BDH). After drying, the slide is stained for a few seconds in 0·25% crystal violet (BDH). The cytoplasm is stained but not the cell wall or the cell plate. During this treatment cells shrink by about 15%.

5 *Optical density.* When removing the 0·5 ml samples used above remove a further 3 ml sample, rapidly measure its optical density at 600 nm against a growth medium blank and return it to the incubating culture. The optical density is a relative measure of light scattered by the culture but its actual value depends on many other factors including the geometry of the spectrophotometer used. Comparable results are, therefore, only obtained when measurements are made under identical conditions on the same instrument.

The synchronous culture
Recently divided cells have been selected from an exponential culture by gradient centrifugation. The growth of this culture can be followed by exactly the same methods as have been used for the exponential culture. Start the measurements as soon as the cells have been removed from the gradient and continue at 15 min intervals for 2 to 2·5 hr.

Results
Record the results in tabular form as exemplified below. The data can then be presented graphically in the way you yourselves find most suitable.

The growth of a synchronous culture of *S. pombe*

Time (hr)	Cell No. x 10^6	Cell plate index	OD 600 nm
0·25	2·14	0	0·118
0·50	2·17	0·2	0·124
0·75	2·06	0	0·129
1·00	2·05	0·4	0·135
1·25	2·11	0	0·143
1·50	2·26	1·2	0·156
1·75	2·09	1·2	0·165
2·00	2·26	6·6	0·177
2·25	2·53	15·4	0·190
2·50	3·15	23·0	0·206
2·75	3·85	17·8	0·215
3·0	3·88	8·0	0·230
3·25	3·89	1·2	0·239

Discussion

Calculate the mean generation time of the cells in the exponential culture and compare it with the value observed in the synchronous culture.

Is the distribution of cell sizes a normal distribution? If not, why not? Consider both cultures.

All the cells in the synchronous culture do not divide at precisely the same time. The percentage of cells dividing together is termed the degree of synchrony. How would you estimate the degree of synchrony from your results?

Extensions and applications

A variety of methods exists for the preparation of synchronous cultures. As well as selecting cells of a similar age, it is also possible to induce all of the cells in a culture to divide synchronously. For example, if DNA synthesis is continuously blocked in an exponentially growing culture, without hindering cell growth, then all of the cells in the culture will grow up to the point where further division is held up through lack of DNA. If the culture is released from the DNA block, the cells then proceed to make DNA synchronously and subsequently undergo a synchronous division. Examples of other methods of induction synchrony can be found in the literature. Which events might be blocked temporarily to give rise to a synchronous culture?

The sequence of events that constitute the cell cycle may also be analysed by studying single cells and many measurements can be made on single cells. These include length, generation time, u/v sensitivity, respiration using a Cartesian diver, and dry mass using an interference microscope. Find examples of these

and of other measurements on single cells, in the literature.

Discuss the relative merits of synchronous culture and single-cell studies.

Further reading

Methods for *S. pombe:*

Mitchison, J.M.
1970 In 'Methods in Cell Physiology', Vol. 4, ed. D.M. Prescott, Academic Press, New York pp. 131–65

Synchronisation of Mammalian cells with reference to the mitotic cycle:

Nias, A.H.W. and Fox, M.
1971 'Cell Tissue Kinetics', 4, 351–74

Methods for studying the microbial division cycle:

Helmstetter, C.E.
1969 In 'Methods in Microbiology', Vol. 1, eds Norris, J.R. and Robbins, D.W., Academic Press New York, pp. 327–63

Cameron, I.L. and Padilla, G.M. (eds.)
1966 'Cell Synchrony Studies in Biosynthetic Regulation', Academic Press, New York

Mitchison, J.M.
1971 'The Biology of the Cell Cycle', Cambridge University Press, Cambridge

Hints for the teacher

The cells for an exponential culture are prepared by inoculating 4 litres of malt extract broth with 5 ml stationary culture and growing up overnight at 32°C. This should result in a culture of 4 x 10^6 cells per ml.

To prepare a synchronous culture, the cells are harvested at a concentration of about 4 x 10^6 per ml by centrifuging at 1 000 g for 5 min. After collection, the cells are resuspended in 4 ml of medium and layered on 80 ml of a 10–40% sucrose gradient made up in malt extract broth. The gradient is spun at room temperature for 5 min at 500 g in a swinging bucket head. The top of the cell layer will travel about halfway down the gradient. 1–3 ml of cells are then removed from the top of the cell layer using a 5 ml syringe with a long needle and are added to 50–200 ml of warm medium. The concentration of cells in the synchronous culture should be about 2 x 10^6 per ml. Since the smallest cells are selected from the gradient, about 2 hr elapse before the first synchronous division. For a 3 hr practical the preparation can therefore be made either immediately before the class, or about 1·5 hr earlier, in which case the students would follow the culture through the first division and into the second. The synchronous culture may be presented as an unknown and

tudents asked to investigate its growth and interpret
heir results.

We wish to acknowledge the involvement of many
olleagues during the evolution of these practicals in
his department.

J. CREANOR
A. H. MADDY
Edinburgh.

8 Survival of cell reproductive capacity of mammalian cells *in vitro* after X-irradiation

Time required

1 Trypsinization of stock cell culture, counting, dilution, seeding into culture vessels for experiment; approx. 1 hr.
2 Irradiation of cultures; depends upon available radiation source(s), can be done for students so that no class time is involved but should certainly take less than 1 hr.
3 Fixing and staining of fully grown cultures; approx. 30 min.
4 Counting colonies of cells and preparing graphs of results; depends upon number of individual culture vessels per student, can be as little as 15–20 min or as much as 1 hr or more.

Assumed knowledge

As detailed here, the technique requires no special knowledge on the part of the student bar the ability to read a pipette, and on the part of the instructor the ability to use a microscope. Because of the high risk of bacterial contamination of cultures of mammalian cells, however, a higher rate of success will be achieved with students who have previous experience of aseptic or 'clean' techniques and bacteriology.

Theoretical background

In the mid 1950's, Dr T.T. Puck at the University of Colorado was the first to obtain successful clonal (colony) growth from mammalian cells *in vitro*. Since differentiated mammalian cells have a finite life-span (longer or shorter depending upon the tissue), they can only be replaced by the proliferation of precursor cells which then differentiate into functional tissue cells. Failure of cell proliferation in the face of continuing loss of functional cells eventually must result in failure of function of the whole tissue, with consequences which may be disastrous for the animal.

With the exception of a few cell types, the majority of mammalian cells die after irradiation when they attempt to divide; whether at the first post-irradiation mitosis or after several divisions depends to a certain extent upon the size of the radiation dose. This *reproductive death* of the cell occurs at doses only a small fraction of the size of those necessary to disturb the biochemical function of the differentiated cell. Hence, survival of cell reproductive capacity is not only the most important, but also the most radiosensitive functional parameter which can be measured in the mammalian cell. In addition, malignant tumours can only grow and pose a threat to the life of the host animal (or man) if the cells of these tumours can proliferate more rapidly than tumour cells are being lost by random cell death, shedding (desquamation), necrosis due to bad tumour blood supply, etc. The survival of cell reproductive capacity *in vitro* can offer a first-approximation estimate of the potential cure or failure to cure a localised tumour *in vivo*. The HeLa cells used in this experiment were first taken from a carcinoma of the *cervix uteri* in a patient in the USA; but they have now been maintained in continuous culture *in vitro* for approximately 20 years.

Necessary equipment

The numbers given below assume that the experiment is being performed by a class of 20, in which each student inoculates one tissue-culture vessel.

Major equipment

Clean work bench with Formica, glass or similar top which can be sterilised by washing down with methylated spirit. It is best if this bench is in a relatively non-dusty area; a laminar air-flow 'clean bench' as is used in bacteriology and in some high-quality industrial assembly work is ideal.

Incubator capable of maintaining a temperature of $37°C$, either humidified with CO_2 flow (in which case the cells can be cultured in the relatively less expensive Petri dishes) or anhydric (in which case the cells will be grown in sealed flasks or Petri dishes sealed into gas-tight plastic boxes).

Access to an appropriate source of X-rays of 140 kV energy or above, or to an isotopic source of gamma rays such as cobalt-60 or caesium-137. Such equipment is normally present in all hospital radiotherapy departments, but *not* in most X-ray departments. In addition, many university and industrial laboratories have small self-contained gamma irradiators which would be more than adequate for the requirements of this experiment. As the use of all equipment which produces ionising radiations is strictly controlled by statutory regulations, the first approach should be made to the Hospital or University Radiation Protection Officer, whose responsibility it will be to know of the existence and capabilities of all

such equipment. He will be well placed to give further guidance in getting the irradiations carried out.

Initial cell culture

HeLa S-3 clone (Flow Laboratories Cat. No. 0-072 o Bio-Cult Laboratories Cat. No. BCL-643)

Disposable plastic or glass labware

1 sterile 250 ml glass bottle for preparation of media

10 x 30 ml sterile universal containers (Sterilin, Ltd. Cat. No. P128/A minimum order quantity 400 but can be stored indefinitely)

10 x 10 ml sterile disposable pipettes (Sterilin, Cat No. P133, packed 10 per sleeve, but 500 per case)

30 x 5 ml sterile disposable pipettes (Sterilin, Cat No. P134, packed 10 per sleeve, but 200 per case

10 x 1 ml sterile disposable pipettes (Sterilin, Cat No. P106, packing as 10 ml pipettes)

20 Falcon TC Flask, 30 ml with screw cap (Biocul 3012, packed 20 per sleeve, minimum order 5 sleeves)

or:

20 Nunclon TC Petri dishes, 50 x 13 mm (Sterilin P122/TC/V, packed 10 per sleeve, 700 per case)

Media and chemicals

200 ml basal medium (Eagle) for Flow HeLa cell (Flow 1-010D, 100 ml) or minimum essentia medium for Biocult HeLa cells (Biocult BCL-100c 100 ml)

100 ml calf serum (Flow 4-020D, 100 ml, BDH 59001)

20 ml glutamine (Flow 6-134B or Biocult BCL-405a

20 ml penicillin/streptomycin (Flow 7-010B or Bio cult BCL-265a)

100 ml trypsin solution 1:250 (BDH 59010)

100 ml Leishman's stain (BDH 35022)

100 ml isotonic NaCl ($0·85\%$ w/v),

Tri-sodium citrate, Distilled water

Miscellaneous

Methylated spirit and clean gauze squares for steriliz ing work bench

Small bench centrifuge (e.g. Gallenkamp Junior CF405) with 50 ml buckets

1 Haemocytometer, to BS 748 (e.g. Gallenkamp MD-200) and access to microscope with approx x 100 overall magnification for counting cells

1 'magic marker' felt-tip pen with non-water-soluble ink for identifying culture flasks

pH paper for adjusting pH of distilled water with tri-sodium citrate when staining colonies

1 magnifying glass of the 'reading glass' type for viewing the colonies of cells
Vichy-water tablets

Type of group
The initial handling of the culture (trypsinisation to remove the cells from the culture vessel and cell counting) must be dealt with as a demonstration by the instructor. Thereafter, plating of the cells can be done in smaller groups (e.g. groups of 4 students) using as many 'clean' bench areas as are available. The staining and scoring of colonies of cells can be done individually or again in small groups.

Instructions to the student
Trypsinisation and cell counting
The instructor takes the culture of HeLa cells and, working on the 'clean' bench, tips the growth medium out of the culture bottle into a waste beaker or other vessel. Using a sterile 10 ml pipette, this volume of the trypsin solution is transferred into the culture vessel. The culture is then placed with the trypsin solution in contact with the cells (which will be attached to one side of the vessel) in the $37°C$ incubator for 5 min. The vessel is shaken vigorously to remove all the cells from the glass, and the trypsin solution containing the cells is poured into a disposable universal container which has already had 10 ml of the complete growth medium (cf. below) pipetted into it under sterile conditions. The universal container is then centrifuged at 1500 rev/min for 5 min, and the supernatant poured off to leave a small pellet of cells in the bottom of the container. 10 ml of complete growth medium is pipetted into the container using another sterile pipette, and the container capped and shaken to resuspend the cells. Using a sterile pipette, 1·0 ml of this suspension is withdrawn for counting. This 'stock' cell suspension should then be kept capped and sterile. A drop of the suspension for counting is placed in the two haemocytometer chambers. The haemocytometer is used as though for counting white blood cells; under the microscope all the cells are counted in each of ten large squares (5 of the 9 in each chamber), and the sum is the number of *thousands* of cells per ml in the stock cell suspension. Most fully grown HeLa cell cultures will contain about 5–10 million cells per bottle, so that the cell density in the stock suspension should be around 500 000 to 1 000 000 per ml, and the number of cells counted in ten large squares should be approx. 500–1000. If the cell suspension is more crowded than this it becomes difficult to count, and it may be simpler to make a 1:10 dilution of the stock suspension into isotonic saline before counting (remember, however, to add the extra factor of 10 back when calculating the number of cells in the stock suspension!). A 'diluted stock' suspension should now be prepared under sterile conditions by appropriate dilution of the 'stock' suspension into another universal container, so that the final cell concentration is *40 000 cells per ml*. For example, if the 'stock' suspension contains 1 000 000 cells per ml, a dilution of 1:25 will be needed—this is achieved by adding 0·4 ml of the 'stock' suspension to 9·6 ml of complete growth medium. This 'diluted stock' suspension will be used by the students in preparing the final dilution for plating.

Plating
Each group of 4 students is provided with one universal container, four tissue culture flasks or Petri dishes, one 10 ml, one 1 ml and four 5 ml disposable sterile pipettes. One of the students takes the universal container and the 10 ml pipette to the instructor's 'clean' bench and transfers 19 ml of complete growth medium to the universal container under sterile conditions; another student uses the 1 ml pipette to transfer this volume of 'diluted stock' cell suspension to the universal container after having gently shaken the suspension to ensure its uniform dispersion. The universal container ('plating suspension') is taken back to the students' working bench and gently shaken. Each member of the group then takes it in turn to pipette 4 ml of this plating suspension into a tissue culture flask or Petri dish using a new, sterile pipette. The final cell concentration is 100 ml, or 400 cells per flask. If the tissue culture flasks are used, the student is to exhale repeatedly into the flask *through the pipette* for at least 1–2 min before capping the flask tightly. This leaves a moist atmosphere of approx. 5% CO_2 in air in the flask—ideal for *in vitro* growth of mammalian cells. Do not skimp on this step, if there is no additional CO_2 in the atmosphere in the flask, the growth medium will become too alkaline (deep purple-pink) and the cells will die! If Petri dishes are used, either an incubator with a humidified 5% CO_2–air atmosphere will be required, or the dishes will have to be placed in gas-tight plastic boxes (e.g. 'Tupperware' or plastic food containers) into which a Vichy-water tablet has been placed in a small dish of water to add CO_2 to the atmosphere. The correct pH for cell growth is 7·0–7·2, hence the phenol red indicator in the medium should be faintly salmon pink in colour; if too acid it will be yellow and if too alkaline it will be dark purple-pink. The flasks or

dishes should be gently swirled or rocked to spread the cells evenly over the bottom surface, and then placed in the 37°C incubator overnight for the cells to attach.

Irradiation and incubation

Irradiation should be carried out between 16 and 20 hr after plating; if too early, cells are poorly attached and may not have recovered from the damage done to them by trypsinisation, and if too late the cells will have already divided, so that one is irradiating micro-colonies rather than single mammalian cells. Of the 20 flasks (or Petri dishes) into which cells have been plated, 10 will be used as unirradiated controls, and 2 each should be given doses of 200, 300, 400, 500 and 600 rads, respectively. The Hospital or University Radiation Protection Officer should be able to suggest the best method by which this should be done with the apparatus available. In our laboratory, one simple arrangement uses a 140 kV superficial therapy X-ray unit which is aimed upwards. The treatment applicators have a flat perspex face, so the flasks (or Petri dishes) can be placed directly on the applicator and the attached tissue culture cells irradiated from below.

The cultures should then be returned to the 37°C incubator. If Petri dishes are used for the cultures, remember that they must be re-equilibrated with an atmosphere of approximately 5% CO_2 in air before being sealed for incubation after irradiation. Whether flasks or Petri dishes are used, they should be inspected at intervals during incubation and any in which the pH has become too alkaline (growth medium bright purple-pink) should be re-gassed under sterile conditions exactly as was done when the cells were first plated. The cultures should be incubated at 37°C for 14 days.

Staining and counting

When the cultures are removed from the incubator, the growth medium should be poured off gently. Colonies of cells will be seen as white, opaque spots on the bottom of the culture vessel. Each flask (or dish) should be washed gently with 5 ml of isotonic saline to remove residual growth medium, and 3 ml of Leishman's stain then added *after* the saline has been poured off. Allow to stain at room temperature for 5 min with occasional gentle agitation by rocking the flasks, then add to each flask 3 ml of distilled water which has been brought to pH 7 by the addition of tri-sodium citrate, and leave the flasks for a further 15 min. Discard the stain, rinse the flasks gently in

cold tap water and the blue-purple stained colonies should be clearly visible. When the flasks (or dishes) have dried, the colonies can be counted using a hand magnifier. All colonies which exceed 0·5 mm in diameter should be counted.

Recording results

Each group of 4 students should have 4 stained culture flasks (or dishes), 2 unirradiated (control) and 2 which have been irradiated, and will have counted the number of colonies on each. The *plating efficiency* is calculated by dividing the number of colonies in each flask by 400 (the number of cells initially plated), and then multiplying by 100%. For unirradiated HeLa cells the plating efficiency should be around 50% or better. The *survival of cell reproductive capacity* is calculated by dividing the number of colonies in one irradiated flask by the number of colonies in an unirradiated control flask and multiplying by 100%. The results for all students should be combined on a single graph using 2-cycle

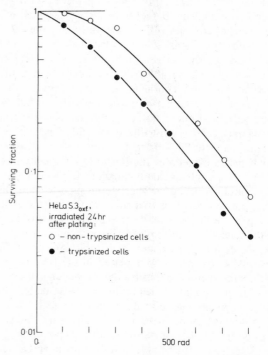

Figure 1 Survival of cell reproductive capacity of HeLa S–3$_{oxf}$. cells irradiated 24 hours after plating. The untrypsinized cells were removed from the culture flask by vigorous shaking only. (From Berry R.J., Evans, H.J. and Robinson, D.M., *Experimental Cell Res.*, 42, 512, 1966.)

semi-logarithmic paper. Radiation dose is plotted linearly along the ordinate, survival using the logarithmic scale along the abscissa. A typical survival curve for HeLa cells is shown in Fig. 1.

Discussion
Note the shape of the survival curve, it has a 'shoulder' at low radiation doses followed by a steeper portion which approximates a straight line. This means that small doses of radiation are less effective in killing mammalian cells than the same small dose *added* to a dose already given. It is believed that much of the damage done by ionising radiation is repaired by mammalian cells so that only after a certain amount of damage is accumulated within the cell does an additional ionising event prove lethal to it. Note that the colonies in the flasks which have been irradiated are smaller than those in unirradiated flasks, and that the larger the radiation dose the smaller the colonies. This is due to two factors; there is a delay in the first cell division after irradiation which is proportional in length to the size of the radiation dose, but as some of these colonies will be the result of up to 12–14 cell divisions this is a minor factor. The major cause of small colonies among radiation survivors is that *heritable non-lethal damage* done to the cells results in their being permanently rendered slow-growers. The proportion of surviving cells showing this permanent change in growth rate is also dependent upon radiation dose and increases as the dose becomes larger.

Extensions and applications
Although the technique of clonal culture of mammalian cells was first devised to study the survival of cell reproductive capacity after exposure to ionising radiation, the method has now been more widely used to study the effects of chemo-therapeutic agents which might be effective in the treatment of human cancer, to study the effects of chemical pollutants present in the environment, and to study the interaction of radiation and chemical damage. The complete human being is a rather more complex organisation than is a flask of cultured mammalian cells, but the *in vitro* culture techniques have developed into a relatively quick and inexpensive first-stage 'screening' method for a number of applications. Only studies which show particular promise are then extended to require the use *in vivo* of animal experimental systems which are costly, often slow to yield the required information, and against which ethical arguments are often raised. *In vitro* tissue-culture techniques will probably never replace the use of animal

experimental systems simply because the complete animal is more than a 'bag of cells' of different types, and because it is well-nigh impossible to simulate *in vitro* all the variables relevant *in vivo*—even if we believed that we knew them all! However, clonal growth of mammalian cells *in vivo* is a valuable tool for biochemistry, pharmacology, toxicology and radiobiology.

Further reading
Cell-culture techniques:
Paul, J.
> 1961 'Cell and Tissue Culture', E. & S. Livingstone, Edinburgh

Biological effects of ionising radiation:
Abraham, E.P. and Berry, R.J.
> 1970 'Some Biological Effects of Radiant Energy' *and* 'Some Effects of Radiation on the Higher Animals', *in* 'General Pathology', 4th edn., Lord Florey, Ed., Lloyd-Luke, London, pp. 781 and 804

Pizzarello, D.J. and Witcofski, R.L.
> 1967 'Basic Radiation Biology', Henry Kimpton, London

Coggle, J.E.
> 1971 'Biological Effects of Radiation', Wykeham Science Series No. 14, Wykeham Publications, London

Hints for the teacher
Ordering of initial cell culture
When ordering the culture of HeLa cells, notify the supplier of the date on which it is proposed to do the experiment, and the approximate number of cells wanted (10 million). This will enable the supplier to deliver the culture at an appropriate stage of growth. The culture should be kept in the 37°C incubator from the time it is received, and should have at least 24 hr to 'settle down' in the instructor's laboratory before it is used for the experiment.

Preparation of complete growth medium
This should be done before the class session. Take the 250 ml glass bottle which has been sterilised and under sterile conditions transfer to it the following:
164 ml 1 x Basal Medium (for Flow HeLa Cells) or 1 x Minimum Essential Medium (for Biocult HeLa Cells)
30 ml calf serum
4 ml penicillin/streptomycin solution (5000 IU/ml)
2 ml glutamine solution (200 mM)
This complete growth medium should be stored in a refrigerator at 4°C and used within a few days of

preparation. The stock solutions should also be stored in a refrigerator and should not be kept longer than one month; Calf serum and trypsin solution can be kept longer if deep-frozen but deteriorate if frozen and thawed repeatedly. If it is intended to keep these, it is best to divide the solutions into aliquots in sterile universal containers and then freeze these, so that only the volume to be actually used in the experiment need be thawed out. Packets of disposable labware from which only a portion of the items have been used should be resealed with sellotape, but the likelihood of items not separately wrapped remaining sterile for prolonged periods is relatively poor.

ROGER J. BERRY
Oxford.

9 Cell killing by ionising radiation as demonstrated by fern spores

Time required
3 hr on day 1; 2 hr on day 2, 3 or 4; 3 hr on day 20, 21 or 22.

Assumed knowledge
How to use the filter pump and the dissecting microscope. Elementary knowledge of fern life-history and growth of prothalli.

Theoretical background
None (but see 'Discussion' and 'Extensions and applications').

Necessary equipment
Spores of *Osmunda regalis* (see 'Hints for the teacher: Collecting fern spores'.)
Filter papers:
 Membrane filters Millipore AAWG 04700, pore size 0·85μm, gridded (Millipore)
  (One filter is required for each spore sample)
 Whatman No. 17, 5 cm filter pads to use in the filter pump.
Petri dishes, 5 cm diameter, glass or plastic. Plastic can be either Escoplastic sterile disposable Petri dishes (2in) in packs of 10 in cartons of 300, grade A (Esco) or Sterlin 13 x 50 mm unvented (Sterlin).
Agar: Ordinary bacteriological Bacto-agar is suitable (Difco).
Graduated pipettes to deliver, 1 to 10 ml
Laboratory, bench or growth chamber where the culture dishes can be placed during growth period. Daylight or ordinary room lighting is sufficient. Direct sunlight should be avoided. If the temperature can be controlled it should be between 19 and 24°C. If the temperature is uncontrolled, large temperature fluctuations should be avoided.
Stereoscopic (dissecting) microscopes or low-powered microscopes giving a magnification between x50 and x400.
Filter pump: (not essential; plating method 2 can be used but is less satisfactory for large scale work). A

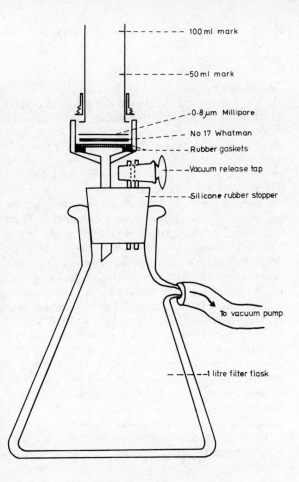

Labels on figure:
- 100 ml mark
- 50 ml mark
- 0·8 μm Millipore
- No 17 Whatman
- Rubber gaskets
- Vacuum release tap
- Silicone rubber stopper
- To vacuum pump
- 1 litre filter flask

Figure 1 Filter-pump set-up.

sketch of the filter pump set-up is shown in Fig. 1. A Gallenkamp membrane filter-holder FD300 accepts 47 mm or 50 mm membranes. A support filter, Whatman No. 17 pad, 50 mm in diameter, is placed over the sintered plate, which is used in the upper position. Any vacuum pump can be used. A good cheap vacuum system is *Mityvac* (Jencons) for about £2·50.

Graph paper (4-cycle semi log) and *squared paper* (about 1 cm squares)

A source of X- or γ-radiation. The dose-rate to a 5 cm Petri dish at a position where the beam is reasonably flat over this area should be measured beforehand by a physicist. A vertical beam is most convenient, and a dose rate between 50 and 500 R/min is suitable. Irradiations should be performed by, or under the close supervision of, a trained operator.

Type of group

A group of 5 could do a complete experiment with controls and 4 radiation doses each in duplicate, and would need one microscope used successively on days 2, 3, or 4 and on days 20, 21 and 22. Alternatively, a group of 8 with two microscopes could do a bigger experiment (controls and 7 radiation doses) and get a better survival curve. Three is the minimum group.

Instructions to the student

Preparation of agar plates

About 18 ml of agar-water gel are required for each 5 cm Petri dish. Add 1·5 g agar to each 100 ml water, allow to soak for 10−15 min, then autoclave at 10 lb for about 20 minutes. Allow to cool to 50−60°C and pour into the Petri dishes to a depth of about 9−10 mm. Cover and allow to cool to room temperature before use.

Plating, Method I (with filter pump)

The vacuum should run continuously and be controlled by the vacuum-release tap. Make sure the vacuum release tap is open and then place a No. 17 Whatman filter pad on the rubber gasket of the filter pump. Place a Millipore filter on the filter pad with the gridded side up. Put about 50 ml of water into the funnel and add 1 ml of the spore suspension prepared as described later.

Plating, Method II (no filter pump required)

Put some spores into a clean 5 cm glass Petri dish, with a cover. Shake. Remove the lid to a dish with agar and a Millipore filter, and tap the lid. Check the number of spores which have fallen on to the filter and repeat until the required density is obtained (about 10 spores per square.)

Preparation and handling of spore suspension for Plating Method I

Weigh out 2−5 mg of spores as required. Put them into a volume of water such that 0·1 mg spores is suspended per 1 ml of water. Soak for 10−30 min with intermittent gentle shaking. Take 1 ml of the suspension while it is being shaken and transfer to about 50 ml of water in the funnel of the filter pump, using a graduated fast-flow blow-out pipette. Rinse the pipette by sucking up and blowing out some water. Close the vacuum release tap and allow the water to be sucked through the filter paper. When the water has almost gone, open the vacuum tap, transfer the filter paper (on which the spores are now deposited) to a Petri dish containing agar. Just before putting the filter down, place two or three drops of

213

distilled water on the agar surface.

Irradiation

Expose samples (on the day of plating) to X- or γ-radiation, under the direction of the operator of the machine. Samples should be covered during exposure. Make certain each dish is identified so that the radiation dose received by it is known.

Culturing

The covered Petri dishes should be placed in a lighted plant culture box or left on a bench in natural daylight or room lighting. Avoid direct sunlight and wide temperature changes.

Counting the spores

1–3 days after plating, count the numbers of spores on each plate, using a dissecting or low-powered microscope at a magnification which allows a whole square of the gridded filter paper to be seen at one time. Make a mark at the edge of the filter paper to indicate the top. Draw a circle of appropriate size (14 squares in diameter) on squared paper and enter the number of spores in each square to correspond with the plate.

Scoring for survival

20 to 22 days after plating and irradiation, count the number of surviving prothalli on each plate. Use a low-powered microscope at the same magnification as for counting, or a higher one for close examination of individual prothalli.

A survivor is defined as a prothallus having 10 or more cells in the apical meristem after 21 days' growth. A normal prothallus of this age is heart-shaped and more or less upright. The meristematic cells are small and form the top edge and notch region of the heart. Examine the control (unirradiated) sample first and note the appearance of the largest prothalli. They will probably each consist of several hundred cells altogether, and the meristem may contain up to 20 cells on each side of the notch.

1 *Controls.* A count of all the spores and sporelings on the control plate will reveal:
(a) The number of spores which have not germinated, i.e. have not opened nor produced a rhizoid
(b) The number which have germinated but remain small and do not qualify as survivors
(c) The number of survivors

The total of these three groups should equal the total counted on the plate as described previously.

2 *Experimental plates.* Score each experimental plate in the same way. Depending on the dose of radiation, there will be fewer survivors and the prothalli that qualify as survivors will have fewer cells.

Recording results

On the control unirradiated plate and on the plates which were exposed to radiation, determine the number of survivors as a percentage of the original number counted on day 2 and 4 minus the number which did not germinate. Put the result for the control plate at 100% and express the results for the irradiated plates as a fraction of the control. Using the lower 3 cycles of 4-cycle semi-log paper, plot the log of each result against the dose of radiation (linear scale) used for that plate. The resulting survival curve will yield an approximately straight line at the higher doses, but will have a 'shoulder' region at the lowest doses. Draw the best line of this shape to fit your results. Extrapolate the straight portion backwards to intersect the zero-dose axis. The point of intersection is known as the *extrapolation number* (N). Where the line you have drawn crosses the 100% survival level gives the value of the *quasi-threshold dose* (D_Q). The slope of the straight-line portion of the curve is described by D_{37}. Its value can be found graphically be determining the radiation dose required to reduce the cell population to 37% of its initial value *on the straight-line portion of the survival curve.*

From the survival curve you have plotted, determine the values of N, D_Q and D_{37} for your spore sample.

Discussion

Mode of cell killing by radiation

Most kinds of cells show little immediate effect of exposure to ionising radiation, but proliferating cell populations always show delayed effects. These appear most obviously when the cells try to divide. Irradiated cells often fail to divide successfully, or if they do, their daughter cells die or fail to divide. This result of irradiation is called *loss of reproductive integrity*: the irradiated cell is killed in the sense that it does not give rise to viable daughter cells. Your experiment is designed to measure this kind of cell killing in fern spores.

Radiobiological considerations

Ionising radiation, when it interacts with matter, deposits its energy in discrete sites along the tracks of ionising particles. In cells or soft tissues energy absorption events are distributed at random. When they occur in a sensitive structure in the cell we can think of them as *hits*. Spores exposed to X-rays (or gamma rays) in your experiment may have received many hits, but whether or not a particular spore has received enough hits to kill it will depend on chance. The chances that it gets killed are of course greater if the radiation dose is greater. But at any radiation dose each spore has a certain probability of survival, not because it was less sensitive to radiation than the others, but simply because, by chance, it did not receive a lethal number of hits.

A useful analogy is to consider a field of corn in a hailstorm. Even if most of the plants are flattened, there is a chance that some will escape, and the proportion which do so will depend on the severity of the hailstorm.

As a result of the quantised manner of energy absorption from ionising radiations, the survival curve is characteristically different from that usually seen after exposure to a toxic chemical agent, or to other damaging physical agents such as heat. It has no true threshold, since even at very low doses there is a chance that some cells will have received a lethal number of hits. At higher doses, more cells will be killed, and those not killed will have received a larger proportion of the hits needed to kill them. At still higher doses, almost all the survivors will need only one more hit to kill them. Therefore at high doses each dose increment of a given size will kill the same proportion of the survivors remaining before that increment was given: in other words, the survivors will decrease exponentially with increasing dose. This gives a straight-line relationship between the logarithm of the surviving fraction and the radiation dose. On this straight-line part of the survival curve, a dose increment such that a fraction equal to 0·37 of the original number survive is known as the D_{37} dose (often called the D_0 dose). Its value gives the reciprocal of the slope of the survival curve. Thus a larger value of D_{37} means a lower sensitivity to killing by radiation.

The 'shoulder' in the low-dose region of the survival curve means that cells can sustain a certain amount of damage without being killed. This damage is called *sublethal* damage. The value of the extrapolation number (N) and of the quasi-threshold dose (D_Q) are measures of the amount of sublethal damage which a cell population can sustain. The parameters D_{37}, N and D_Q define the survival curve.

Question

A population of mammalian cells is found to give a survival curve with $D_{37} = 100$ R and $N = 2$. On semi-log paper, draw a survival curve having these parameters and compare it with the survival curve

you obtained for fern spores. Which cell population is more sensitive to killing by radiation? If you exposed the mammalian cells to 300 R, what proportion of them would survive?

Extensions and applications

The principal therapeutic use of ionising radiations in medicine (usually X- or γ-rays) is in the treatment of cancer. Radiation is an effective agent because it destroys the capacity of cells to go on dividing, i.e. their reproductive integrity. To demonstrate this type of cell killing single cells are irradiated and the number (as a fraction of the original population) which retain the ability to grow into clones is determined and plotted as a survival curve. The most useful form of survival curve is a plot of the logarithm of the surviving fraction on the Y-axis against the radiation dose on the linear X-axis.

Colony formation in Petri dishes has been used for many years to observe the growth of bacteria. It is also possible to grow mammalian cells of many types, including human cancer cells, in Petri dishes, and to score their survival in terms of the ability to form clones. Experiments can be performed to test the sensitivity of cells under various conditions of irradiation and to see how different cell types differ in their sensitivity to radiation killing.

This laboratory experiment with fern spores yields the same type of information as experiments carried out in radiobiological studies with mammalian cells. The prothallus growing from a spore is, in fact, a clone of cells. The resulting survival curves are similar to those which are typical of mammalian cells after irradiation. At low radiation doses there is a shoulder region followed at higher doses by a straight (exponential) response. The slope of this exponential region (D_{37}) is a measure of the sensitivity of the cell population to radiation killing. The width of the shoulder (D_Q) is related to the amount of sub-lethal damage which cells can sustain before being killed by radiation.

Further reading
Lawrence, C.W.
   1971 'Cellular Radiobiology', Institute of Biology's 'Studies in Biology' No. 30. Edward Arnold, London, pp. 15—20 (A good discussion of dose-survival curves)
Coggle, J.E.
   1971 'Biological effects of radia ham Science Series, Wykeha London, Chapters 3 and 4 on cell

Hints for the teacher

Collecting fern spores

Osmunda regalis (the royal fern) is a wild species growing usually in the shade and on wet ground in the United Kingdom, Europe and North America. Specimens can be found in many parks and private gardens in the UK and are especially common in the gardens of Victorian houses. (It was a popular plant in the nineteenth century in England.) The spores are produced in May or June. The fertile fronds should be cut when the majority of sori are greyish-green, and those near the tip show brown edges; a hand lens will confirm that sporangia are opening. Lay the fronds in clean sheets of paper for 5–10 days. The green spores will be shed and can be shaken through a sieve (mesh 240) to remove sporangia and other debris. Store the spores in small stoppered glass vials in a refrigerator or deep freeze. They do not retain their viability for long at room temperature.

Other fern species commonly available are less suitable because of smaller spore size and slower germination.

Miscellaneous notes

1 If class time is limited, agar plates can be prepared beforehand.
2 Plating method I (with a filter pump) gives fairly uniform and reproducible densities of spores on the plate, but method II is satisfactory for class work if clumps of spores can be avoided and the density is such that no square contains more than about 20 spores.
3 If class time is short, a part only of the plate can be used by counting and scoring spores in identifiable squares.
4 The optimal size of student groups should be decided after the constraints of the irradiation facility are known. Spores must be irradiated on the day of plating (otherwise their sensitivity to radiation will be higher and more variable). The number of plates which can be exposed on one day will depend on the dose-rate and the size of the radiation field. Each group of students should have at least 4 plates irradiated with different doses (say 2, 4, 6 and 8 kR) and two or more unirradiated (control) plates. This much material would be enough for a group of 3 or 5 students. With 8 students in a group there should be 7 irradiated samples in the range 1 to 8 kR. For counting and scoring at least one microscope for every 2 students is required unless the work can be spread out over 2 or 3 days for each operation.

Background information about the material

The life history of ferns and the general morphology of their gametophytes (prothalli) are described in any textbook of Botany. If students have no botanical knowledge they should be provided with a simple description or referred to an available textbook.

This work was supported by grants from the Cancer Research Campaign and the Medical Research Council.

A. HOWARD
M.V. HAIGH
F.G. COWIE
Manchester.

Section 12

Enzyme Induction

1 Induction of beta-galactosidase synthesis in bacteria

Time required
3 hr laboratory time.

Assumed knowledge
Accurate use of graduated pipettes.

Theoretical background
The best known operon is the *lac* region of the chromosome of *E. coli*. The region is so termed because it controls the production of enzymes responsible for the utilisation of lactose by the cell. Three genes are involved in the process: the Y gene specifies the structure of a permease enzyme which is responsible for the accumulation of lactose in the cell; the Z gene specifies the structure of the enzyme β-galactosidase which hydrolyses lactose; the third gene, the i gene, is a regulator gene which controls the synthesis of a diffusible product which is thought to act on a fourth gene (o gene) called the operator gene, and thereby regulates the synthesis of the two structural genes, Y and Z.

In this experiment we will utilise four strains of *E. coli* termed EMG 7, 8, 9 and 10 and having the genetic form of $i^+Z^+Y^+$, $i^+Z^-Y^+$, $i^-Z^+Y^+$, and $i^+Z^+Y^-$ respectively. These 4 strains are provided in log phase growing in a minimal medium with glycerol replacing glucose as an energy source. We will expose these bacteria to enzyme inducers and then assay for the presence of β-galactosidase.

The assay involves the use of a colourless substrate, o-nitrophenyl-β-galactopyranoside (ONPG), which is hydrolysed by the enzyme to give a yellow product, o-nitrophenol.

Necessary equipment
For each bench
1 water bath at 37°C
1 water bath at 28°C
For each pair of students
Ice bath i.e. ice in plastic trays
12 plugged sterile test tubes in which to run the experiment. To each will be added 5 ml medium

217

M9 glycerol (plus leucine, threonine and thiamine, as appropriate)

12 test tubes in rack

10 ml 2% glucose

5 ml ONPG solution (Sigma)

10 ml 1M Na_2CO_3 (= 143·07g in 500 ml)

5 ml IPTG solution ($5 \times 10^{-3}M$) isopropylthiogalactoside (Sigma)

12 x 1 ml graduated pipettes (Sterile)

2 x 10 ml graduated pipettes

Toluene in dropping bottles

Sterile distilled water (10 mls)

In the laboratory

2 litres M9 glycerol medium

Marbles for capping test tubes

ONPG solution:

o-nitrophenyl-β-galactoside

Make up 100 ml phosphate buffer (pH 7·0) as follows:

60 ml $Na_2HPO_4 . 2H_2O$ (11·876 g/litre)

40 ml KH_2PO_4 (9·078 g/litre)

In this dissolve the following, to give the required molarity:

0·001M $MgSO_4$ (= 0·247 g/litre)

0·0002M $MnSO_4$ (= 0·447 g/litre)

0·1M mercaptoethanol (= 7·8 g/litre) (BDH)

To every 100 ml of this solution add 0·5 g ONPG, heating as little as possible to dissolve the ONPG, cooling immediately and storing in the refrigerator.

M9 glycerol media

Na_2HPO_4 anhydrous 60 g

KH_2PO_4 anhydrous 30 g

NaCl anhydrous 59 g

NH_4Cl anhydrous 10 g

Water to 1000 ml

Dissolve in order and autoclave.

To each 100 ml of this salt solution add 10 ml of 20% glycerol,

10 ml 0·1M $MgSO_4$

10 ml 0·01M $CaCl_2$ and sterile water to 1000 ml. Each solution should be sterilised separately before final mixing.

To this M9 medium should be added the following to permit growth of the particular strains.

EMG 7, 8, 9—thiamine at 20 μg/ml (BDH)

EMG 10—thiamine, threonine and leucine at 20 μg each/ml (BDH).

Bacteria

E. coli EMG 7, 8, 9, 10. These four strains can be obtained, growing on nutrient agar, from National Collection of Industrial Bacteria, Aberdeen. They must be obtained and subcultured some days before the experiment.

Type of group

Pairs of students in large class, or individual students in small class.

Instructions to the student

You are provided with a set of 12 test tubes, which should be labelled and filled according to the table. Each tube should initially be given 0·2 ml of the appropriate bacterial culture and made up to 5 ml with the minimal medium (note that a different medium is required for 10 than for 7, 8, and 9). Incubate at 37°C for 1 hr. Then to appropriate tubes add either 0·6 ml IPTG plus 0·6 ml sterile distilled water, or 0·6 ml IPTG plus 0·6 ml glucose or 1·2 ml water. Place the tubes in the water bath and incubate for a further 45 min to allow induction to take place. A marked set of 12 test tubes in an ice bath should now be ready, each containing 1 drop of toluene. Add to each tube 1 ml of the bacterial culture, shake vigorously, and return the tube to the ice bath. When all the samples have been collected, the tubes are incubated in the water bath at 37°C for 30 minutes precisely, with occasional shaking. They are then transferred to a 28°C bath, and 0·2 ml of ONPG reagent added to each. Add the ONPG in an ordered sequence of tubes so that the reaction can be stopped in the same order. Allow to incubate for 30 min and then stop by adding 0·5 ml of 1M Na_2CO_3 solution to each tube. The intensity of the yellow colour, indicating the phenol concentration, can be measured at 420 nm.

N.B. IPTG is isopropylthiogalactoside, which is a potent inducer of the operon; glucose should inhibit the reaction. Why?

Note that strict sterility must be maintained if the exercise is to work. Contamination of one strain by another, or of one solution by another, is fatal to the result.

| | Mutant | | | |
Supplement	EMG7	EMG8	EMG9	EMG10
H_2O	1	2	3	4
Glucose and IPTG	5	6	7	8
IPTG	9	10	11	12

Recording results

Tabulate class results and account for any individually aberrant experiment.

Discussion

How does the inducer work, and what is the effect of added glucose on the induction process. Explain the performance of each strain of bacterium.

This experiment has formed the theoretical basis for much of our understanding of genetic regulation in bacteria. It is the classical operon.

Further reading

Watson, J.
1965 'Molecular Biology of the Gene', W.A. Benjamin, New York
Jacob, F., and Monod, J.
1961 J. Molec. Biol, 3, 318

Hints for the teacher

It is imperative to grow the strains of bacteria in the minimal media for some two days, shaken, at 37°C, to ensure that they are adapted to the media. They should be subcultured in the media at least twice to prevent any carry over of other substances from the nutrient agar in which they will originally be obtained. The colour reaction in this experiment is normally sufficiently convincing without recourse to the use of the colorimeter at 420 nm.

This experiment was adapted from the procedure suggested by Hayes, W., in 'Experiments in Microbial Genetics', Blackwell, London 1968.

N. MACLEAN
Southampton.

2 Gibberellin-induced amylase activity in barley endosperm.

Time required

This experiment involves a sequence of operations (see 'Instructions to the student') which are set out below in order together with an indication of the time each operation requires:

	Operation	Time required
1	Cutting of half-fruits	15 min
2	Sterilisation of half-fruits and transfer to moist sand	24 hr
3	Pre-incubation in sand	24 hr
4	Preparation of concentration range of Gibberellic acid (GA)—this can be carried out during operation (3)	45 min
5	Transfer of half-fruits from sand to GA concentration range	15 min
6	Incubation of half-fruits in GA concentration range	24 hr
7	Assay of amylase activity	1 hr

The whole experiment runs for approximately 50 hr and the action is best performed in three one hour periods on successive days. It is possible to fit the experiment into a one hour period separated by any convenient long period from a 15 minute period separated by 24 hr from a final 1 hr period. To do this the students should first perform operations (1) and (4); operations (2) and (3) can then be performed by technical staff so that the students continue from the end of operation (3) at an appropriate time for the students to complete the experiment.

Assumed knowledge

Simple principles of aseptic transfer technique and use of colorimeter.

Theoretical background

Early in the mobilisation of the carbohydrate reserves of the barley endosperm there appear hydrolytic enzymes (e.g. α-amylase) which are not present in any significant amount in the dry fruit. The site of synthesis of these enzymes is the aleurone layer of the endosperm and this experiment provides evidence that it is stimulated to perform this function by a chemical messenger of the 'gibberellin' type which has its origin in the embryo. This experiment shows that an exogenous supply of GA is required to induce the synthesis of α-amylase in fruits from which the embryo has been removed.

The amylase assay used in the experiment involves the incubation of a standard amount of starch for a standard time with the solution containing amylase activity. The amount of starch remaining at the end of the incubation period, which can be measured colorimetrically utilising the starch-iodine reaction, is inversely related to the activity of the enzyme.

Necessary equipment

Barley, *Hordeum vulgare*. The variety 'Graupenkorn' has been used in these experiments because of the ease with which it can be dehusked, and small amounts of this variety can be obtained from the Department of Botany, School of Biological Sciences, University of Leicester, in order to build up a stock. Other more readily available varieties of barley can be used although they may be a little less convenient to handle and may respond to different concentration ranges of GA.

For each student: 80 barley fruits ('seeds')
1 solid backed razor blade
2 containers for halved fruits (e.g. plastic Petri dishes)
2 x 250 ml conical flasks each containing 100 ml distilled water
'Parafilm' to cover flasks
2 x 1g portions of sodium hypochlorite (BDH)
500 ml sterile distilled water
Forceps
Bunsen burner
5 Petri dishes containing sterile, moist, washed sand
7 screw top vials—10 ml capacity
Some means of shaking vials at 25°
10 ml 2×10^{-6}M Gibberellic acid GA$_3$ (BDH)
10 ml 1×10^{-3}M acetate buffer, pH 4·8, containing 1·0 mg/ml streptomycin sulphate (Glaxo)
40 ml distilled water
1 x 10 ml graduated pipette
4 x 1·0 ml graduated pipettes
2 x 0·1 ml graduated pipettes
12 test tubes to contain at least 10 ml
15 ml starch solution (6·7 g soluble starch and 8·16 g KH$_2$PO$_4$ per 1000 ml)
15 ml iodine solution (0·6 g Kl and 0·06 g I$_2$ per 1000 ml of 0·05M HCl)
Colorimeter with red filter.

Type of group
Individual students in classes of any size.

Instructions to the student
Preparation, Sterilisation and Pre-incubation of 'Half-fruits'.
Use the razor blade to cut 75 barley fruits in half as indicated in Fig. 1.

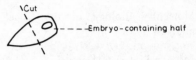

Figure 1

Discard all but 12 of the embryo-containing half-fruits and place them in the container provided. Retain all the embryo-less half-fruits and place them in the container provided.

Sterilise both kinds of half-fruits by soaking for 15 minutes in freshly prepared 1% sodium hypochlorite solution, then rinse them thoroughly with sterile distilled water. Using aseptic technique transfer all twelve embryo-containing half-fruits to one Petri dish containing moist sterile sand and distribute the embryo-less half-fruits amongst the remaining 4 dishes. These Petri dishes will then be incubated for 24 hr, after which the half-fruits should be transferred—again using sterile technique—to screw-topped vials containing various concentrations of GA prepared as detailed below.

Preparation of GA solutions
Using the stock solution of GA (2×10^{-6}M), acetate buffer (1×10^{-3}M containing 1·0 mg/ml streptomycin sulphate) and distilled water, prepare vials containing the following concentrations of GA in 2·0 ml 5×10^{-4}M acetate buffer. Each vial should also contain 0·5 mg streptomycin sulphate per ml.

Vial	GA concentration(M)
1	0
2	0
3	1×10^{-6}
4	5×10^{-7}
5	1×10^{-7}
6	5×10^{-8}
7	1×10^{-8}

It will be necessary to dilute the GA stock solution tenfold in order that the volumes required to give the lower concentrations can be accurately measured.

Incubation of half-fruits in GA solutions

Using sterile technique transfer 10 embryo-containing half-fruits from the sand in which they have been incubated to vial 1. Transfer 10 embryo-less half-fruits to each of the remaining 6 vials, screw caps loosely on the vials and place them on a shaker where they will be incubated for 24 hr. When the incubation is complete, assay the solutions surrounding the half-fruits for amylase activity.

Assay for amylase activity

Withdraw 0·1 ml of each of the solutions surrounding the half-fruits and transfer to test tubes containing 1·0 ml of starch solution in phosphate buffer and 0·9 ml water. Mix thoroughly then incubate at room temperature for *exactly* 10 min, then add 1·0 ml of iodine solution. Add 1·0 ml of water to each tube, mix and measure the intensity of colour in the colorimeter with a red filter using a water blank to set the instrument to zero.

Recording results

Draw a histogram of colorimeter reading against GA concentrations.

Discussion

In the course of interpreting your results comment on the following:

1 The relationship between GA concentration and amylase activity
2 The role of the embryo in the development of amylase activity
3 Experimental methods which could be used to further locate the site of enzyme production
4 The preliminary incubation in moist sand
5 The presence of streptomycin sulphate in the incubation media
6 Use of a red filter in the colorimeter
7 Any control assays you performed.

Extensions and applications

Experiments of this kind have shown that the site of synthesis of α-amylase is in the 'aleurone layer' of the fruit and the system has been further studied with the aim of elucidating the way in which the hormone GA interacts with the aleurone layer to induce enzyme synthesis.

The present experiment also illustrates that there is a quantitative relationship between the amount of enzyme synthesis induced and the concentration of gibberellic acid to which the fruits are exposed, and in fact the induction of amylase activity has been used as the basis of a 'bioassay' for gibberellins.

Further reading

Paleg, L.

1964 'Cellular Location of the Gibberellin-Induced Response of Barley Endosperm', Régulateurs de la croissance végétale', Colloques internationaux du centre national de la recherche scientifique; No. 123, 302–17.

Jones, R.L. and Varner, J.E.

1967 'Bioassay of Gibberellins', Planta (Berl.) 72, 155–61.

Hints for the teacher

This experiment should give no difficulty, although occasionally samples may be lost through massive microbial contamination during the incubation of the half-fruits in the GA solutions.

W. COCKBURN
Leicester.

Section 13

Differentiation

1 Embryogenesis in cultures of carrot

Time required

As performed in our laboratory where students are presented with a suspension culture of carrot cells the experiment is set up in 1–1·5 hr and the cultures are examined weekly (each examination takes 20–45 minutes) for the next three weeks.

The exercise can be converted into an extended project if the students are to initiate the callus culture, grow up the callus, initiate and grow up the cell suspension prior to the actual experiment on embryogenesis.

Assumed knowledge

The student must be used to preparing solutions, pouring agar plates and should have done previously simple aseptic manipulations.

Theoretical background

The only necessary background is (1) an elementary knowledge of sexual reproduction—normally the formation of an embryo requires the fusion of male and female gametes to produce a zygote which subsequently develops into an embryo; (2) a knowledge of the morphology of dictoyledonous plant embryos (embryo axis with shoot and root tip, presence of two cotyledons). In certain flowering plants (carrot is the classic example) cells other than the zygote may be induced to undergo the organised sequence of cell divisions followed in embryology to give rise to *adventive embryos* (embryoids). The experiment is to demonstrate this phenomenon and follow stages in the development of the adventive embryos.

Necessary equipment

Each student or pair of students will require the following.

2 tubes containing 10 ml sterile Murashige and Skoog (MS) medium solidified with 0·5% agar and containing 0·1 mg/1 2, 4-D and 2 tubes of the same medium with no 2, 4-D (2, 4-D = 2, 4-dichlorophenoxyacetic acid—a synthetic auxin or growth hormone)

A boiling water bath and a water bath at 40°C

4 sterile serum pipettes (1·0 ml) of sufficiently wide

bore to not block when pipetting the cell suspension

4 sterile plastic Petri dishes (9 cm)

Glass marking pencil to label dishes

Parafilm (Gallenkamp) to seal the dishes

A hand-lens (x10) and a low-power binocular microscope

Incubator at 25°C

A sterile suspension culture of carrot cells prepared as detailed below

A bunsen burner

High-power microscopes, preferably incorporating phase optics, should be available. One microscope per 4 students if they are working individually (one per 8 students if working in pairs).

The *technical staff* require to prepare in advance the suspension culture of carrot cells and the culture medium required to pour the Petri dishes (as above). The experiment is most successful if a newly initiated callus culture is prepared, but alternatively a culture of high embryogenic potential may be obtainable from a laboratory specialising in plant tissue and cell culture.

The technical staff must have the following.

Facilities for culture media preparation and sterilisation

Facilities for and training in aseptic manipulations

Incubator space at 25°C

A horizontal platform shaker fitted with clips for 100 ml wide-mouthed Erlenmeyer flasks (such as the Gallenkamp incubator shaker).

The basic culture medium (MS medium) (Murashige, T. and Skoog, F. (1962), 'A Revised Medium for Rapid Growth and Bioassays with Tobacco Tissue Cultures', Physiologia Plantarum, 15, 473) is as follows:

Constituent	Final concentration per litre
$CaCl_2.2H_2O$	0·44 g
KNO_3	1·9 g
NH_4NO_3	1·65 g
KH_2PO_4	0·15 g
$MgSO_4.7H_2O$	0·37 g
$MnSO_4.4H_2O$	22·3 mg
$ZnSO_4.7H_2O$	11·5 mg
H_3BO_3	6·2 mg
KI	0·83 mg
$CuSO_4.5H_2O$	0·025 mg
$Na_2MoO_4.2H_2O$	0·25 mg
$CoCl_2.6H_2O$	0·025 mg
$FeSO_4.7H_2O$	27·8 mg
Na_2 EDTA	37·3 mg
Nicotinic acid	0·5 mg
Thiamine HCl	0·1 mg
Pyridoxine HCl	0·1 mg
Glycine	3·0 mg
meso-inositol	100·0 mg
Sucrose	20·0 g

Adjust to pH 5·5 with 0·1 N NaOH immediately before autoclaving at 15 lb/in² for 15 min. 0·5% Difco Ion agar is incorporated for pouring plates, 0·8% for callus initiation and maintenance media.

For *callus initiation* and *maintenance* this medium is supplemented by 1·0 mg/l 2,4-D (2,4-dichlorophenoxyacetic acid) (BDH), dispensed into 100 ml wide-mouthed Erlenmeyer flasks (25 ml medium), closed by a double layer of domestic aluminium foil, protected by an inverted small beaker and autoclaved. Callus is most easily initiated from commercially available carrot seed (*Daucus carota*), surface sterilised by 15 min shaking in hypochlorite solution (10 g calcium hypochlorite shaken with 100 ml water and filtered), washed with three rinses of sterile water and germinated in sterile Petri dishes containing two No. 1 Whatman filter circles moistened with water. After 7 days incubation at 25°C, 10 mm segments of the seedling roots (apical 10 mm not used) are excised with a razor blade and transferred individually to the surface of the MS medium + 1·0 mg/l 2,4–D and incubated in darkness or low light intensity (intensity *not* critical—say 50–100 lumens/ft²). After 2 to 3 weeks, portions (each portion should be approx. 200 mg fresh weight) of the callus which develops are separated from the root explant using a suitable spatula and transferred individually to new flasks of medium. One such culture should, after 3–4 weeks incubation, give sufficient callus to initiate 6, 8 or more new cultures. The stock of callus is maintained and multiplied in this way. A high level of asepsis is important at subculture. The student experiment should be performed with a cell-suspension culture (see below) recently derived from a callus which has been maintained in culture for only a limited period (for safety a callus which has not been through more than 3 or 4 transfers) because callus gradually declines in embryogenic potential with its continued growth in culture.

The suspension cultures needed for the student experiment are prepared by transferring a fragment (approx. 200 mg fresh weight) of callus to 25 ml liquid MS medium containing 0·1 mg/l 2,4-D in a

100 ml flask and incubating on a platform shaker (speed 100–130 rev/min) at 25°C in continuous low light intensity. After 24–28 days incubation 2·5 ml of the resulting suspension should be transferred to a new flask of medium and this 2nd passage suspension after 24–28 days incubation should be used by the students.

The subculture of these suspensions is facilitated by using an automatic pipetting unit (Horwell) fitted with a cannula of 1·5 mm bore to exclude any large cell aggregates.

Type of group
Experiment can be given to a large class working individually or in pairs.

Instructions to the student
Procedure to be followed by the student: Melt the agar medium in the tubes by immersing them for the minimum amount of time in the boiling water-bath. Transfer and keep molten in the 40°C bath. It is desirable that the medium should not be above 40°C when the tubes are inoculated with cell suspension.

Using aseptic technique proceed as follows: Remove a tube of medium from the 40°C water-bath. Pipette into it 1·0 ml of cell suspension using one of the serum pipettes provided. Swirl the tube gently to distribute the inoculum. Immediately pour the contents of the tube into a sterile Petri dish again using aseptic precautions. Move the dish by gently rocking about the horizontal to form an even layer of medium. Set the dish aside until the medium has solidified. Seal the Petri dish with a strip of parafilm (this is to prevent drying out during incubation). Label the Petri dish with date and whether + 2,4-D or – 2,4-D. Let the labelling not be conspicuous so that it will not interfere with subsequent observation of the cells. Examine the agar medium with the binocular microscope. Make a drawing to show how the cells and cell aggregates are distributed in a specimen area of the dish (dishes examined without removing the lid). This process is repeated for all 4 dishes. Examine some of the remaining cell suspension in the high-power microscope and make simple outline drawings of the cells and cell aggregates.

The dishes will be examined as described above after 7, 14 and 21 days' incubation in the dark at 25°C. On the last occasion open dishes containing adventive embryos for more detailed examination. On each occasion make drawings of embryos observed—try to depict successive stages in their development.

Recording results
Results will include drawings of stages in embryo development and counts of numbers of embryos per dish.

Discussion
Does the presence or absence of 2,4-D in the plating medium influence embryogenesis? Are the embryos quite discretely distributed or are they associated in clusters? Do you think the embryos arise from isolated cells, very small aggregates of cells, or the larger cell aggregates present, or from all of these? Do the embryos arise simultaneously or do they continue to arise during incubation? Are there size differences in the cells of the original cell suspensions—is there a normal distribution of cell sizes or are there two classes of cells? Is there any relationship between cell size and isolated versus associated cells (cells in the aggregates)? If somatic cells (as here) are to behave like zygotes, do they have to be polarised? How could such polarisation arise in the plated suspension culture you have been using?

Extensions and applications
If tissue cells derived originally from a carrot seedling or a storage root of carrot (embryogenic cultures can be derived in both these ways) are to behave as zygotes, what changes have to occur in these cells? Does this experiment tell us anything about the process of differentiation in plant cells? What do you think is implied by the statement that 'plant cells can retain their totipotency during differentiation'? Would you expect all the embryos produced in this way to be identical (as identical as identical twins)? Can you see any practical application of this phenomenon of embryogenesis from somatic cells?

Further reading
The following references are given as selected further reading:

Steward, F.C., Mapes, M.O. and Mears, K.
    1958 'Growth and Organized Development of Cultured Cells. II. Organisation in cultures grown from free suspended cells', Am.J.Bot., 45, 705
Steward, F.C., Kent, A.E. and Mapes, M.O.
    1966 'The Culture of Free Plant Cells and its Significance for Embryology and Morphogenesis', in 'Current Topics in Developmental Biology', Academic Press, New York
Reinert, J.
    1973 'Aspects of organisation—organogenesis and embryogenesis' pp. 338-354, in 'Plant Tissue and Cell Culture,' Ed. H.E. Street, Blackwells, Oxford

Halperin, W. and Jensen, W.A.
    1967 'Ultrastructural Changes during Growth and
    Embryogenesis in Carrot Cell Cultures', J. Ultra-
    struct. Res., 18, 428

Hints for the teacher
This experiment as presented to the student is simple
and yields consistent results. The success depends on
obtaining an actively growing cell suspension culture
free from contamination. Care must be exercised in
the preparation of the culture medium, in the
cleanliness of all glassware and in rigorous asepsis.
Incubator temperature must be controlled at the
temperature specified, even at times of high ambient
temperature.

H.E. STREET
Leicester.

2 Hormone-controlled induction of organogenesis in a tobacco callus.

Time required
The experiment is in two parts:
1 the induction of the callus from a segment of
   tobacco stem;
2 the induction of organogenesis using the resulting
   cultured callus.
Part (2) can be performed by the students alone if a
stock callus is provided to them. In calculating the
student time, it is assumed that students are provided
with tubes of sterile medium—if this is done the
experiments can be given to a large class (we have
carried them out with a class of 100 students). The
minimum time for Part (1) is one period of about
1 hr followed in 14 days time by a second period of
½ hr. If the callus initiated in Part (1) is used in
Part (2) there must now be an interval of approx. 28
days. Part (2) requires one period of 1 hr with an
opportunity to look at the cultures every 7 days for a
few minutes until on the 4th occasion 0·75 hr is
required to make drawings. By providing a stock
callus for Part (2) both Part (1) and Part (2) can
proceed simultaneously and the whole experiment is
completed in 4 weeks.

Assumed knowledge
Simple principles of aseptic transfer technique and
some practice in carrying out such transfers with
dexterity.

Theoretical background
The students are repeating a 'classical' experiment
which was the key to the discovery of the cytokinins.
They should know the history of this experiment and
have been introduced to the concept that organogenesis is controlled by an interaction between
natural plant growth-regulating hormones and that
this interaction also controls whether it is root or
shoot primordia that are initiated (basic reference:
Skoog, F. and Miller, C.O. (1957), 'Chemical Regulation of Growth and Organ Formation in Plant Tissue
Cultures *in vitro*', Symp. Soc. Exp. Biol., 11, 118).

225

Materials and equipment

Part (1): Initiation of a callus from a piece of tobacco stem

Each student will need the following.

A sterile Petri dish containing 2 x 10 mm long cylinders of surface sterilised tobacco stem (*Nicotiana tabacum* var. Wisconsin 38)

3 x 2·5 cm diameter test tubes containing 20 ml sterile solidified culture medium (MSI medium)

A sharp metal scalpel and appropriate metal forceps

Spirit lamp or bunsen burner

Bottle of ethanol

Small beaker of boiling water

A disposable face mask

Glass writing pencil

A rack for the culture tubes

Optional: a small bench mounted glass or perspex inoculation hood to facilitate the aseptic manipulations.

Tobacco (Wisconsin 38) plants (3–4 ft tall) should be displayed in the laboratory so that students can see the source of their stem cylinders.

14 days later each student requires the same basic equipment plus:

A small long handled metal spatula

Two empty sterile Petri dishes

2 tubes of solidified sterile culture medium (MSM medium).

Part (2): Induction of organogenesis in tobacco callus

Each student will need the following.

A callus culture of tobacco (Wisconsin 38)—either the callus grown up in Part (1) or a stock callus

2 tubes of culture medium—MS + 1·5 mg/l IAA (BDH) + 0·5 mg/1 kinetin (BDH)

2 tubes of culture media MS + 0·5 mg/l kinetin

2 tubes of culture medium MS + 2·0 mg/l IAA + 0·2 mg/l kinetin

The metal spatula and basic equipment (including tube rack) for aseptic transfer used in Part (1). There should be in the laboratory several Petri dishes containing fragments of callus of approx. 200 mg fresh weight so that the students can judge the size of callus piece they are to transfer from the stock culture to the new media provided.

The technical staff are to prepare the surface sterilised stem cylinders for Part (1). Seed of *Nicotiana tabacum* (Wisconsin No. 38) can be obtained from our laboratory or from Professor F. Skoog, Department of Botany, University of Wisconsin, Madison, Wisconsin, USA. Plants should be grown to the required size in the greenhouse. The upper part of the main stem is used. Wash surface with water, followed by ethanol, and then using suitable sterile secateurs cut 2 cm lengths into a clean wide-mouthed bottle. Cover the stem cylinders for 5 min with hypochlorite solution (10 g calcium hypochlorite in 100 ml water, filtered), pour off the hypochlorite and rinse three times with sterile water. Using aseptic technique cut off the terminal 0·5 cm from each end of each segment (sharp scalpel) to obtain 1·0 cm cylinders and place these in sterile Petri dishes (2 per dish).

The technical staff are also to prepare the culture media required. The basic medium (MS) is according to Murashige, T. and Skoog, F. 'A Revised Medium for Rapid Growth and Bioassays with Tobacco Tissue Cultures', *Physiologia Plantarum*, 15, 473, 1962. Its composition and sterilisation is detailed under *Embryogenesis in cultures of carrot (Daucus carota)* (see Section 13.1). For the work here it is solidified with 1% Difco Ion agar.

The callus initiation medium (MSI) differs from the basic MS medium in containing 10 g/litre sucrose, 1 g/litre Difco casein hydrolysate and 2·0 mg/litre IAA (indol-3yl-acetic acid) in addition to the listed components of the MS medium.

The callus maintenance medium (MSM) differs from the basic MS medium in containing additionally 1 g/litre Difco casein hydrolysate, 1·5 mg/litre IAA and 0·2 mg/litre kinetin (K) (6-furfurylaminopurine).

For Part (2) the students receive MSM medium labelled 'Medium containing 1·5 mg/litre IAA + 0·2 mg/litre K' and this medium with the following differences as indicated in the labelling 'Medium containing 0·5 mg/litre K' and 'Medium containing 2 mg/litre IAA and 0·02 mg/litre K'. Preparation of these media is facilitated by preparing (a) stock solution IAA—0·05 g IAA dissolved in 5 ml ethanol and solution diluted to 50 ml with water, (b) stock solution of K—0·025 g K added to 500 ml water in litre conical flask—steam to dissolve, cool and dilute to 1000 ml.

The medium is dispensed in tubes as indicated, each containing 20 ml medium and closed with a gauze-covered plug of non-absorbent cotton protected from drip in the autoclave and dust by a cap of aluminium foil.

Type of group

Students can work individually and the experiment can be performed by a large class.

Instructions to the student

Part (1): Initiation of a callus culture of tobacco

You are provided with a Petri dish containing two surface-sterilised segments of tobacco stem (var. Wisconsin 38). Keeping the segments in the Petri dish and using aseptic precautions, remove the layers external to the cambium from each segment. The outer layers are no longer required, therefore put to one side in the Petri dish. Holding one of the central cylinders in the forceps, cut three tangential sections of thickness approximately 2 mm and of width approximately 6 mm (see Fig. 1A) (the second stem cylinder is a reserve if you feel that one or more of your sections has not been properly cut). Transfer one section to each of the tubes of media provided;

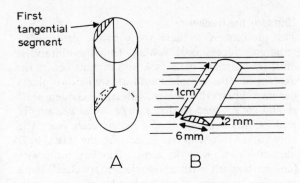

First tangential segment

1 cm

2 mm

6 mm

A   B

Figure 1 A. Diagram of central cylinder of tobacco stem showing the plane of cut of the tangential sections..
B. Tangential section showing dimensions and how placed on the surface of the agar-solidified media (parallel lines indicate medium surface).

the exposed pith should be face downwards on the surface of the medium (see Fig. 1B). All these manipulations should be performed aseptically. The operating end of instruments should be dipped in the ethanol, flamed off, and the hot instrument tips quenched in the boiling water. Only raise the Petri dish lid just sufficiently to operate on the stem cylinders until you come to cut the tangential sections, then lay the Petri dish lid top downwards and work quickly. Replace the lid as soon as the sections are cut. Transfer the sections neatly to the tubes of media holding the plug in the correct manner between the little finger and next finger of the right hand as shown to you by the demonstrator. Flame the neck of the tube before and after inserting the stem section. Take care not to catch the plug in fire. Wear the disposable mask and be careful at no time to touch a sterile surface with your hands.

Label the inoculated tubes of media with your initials or class number. The tubes will be incubated at 25°C in the light. They will be available for inspection in 7 days–note, if necessary making sketches, where callus is developing. After a further 14 days the tubes will again be available and at this stage proceed as follows: Choose the two tubes in which callus development has proceeded best (reject any tubes which show evidence of infection with fungus). Remove the stem sections–one to each of the available sterile empty Petri dishes–using sterilised forceps or a sterile spatula. Using a sterile scalpel excise as large a piece of callus as possible clearly from the stem section. Using the sterile spatula transfer this to one of the tubes of media provided. Repeat for the second stem section. If only one of your stem segments has given rise to sufficient callus for excision or if only one is free from obvious contamination, endeavour to cut two callus fragments so that both tubes of media provided can be inoculated. Label the tubes with your initials or class number. The culture tubes will now be incubated at 25°C in the light and will be available for inspection so that you can monitor the growth of the callus. If you do not use one of these callus cultures for Part (2) of the experiment, you should, after 21 or 28 days, carefully remove the calluses and weigh them and also weigh a piece of callus as near in size as possible to the original callus fragment you excised from the stem section (do this with the spare callus if one is used for Part (2)). Now tease out some of the callus and examine first in a low-power binocular microscope and then in the high-power microscope mounting a small portion in water and gently pressing under a cover-slip.

Part (2) Organogenesis in tobacco callus
You are provided with either a stock callus of tobacco or you will use one of the calluses you have produced in Part (1). Note the composition of the culture medium in each of the 6 tubes provided. MS describes the composition of the basic medium; IAA = indol-3yl-acetic acid; K = kinetin. On the demonstration table are some Petri dishes containing callus fragments of about 200 mg fresh weight. You are to obtain similar size fragments from your callus culture to inoculate the 6 tubes of culture medium (one fragment per tube). The spatula is the instrument to be used. The sterile moist spatula is used to cut away each fragment and transfer it *aseptically* to a new tube of medium. Think out carefully how you are going to organise this. (It is suggested that students work in pairs–you actually do the transfers

but your neighbour unplugs, flames and replugs the tubes which are side by side in a test-tube rack. The tube containing your callus culture will need to be temporarily removed from the rack by yourself to assist in transferring the fragment to the spatula. You then reciprocate the assistance.) Remember the spatula must be resterilised each time a new fragment is transferred. The inoculated tubes will now be incubated at 25°C in the light. Label with your initials or class number. You will inspect the 6 tubes each 7 days for 3 or if possible 4 weeks. Make notes on the development of the culture in each tube. Make a sketch of each culture at the time of the final inspection.

Recording results
Your results should include a measure of the rate of growth of the callus initiated in Part (1), and drawings clearly depicting any organ formation from the calluses which occurs in Part (2).

Discussion
Say how far you have reached any conclusion regarding the site of origin of callus tissue on the stem segments. Is there any tissue in the segment which you might expect to give rise to callus? What is it in the procedure you have followed which causes callus to form? The culture medium used for initiation of the callus contained 2·0 mg/litre IAA and no kinetin, that used to grow the detached callus contained 1·5 mg/litre IAA and 0·2 mg/litre kinetin, why do you think two different media were used for the two stages of Part (1)? What can you say about the growth *rate* of the callus?

Can you reach any conclusion regarding the importance of growth hormone balance in organogenesis from Part (2) of the experiment? How would you proceed to define more precisely the hormone balance promotive of root initiation and that promotive of shoot initiation? Kinetin is not a naturally occurring cytokinin—how therefore do you account for its physiological activity? How would you explain the observation that induction of root and shoot formation from callus cultures of many other species *cannot* be achieved by an appropriate supply of auxin and cytokinin in the culture medium? What technique would you use to study in greater detail the change in organisation within the callus which leads to the formation of organ primordia?

Extensions and applications
It was in studies on the culture of tobacco callus that the cytokinin hormones were discovered—follow this story from the 'Further reading' and consider whether this approach might be used to search for other and as yet unidentified plant-growth-regulating hormones.

Further reading
Skoog, F. and Miller, C.O.
  1957 'Chemical Regulation of Growth and Organ Formation in Plant Tissues Cultured *in vitro*', Symp. Soc. Exp. Biol., 11, 118
Street, H.E.
  1969 'Growth in Organized and Unorganized Systems', in 'Plant Physiology', Vol. VB, Chap. 6 Ed. F.C. Steward, Academic Press, New York

Hints for the teacher
The problem is to have the tobacco plants when required. Successional sowings (monthly intervals) must be started as early as 4 to 5 months before the experiment (the actual time to get well-grown plants depends upon the season). Botanic Garden staff should collect seed so that a single initial seed supply enables this experiment to be done each year. The only other problem is to school the class in a 'dummy' run on good aseptic handling, otherwise (and particularly with a large elementary class!) there may be a high rate of contamination of the cultures.

H.E. STREET
Leicester

3 Induction and functioning of a vascular cambium in cultured excised roots.

Time required
The experiment as described under 'Instructions to the student' requires about 1 hr, 1·5 hr, 0·5 hr, and 3 hr at weekly intervals (4 occasions spread over 3 weeks).

Assumed knowledge
The limited conducting tissues (xylem and phloem) of the seedling root of dicotyledonous plants are supplemented as the plant grows by secondary conducting tissues formed from a secondary meristem (the vascular cambium). The student should have a basic knowledge of root anatomy—primary and secondary. Root cultures initiated from seedling root tips, although capable of continuing growth in culture, retain the primary structure of the initial embryonic root; they do not develop a vascular cambium and secondary xylem and phloem. The experiment is an attempt to induce secondary vascular tissue formation in such cultured roots.

Theoretical background
The conventional root-culture technique exposes the whole surface of the root to a medium from which it absorbs all its nutritional requirements. In contrast, the root in the intact plant obtains only mineral ions via its surface, the remainder of its nutritional requirements (sugar, vitamins, growth hormones) are either synthesised within the root or imported from the photosynthetic shoot system. In the intact plant not only water but inorganic ions and metabolites (e.g. amino acids) are passed from the root to the shoot. Is the development of secondary vascular tissues in the root a consequence of this root-shoot interaction? The root-culture technique developed by Raggio and Raggio[1] is an attempt to approach nearer to the natural situation by making it possible to supply part of the nutritional requirements of the root to the apical region and the remainder via the basal cut surface and adjacent root surface. Essential materials supplied basally will have to be transported to the apical meristem if the culture is to grow. In the

experiment we use a Petri dish version of the Raggio and Raggio technique[2] to study how far the site of application of an auxin (the natural auxin—indol-3yl-acetic acid—IAA) influences whether or not a vascular cambium is formed and, if so, how far it functions to produce secondary vascular tissues.

Materials and equipment
Each *pair of students* will need the following at the beginning of the experiment.
Petri dishes (5) each containing 6 sterile, germinating (36—48 hr from moistening) seedlings of pea (*Pisum sativum* e.g. Meteor variety)
Petri dishes (20) each containing 10 ml solid root culture medium (composition given below)
Small bottle of ethanol
Beaker of boiling water
Scalpel and simple rack to rest scalpel free from the bench
Bunsen burner or spirit lamp
Glass writing pencil.

After 7 days each pair will need:
12 Raggio—Raggio Petri dishes (see Fig. 1). These are of 4 kinds (3 of each):

Label	Petri dish medium	Vial medium
P+V+	+	+
P+	+	−
V+	−	+
P−V−	−	−

where + indicates presence of 10^{-5}M IAA and − indicates no IAA present.
The 20 Petri dishes inoculated 7 days earlier and which have been incubated at 25°C in an incubator
A scalpel and a pair of blunt pointed forceps and the

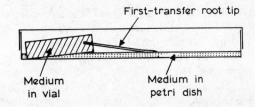

First-transfer root tip

Medium in vial

Medium in petri dish

Figure 1 Diagrammatic section through a Raggio-Raggio Petri dish to show positioning of vial and way root tip is introduced with base immersed in medium in the vial and root tip making good contact with the medium in the dish.

accessories for their sterilisation as at commencement

After 14 days each pair will need:

The Raggio-Raggio Petri dishes for inspection

After 21 days:

The Raggio-Raggio Petri dishes for inspection

Razor blade and pith to assist section cutting

Compound microscope, microscope slides and cover-slips

The reagents phloroglucinol (BDH) and HCl for staining for lignin.

The *technical staff* are required to prepare a modified Bonner and Devirian root-culture medium (as detailed in Torrey, J.G., 'The Role of Vitamins and Micronutrient Elements in the Nutrition of the Apical Meristem of Pea Roots', Plant Physiology, 29, 279, 1954).

Modified Bonner—Devirian root-culture medium (Torrey, 1954)

Constituent	Weight per litre of final medium
$Ca(NO_3)_2 4H_2O$	242 mg
$MgSO_4.7H_2O$	42 mg
KNO_3	85 mg
KCl	61 mg
KH_2PO_4	20 mg
$FeCl_3$	1·5 mg
$MnSO_4.4H_2O$	4·5 mg
$ZnSO_4.7H_2O$	1·5 mg
$CuSO_4.5H_2O$	0·04 mg
$Na_2MoO_4.2H_2O$	0·25 mg
H_3BO_3	1·5 mg
Thiamin HCl	1·0 mg
Nicotinic acid	5·0 mg
Sucrose	60 g
Difco purified agar	5·0 g

IAA (BDH) at 10^{-5}M (MW = 175) is incorporated as required before sterilising. The medium can be sterilised by autoclaving at 15 lb/in^2 for 20 min. It is sterilised in plugged tubes (10 ml medium per tube) and these used to pour the Petri dishes (9 cm glass dishes). The vials (3·5 x 1·2 cm external dimensions) are charged each with 3·5 ml of medium and sterilised upright in crystallizing dishes. These vials are inserted (see Fig. 1) in the Petri dishes after the agar has begun to set in the dishes. Aseptic precautions are observed in pouring the plates and inserting the vials.

Dry pea seeds are sterilised by immersion (usually for 10–15 min) in 0·1% w/v mercuric chloride containing a small amount of a detergent, rinsed three times with sterile distilled water, and transferred to sterile 9 cm Petri dishes containing two No. 1 Whatman circles moistened with 3–4 ml of sterile water. Six seeds are plated per dish (a check should however be made on the per cent germination and normality of germination after the sterilising treatment). Commercial seeds vary in the ease with which they can be effectively surface-sterilised and root tips should be tested for sterility by transfer to root-culture medium and incubation. Seeds are germinated 36–48 hr at 25°C; the radicle should be emergent to a length of 10 mm when given to the students.

Type of group

Applicable to a large group working in pairs. Class results can be compared to study consistency of observations and extent to which Raggio-Raggio dishes showed fungal or bacterial contamination (standard of aseptic technique).

Instructions to the student

You are provided with 20 Petri dishes containing a root-culture medium solidified with agar and dishes containing sterile, germinating pea seedlings. Using a sterile, sharp scalpel (immerse blade in alcohol, burn off, quench in the boiling water) excise root tips (approx. 5·0 mm long) from the seedlings and transfer singly to the surface of the agar medium (1 root tip per dish) using aseptic precautions. The dishes will be removed and incubated at 25°C in darkness until the class next week.

Now (7 days' later) your dishes containing root tips are returned. Select 12 which are free from contamination and in which the root tips have grown. Excise a 15 mm root tip from one of the selected dishes and, using the forceps provided, transfer carefully and neatly to one of the Raggio-Raggio dishes (Figure 1). The basal 5 mm of the tip segment is inserted in the vial and the apical portion of the tip should rest on the agar in the dish. Sterilise the scalpel and forceps and achieve the transfer of the tip aseptically. Repeat this operation until all 12 of the Raggio-Raggio dishes provided have been inoculated and labelled (see 'Materials and equipment'). These dishes will now be incubated for 14 days.

After the Raggio-Raggio dishes have been incubated for 7 days inspect them. Reject any showing obvious contamination. Measure the length of the roots by sight through the lid of the Petri dish (do *not* remove lids). Return to the incubator. When the dishes have been incubated for 14 days, again measure the roots and remove them and cut transverse sections at various points along their length.

Stain the sections in phloroglucinol and HCl (lignified cell walls stain red) to better observe the conducting elements of the xylem. Examine the sections in the microscope and make clear outline drawings to illustrate the anatomy of the roots, paying particular attention to the presence or absence of a vascular cambium and the presence (and number of) or absence of secondary xylem elements.

Recording results

In addition to your anatomical drawings, prepare tables to show root growth and extent of secondary thickening observed in the 4 treatments.

Discussion

What are the general functions of the individual constituents of the basic culture medium used? Why should cultured roots require thiamin and nicotinic acid? Do your results enable you to draw any conclusions regarding the influence of IAA on root growth and on the production of secondary vascular tissues in the root? Since sugar is normally received by the root from the photosynthetic shoot, can you suggest a follow-up experiment to the one you have just completed? Why might one expect a different effect if IAA entering over the whole root surface to IAA being moved within the root from base to apex? What factors do you think are responsible for the variation in growth and vascular cambium activity between the replicate roots submitted to the same treatment in the Raggio-Raggio dishes?

Extensions and applications

Can you think of an improved version of the Raggio-Raggio apparatus you have used? How could the experiments be extended to develop our knowledge of the factors which lead to the induction of division in the cells which form the vascular cambium and which control the differentiation of cells derived from this cambium into xylem and phloem cells? How good a model system is this for studying in plants the hormonal and nutritional control of cell division and cell differentiation?

Further reading

1 Raggio, M. and Raggio, N.
        1956 'A New Method of Cultivation of Isolated Roots', Physiol. Plantarum, 9, 466
2 Torrey, J.G.·
        1963 'Cellular Patterns in Developing Roots', Symp. Soc. Exp. Biol., 17, 285

Hints for the teacher

The snags which have to be watched for are:

1 failure to sterilise the pea seeds or their over sterilisation so that few germinate and germination is abnormal;
2 need for some skill and finesse in transferring the root tips (15 mm) to the Raggio-Raggio dishes, particularly to avoid injury to the tips, bad contact between root and the medium, and contamination.

H.E. STREET
Leicester.

4 Mitochondrial development in yeast

Time required
About 3 to 4 hr.

Assumed knowledge
Use of an oxygen electrode. (Section 10.2)

Theoretical background
This experiment is similar to that entitled 'Respiration and fermentation in yeast' (Experiment 10.1), which should be referred to. However, here an oxygen electrode is used to assay respiration, and an inhibitor of mitochondrial development will be studied.

Under anaerobic conditions yeasts are able to ferment sugars as their energy source and do not develop mitochondria. When transferred to aerobic conditions, mitochondria develop rapidly if the concentration of fermentable substrate is not so high as to repress mitochondrial development. Electron micrographs, cytochrome analyses and oxidative phosphorylation measurements all substantiate the above scheme.

In this experiment you will measure respiratory activity using an O_2 electrode. Yeast cultures will be provided which have been grown under aerobic and anaerobic conditions; the latter cultures will then be aerated in the presence and absence of the protein synthesis inhibitor, chloramphenicol.

Necessary equipment
Growth medium: amounts per litre

Difco yeast extract	5 g
NaCl	0·5 g
$MgCl_2$ $6H_2O$	0·7 g
$(NH_4)_2$ SO_4	1·2 g
KH_2 PO_4	1·0 g
$FeCl_3$	5 mg
Glucose	50 g

Autoclave at
10 lb/in^2 for 5 min

1 Inoculate 250 ml of medium in a litre flask (from a yeast slope of *Saccharomyces cerevisiae* (Natl. Coll. Yeast Cultures; or by plating out baker's yeast), grow at 28°C under forced aeration or with rapid shaking

2 For anaerobic growth, the inoculum is taken from an aerobic, shake culture (at stationary phase) and flushed with O_2-free N_2. Grow with continuous stirring by means of a magnetic stirrer at 25°C, until stationary phase is reached (approx. 16 hr).

Requirements per pair of students
For aerobic growth: Inoculate 100 ml of growth medium with 100 ml of aerobically grown yeast culture, grow at 25°C on shaker
For anaerobic growth: Inoculate 100 ml of growth medium with 100 ml of anaerobically grown yeast culture, grow at 25°C
Inhibitor chloramphenicol: Dilute 100 ml of culture medium (containing 2 mg/ml of chloramphenicol; Parke-Davis) with 100 ml of anaerobically grown yeast culture, grow aerobically at 25°C on shaker
1 complete set of O_2 electrode and recorder
0·1M phosphate buffer pH 7·0 containing 0·1M glucose → 25 ml
10 pipettes
Sodium hydrosulphite (dithionite) (BDH)
Spectrophotometer.

Type of group
Pairs of students seems to be quite suitable.

Instructions to the student
You should measure the rates of O_2 uptake as a function of time aeration of the cultures e.g. at half-hour intervals.
1 Aerobic culture—5% glucose (initially)
2 Anaerobic culture—5% glucose (initially)
3 Anaerobic cultures—5% glucose (initially)
   (a) Exposed to air—glucose diluted 1:1 after 24 hr anaerobic growth
   (b) Exposed to air—glucose (as in (a) + 2·5 mg/ ml chloramphenicol).
Calibrate and check the O_2 electrode before starting the experiments according to instructions in Section 10.2. Measure the density of the yeast suspensions at OD_{540}. Establish the conditions for measuring respiration with the aerobic culture first and then follow identical procedures for the other three cultures at regular intervals. The volume of the reaction vessel is 3–4 ml. Add about 1·5 ml of yeast suspension and 1·5 ml of 0·01M glucose in 0·1M phosphate pH 7, to a final volume of 3·0 ml. You may need to vary these conditions in order to obtain reasonable rates of O_2 uptake; establish your procedure initially. Inhibitors of electron transport, e.g. azide at 10^{-3}M or HCN at 10^{-2}M, can be used to

check the mitochondrial contribution to oxygen uptake.

Recording results
Express data as μmoles oxygen taken up per standard OD_{540} reading per minute. Representative recorder traces (well labelled) and method of calculation of rates should be included in the report.

Discussion
The studies on the development of mitochondria in yeast and *Neurospora* have been significant in establishing many important structural and functional aspects of mitochondria. In addition, cytoplasmic and nuclear control mechanisms of organelle development and protein synthesis have been elucidated using these fungi. Yeast is particularly useful because mitochondrial synthesis can be 'switched on' quickly and easily and its control studied using genetic variants, inhibitors of protein synthesis, uncouplers and inhibitors of oxidative phosphorylation, etc. The student should try to design a more comprehensive experiment to answer these problems, remembering the time and difficulties involved in various approaches.

Extensions and applications
In addition to fungi and bacteria, the technique is generally used in following the development of chloroplasts (from non-green to green state) and the regeneration of oxidative capacity in parasites as they change their environment in their host and thereby their sensitivity to various drugs.

Further reading
Roodyn, D. and Wilkie, D.R.
    1968 'The Biogenesis of Mitochondria', Methuen, London
Schatz, G.
    1970 In 'Membranes of Mitochondria and Chloroplasts', ed. Racker, E., Van Nostrand Reinhold, New York
Smith, D.G., and Marchant, R.
    1968 Biol. Rev., 43, 359
Ashwell, M. and Work, T.S.
    1970 Ann. Rev. Biochem., 39, 251
Linnane, A.W.
    1972 Ann. Rev. Microbiol., 26, 163

Hints for the teacher
The aerobic culture is very easily grown up on the morning of the practical (about 6 hr before required). The anaerobic culture takes a little more effort and care, and should be started about 28 hr before required. The recording oxygen electrodes should be fully checked before the start of the practical. The students should then do a quick calibration of the electrode before starting to use the aerobic yeast.

D.O. HALL
King's College
London.

5 Effect of ions on the transformation of *Naegleria gruberi* from amoeba to flagellate

Time required

The time required for this practical depends on the number of salt solutions investigated by the student. If a number of salts are attempted, 4 hr will be required, but 3 hours is adequate if there is class co-operation. Time required can also be reduced by supplying salts already diluted.

Assumed knowledge

The student should be able to use a phase-contrast microscope (although standard optics can be used) and a haemocytometer. Since the practical requires rapid sampling and systematic working, it is more suitable for experienced students.

Theoretical background

Naegleria gruberi is normally an amoeba which lives in the soil and feeds phagocytically on bacteria. It moves by the production of lobose pseudopodia. If the environment is flooded, and presumably food is scarce, the amoeba can, in a fairly short time (1–2 hr) assume a flagellate form and become motile. These flagellates possess two to four anteriorly directed flagella and are temporary; they do not feed or reproduce.

Willmer[5] investigated the effects of dilute salt solutions on the transformation and proposed that loss of cations from the cell was involved. Results of experiments by Fulton[4] and Jeffery and Hawkins (unpublished) are not in agreement with some of the earlier work.

The amoebae are grown on nutrient agar together with bacteria and can be washed off the surface, concentrated by centrifugation and then the effects of varying dilutions of salt solutions investigated. In order to distinguish between ionic and osmotic effects, solutions of sucrose can be used.

Equipment

Each student will require the following.
1 culture of *Naegleria gruberi* growing on nutrient agar

20 ml of each of the following solutions, all M/40: KCl, NaCl, LiCl, $MgCl_2$, $CaCl_2$
20 ml of M/5 sucrose
100 ml glass distilled water
Test tubes, 4 per solution studied (i.e. 28 for complete experiment)
Graduated pipettes: 6 x 2 ml; 1 x 5 ml; 1 x 10 ml
8 Pasteur pipettes and 8 teats (minimum)
Marking pencil
Water bath running at 25°C with test-tube rack
Bench centrifuge and plastic tubes
Haemocytometer
Phase-contrast (or standard-optics) microscope
Slides and cover slips (at least 24 slides will be required)
Graph paper
Stopclock
10 ml Lugols iodine (4 g iodine, 6 g potassium iodide dissolved in a few drops of water, add water to 100 ml) + Pasteur pipette and teat

Type of group

The experiment should be run by individual students, in large or small classes. If the salt solutions are shared among the class and results collected, all the suggested salts can be investigated.

Instructions to the student

Work carefully: some strains of *Naegleria* have been reported to be pathogenic.

1 If a large class is involved, divide up the salts between batches of students—say 2 each, but *each* student must run a control in distilled water. Using M/40 solutions of the salts, NaCl, KCl, LiCl, $MgCl_2$ and $CaCl_2$, make up a small volume of each (2 ml/test tube) at the following dilutions: M/80, M/320 and M/1280, retaining some salt solution at M/40. The M/5 sucrose solution should be diluted to give M/20 and M/80 samples (glass-distilled water should be used for all dilutions). Set up one tube with 2 ml distilled water only. Remember to label all tubes.

2 When solutions are ready, and only then, the amoebae should be washed off the agar plate with distilled water and centrifuged at 500 g for 5 min. Note the time. Pour off and keep the supernatant (in case the cells have not sedimented), add more distilled water, shake and centrifuge again. Pour off and keep supernatant. Add 1 ml of distilled water to the pellet, shake and take a small drop using a Pasteur pipette for examination under a phase-contrast microscope (x40). If a high density of bacteria still remains,

add more distilled water and centrifuge again. Add 1 ml of distilled water to the pellet, shake and re-examine. Some cysts may be present but will not interfere with the experiment.

3 Take the distilled water control and add a known number of drops (say 4) of amoeba suspension to the test tube using a Pasteur pipette. Sample with another Pasteur pipette (1 drop) and check that sufficient cells are present in the drop (say 30 cells when viewed x 10); if not, add some more cells. Add a standard number of drops of amoebae to each of the diluted salt solutions and place all test tubes in a water bath at 25°C and note time.

4 At known time intervals (15 min) sample all test tubes, preferably using a separate Pasteur pipette for each tube, but with care, one for each salt will suffice. Place drops of suspension on a marked slide and add a drop of Lugol's iodine. The cells are now fixed and can be counted as time permits. The cells may be counted directly or the sample placed on a haemocytometer slide. The numbers of amoebae and numbers of flagellates present in the sample are counted, preferably using a phase-contrast microscope, through which the fixed cells appear orange and flagella are distinct and black. (If phase-contrast microscopes are not available, use dark-field illumination since flagella are not very distinct with bright field. However, the shape of the cells is very distinctive, the flagellates being ovoid or triangular, and if the sample is viewed before fixation, swiftly moving flagellates are easily seen). Continue sampling experiment for at least 90 min. Make drawings of amoebae and flagellates during the experiment.

Recording results

Data should be expressed as the % flagellates in your sample at a known time and known solution. Then graphs can be plotted for each set of salt solutions, compared with the % flagellates in the control, distilled water. In order to compare your results with others in the class, some account must be taken of the varying % of transformation which will occur in the controls. This variation is the result of many factors, e.g. extent to which bacteria were removed from the initial culture, efficiency of sampling, etc. Results can be compared by selecting the time at which the majority of the class obtained transformation in their control experiments, about 75–90 min, and then expressing the % flagellates in salt solutions at this time as a % of those in the control. This latter

% can then be plotted against molarity for each salt solution, all on one graph.

Discussion and extensions

Are there any differences in the molarities at which the amoebae transform, if you compare monovalent and divalent cations? Do your results with sucrose suggest that osmotic effects may be important, and how might this be tested further? Did you notice any tendency to reversion towards the end of your experiment? Try and make a list of the events which must occur during this transformation on receipt of the initial stimulus.

Until recently the transformation of *Naegleria* amoebae into flagellates had been neglected, and yet this cell provides a means of looking at differentiation after the receipt of an initial stimulus. The nature of this stimulus is apparently complex, as are the events which follow. One feature of considerable interest is that the amoebae lack any centrioles or basal granules, and yet the flagellate possesses one basal granule/flagellum (Fulton[2]). Further, the number of flagella produced by the cell may be altered by temperature shock (Dingle[1]). *Naegleria* can be grown on dead bacteria and methods are being developed for their axenic culture, so that there is no doubt that in the future many workers will follow Fulton[2] and select these amoebo-flagellates as research partners.

Further reading

1 Dingle, A.D.
1970 'Control of Flagellum Number in *Naegleria*. Temperature-Shock Induction of Multiflagellate Cells', J. Cell. Sci., 7, 463

2 Fulton, C.
1970 'Amoebo-flagellates as Research Partners: the Laboratory Biology of *Naegleria* and *Tetramitus*', in 'Methods in Cell Physiology', Vol. 4, ed. Prescott, D.M., Academic Press, New York, p. 341

3 Fulton, C.
1971 'Centrioles', in 'Origin and Continuity of Cell Organelles', ed. Reinert, J. and Ursprung, H., Springer-Verlag, Berlin, p. 170

4 Fulton, C.
1972 'Early Events of Cell Differentiation in *Naegleria*. Synergistic Control by Electrolytes and a Factor from Yeast Extract', Develop. Biol., 28, 603

5 Willmer, E.N.
1956 'Factors which Influence the Acquisition of Flagella by the Amoeba, *Naegleria gruberi*', J. Exp. Biol., 33, 583

Hints for the teacher

1 It is important to ensure that each student has an adequate supply of *Naegleria*. Cultures can be obtained from the Cambridge Culture Collection of Algae and Protozoa, 36 Storey's Way, Cambridge, and grown on nutrient agar in association with bacteria. We use one culture from Cambridge to inoculate 2 x 10 cm Petri dishes. The nutrient agar contains, per litre of distilled water, 2 g Difco Bacto-peptone; 2 g dextrose; 1·5 g K_2HPO_4 and 20 g Difco Bacto-agar. The agar is poured into 10 cm glass Petri dishes, using about 12 ml per dish. It is allowed to set and then the Petri dish is wrapped in aluminium foil and autoclaved for 15 min at 15 lb/in^2. The dishes are allowed to cool on a flat surface and may be used when cool, or kept for a short time. [If plastic Petri dishes are used, the medium should be autoclaved first and then poured]. The culture obtained from Cambridge contains bacteria together with *Naegleria* cysts. A little distilled water should be added, the cysts dislodged from the agar slope and pipetted using a Pasteur pipette on to the surface of the nutrient agar. A little liquid should be present on the surface of the plate; they should not be dry. The plates should be allowed to grow at 25°C and are ready for sub-culturing or experimentation after 2 days. One plate should provide enough *Naegleria* plus bacteria to inoculate a further 5 plates. The plates will mature in 1 day at 33°C, but the lower temperature has been chosen deliberately in order to reduce the risk of a pathogenic strain developing. Evidence is accumulating that *Naegleria* can be pathogenic, death occurring after intranasal infection following swimming in tepid water in Australia (Carter, R.F., Trans. Roy. Soc. Trop. Med. and Hygiene, 66, 193, 1972). Most strains grow readily at 37°C but tests carried out by Fulton failed to produce any pathogenic effects in mice. Observations suggest that virulent strains may exist, so that the need for caution when handling these organisms should be emphasised. We have done this experiment for 8 years.

2 *Naegleria* will start to transform as soon as they are removed from the agar plate and washed free of bacteria with distilled water. Thus once the experiment has started, students must spend 2 hours sampling. However, the samples, when fixed in Lugol's iodine, will last 24 hr without damage, providing they are kept in the dark; thus if samples are collected in small tubes and sealed, rather than on slides, they can be counted the following day.

SHIRLEY E. HAWKINS
King's College, London

6 Isozymes of lactic dehydrogenase: electrophoretic separation and colorimetric assay

Time required
This experiment requires at least 3 hr in the laboratory; an additional hour is helpful. The first part of the experiment requires about ½ hr; the electrophoresis apparatus can then be left running for 1 hr, during which time the students can be elsewhere if necessary.

Assumed knowledge
Students should be familiar with the use of a spectrophotometer, while previous use of electrophoresis equipment is helpful but not essential.

Theoretical background
Many proteins have been shown to exist in multiple forms or isozymes, which may vary from tissue to tissue and with age. Lactic dehydrogenase (LDH) catalyses the conversion of pyruvate to lactate in the presence of NADH. Five isozymes of LDH can be separated electrophoretically from animal tissues. These five isozymes are made up of tetramers of two different polypeptide chains, A and B viz. A_4, A_3B, A_2B_2, AB_3 and B_4. In adult heart muscle LDH1 (B_4) predominates, while in skeletal muscle it is LDH5 (A_4) which predominates. All proteins in an alkaline pH will bear a net negative charge and migrate to the anode; LDH1 has a greater net negative charge than has LDH5.

LDH can be assayed colorimetrically. In the presence of NADH, pyruvate will be converted to lactate at a speed depending on the amount of LDH. After a known period of time, a colour reagent, dinitrophenyl-hydrazine, is added and the remaining pyruvate will react with it, forming a complex, pyruvate-dinitrophenylhydrazone. Upon the addition of a base, a coloured compound is formed, the intensity of which is *inversely* related to the amount of LDH.

Necessary equipment
Ice bucket containing supernatant from homogenised rat heart muscle and rat skeletal muscle, 50 μl of each per 2 students
Pair of forceps and scissors
Cellulose acetate papers (Cat. No. SAE 2662) (6 per student pair) (allow a few spare) (Shandon)
Whatman filter paper
Veronal (Barbitone, BDH) buffer 0·05 M, pH 8·6 (amount will depend on volume of electrophoresis tank)
Dish to soak cellulose acetate paper without folding (1 per group)
1 Shandon electrophoresis tank per student pair
1 Shandon Vokam power pack per two pairs of students
Filter paper to make 'wicks'
1 micropipette to deliver 10 μl or 'Microcaps' (Shandon)
Purified rabbit muscle LDH and beef heart LDH (Sigma) (buy for class, allowing 10 μl of each per student)
Ponceau S stain (BDH) } volumes will depend on
5% acetic acid } size of staining dishes
Dish to stain and one to destain acetate strips
Reagents for LDH assay based on Sigma assay kit as follows:
 each student pair requires 4 vials containing preweighed NADH (β-diphosphopyridine nucleotide, reduced form, preweighed vial containing 1 mg–Sigma no. 340–101)
 7 vials Pyruvate substrate (Sigma standardised No. 500 L–1, available in 25 and 100 ml quantities)
 7 ml Sigma colour reagent (Sigma No. 505–2, available in 25 and 100 ml quantities)
 (With large numbers of students, cost of pre-weighed vials becomes prohibitive, so buy NADH (β-DPNH-Sigma) and *accurately* weigh out enough to make a 1 mg/ml solution in pyruvate substrate just before use–*do not store*. In this case, place NADH and pyruvate in 1 ml aliquots in tubes large enough to take a final volume of 12 ml)
70 ml 0·4N NaOH
3 tubes large enough to take 12 ml (to run controls)
Water bath at 37°C and racks
Cuvettes or colorimeter tubes
Spectrophotometer
Stopclock.

Type of group
Experiment can be run by pairs of students, or individual students, the number at any one time

237

depending on the availability of electrophoresis equipment.

Instructions to the student

Separation of LDH

The crude preparations of LDH from rat heart and rat skeletal muscle are provided for you. They are obtained by homogenising heart and muscle in ice-cold 0·25 M sucrose, centrifuging at 800 g for 10 min, the supernatant decanted and kept on ice.

1 Check that tank is *disconnected* from power supply and fill electrophoresis tank with veronal buffer.

2 Take cellulose acetate strips and soak in veronal buffer in dish provided for 2 min. Place strip gently on surface of buffer with forceps and allow to sink—do not push under. Using forceps remove from bottom of dish and blot gently, removing excess moisture, *but* do not dry (remove these strips one at a time and use, do not leave out).

3 Apply 10 µl crude LDH from rat heart towards one end of the strip as a thin, horizontal line—keep away from the edges of the strip. Apply in aliquots, not all at once. Label strip. Place in electrophoresis tank with ends of strip dipping into buffer and with a piece of buffer-soaked filter paper under the end of the strip, lying on the support bar and dipping into the buffer as a 'wick' to ensure a good contact. An additional piece of filter paper soaked in buffer can be laid on top of the end of the strip if necessary. Which way should the strips be orientated with respect to the migration of a protein in an alkaline pH? Put lid on tank.

4 Prepare another strip with LDH from rat heart, label and place in tank. Prepare 2 strips with crude rat skeletal muscle, label and place in tank. Set up another strip with samples of *purified* LDH side by side, keeping the size of the spot small; label and place in tank. Another strip can be set up in this way if required. Each member of the student pair will now have 1 strip with crude heart LDH, 1 strip with crude muscle LDH and 1 strip with purified LDH from heart and skeletal muscle.

5 Make sure that lid is on the tank. *Now and only now*, connect to the power supply. Run at about 1·5 mA per strip for 50–60 min (check that power pack is set on 'constant current'). *Do not touch tank while running.*

6 *Disconnect* power supply.

7 Take each acetate strip with crude samples of LDH and cut in half lengthwise. Take *one* half and stain in Ponceau S protein stain for about 5 min. Rinse in 5% acetic acid, remembering that all stain will come out in time, so watch. Stain whole strip containing purified LDH.

8 Select the best separation for crude heart muscle and best for crude skeletal muscle from each pair of strips. Using stained half as a guide, mark the areas of the other, unstained half of the strip where material is present. Cut out the area of highest negative and lowest negative mobility for heart and skeletal muscle, so that you have four pieces of unstained cellulose acetate to use in the assay.

Assay of LDH

1 Pipette 1·0 ml pyruvate substrate into each of 4 vials containing NADH and place in water-bath at 37°C for a few minutes (or take 4 tubes already containing 1 ml of NADH and pyruvate if pre-weighed vials are not being used and put in water-bath at 37°C).

2 Place 1 cellulose acetate sample into each vial, cap, shake gently, replace in water bath. *Commence timing.*

3 *Exactly* 30 min after adding the sample, remove vial from water-bath and add 1 ml 'colour reagent'. Shake well and leave at room temperature for 20 min.

4 Add 10 ml 0·4N NaOH to each vial, cap, mix by inversion and stand for *at least* 5 min and *not more than* 30 min.

5 Decant liquid into a cuvette, and record OD for all samples at 475 nm, using water as a blank.

A calibration curve is prepared by placing 1, 0·5 and 0·25 ml pyruvate substrate and bringing the volume up to 1 ml with water where necessary. 1 ml colour reagent is added and left for 20 min at room temperature. Add 10 ml 0·4 N NaOH, mix by inversion and allow to stand for at least 5 min (not more than 30 min). Transfer to cuvette and read OD at 475 nm, using water as a blank. (NADH does not contribute any appreciable OD and may be omitted from the calibration.)

Recording results

Remember to save the stained cellulose halves and put in your report. Plot calibration curve as OD versus pyruvate substrate in ml. Remember that LDH activity is inversely related to the intensity of colour complex, i.e. highest OD reading will be given by 1 ml of pyruvate substrate and thus the lowest amount of enzyme, i.e. nil. Express results from cellulose acetate assay in terms of pyruvate substrate used.

Discussion and extensions

Using the results obtained from electrophoresis and assay,

1 can you detect the isozymes present in heart and skeletal muscle, and is there any difference in amount or type, i.e. LDH1 or LDH5?
2 can you think of any physiological advantage for the differences noted, e.g. enzyme activity and substrate concentration?
3 on the basis of one gene = one enzyme hypothesis, how can results be explained? What sort of experiments would be required to determine the genetic basis of isozymes?

The technique of electrophoretic separation of proteins has wide application, both in research and medicine, and use of cellulose acetate enables the separation to be carried out relatively quickly. Changing the solid matrix of cellulose acetate to polyacrylamide gel and subjecting the sample to an electrical field and to a pH gradient enables high-resolution analysis of complex protein mixtures, and is used for detecting mutant forms of protein such as haemoglobin. The colorimetric analysis of serum LDH may be used clinically to look for elevated levels of isozymes indicative of tissue damage, e.g. liver or heart.

Further reading

Markert, C.L.
  1963 'Epigenetic Control of Specific Protein Synthesis in Differentiating Cells', in 'Cytodifferentiation and Macromolecular synthesis', ed. Locke, M., Academic Press, New York, p. 65

Hints for the teacher

1 Crude LDH from heart muscle and skeletal muscle of rat should be prepared just before use. Homogenise tissue in 0·25M ice-cold sucrose, using 1 ml sucrose to 1 g of tissue. Centrifuge in the cold at 800 g for 10 min and decant the supernatant into cold containers in an ice bucket.
2 *Emphasize* the importance of care when using electrophoresis equipment. Check that all tanks are disconnected before the start of the class and are not connected until all is ready. Point out that the actual current used will be 1·5 mA × the number of strips in the tank. The proteins will separate in a shorter time if the current per strip is increased but the problem here is overheating and thus denaturation of the enzyme. The proteins separate well, but in these circumstances the assay yields little or no LDH activity.

3 The use of Sigma preweighed vials is very convenient but expensive, especially when large classes are involved. Provided NADH is bought in fresh and made up only just prior to use, then it should be satisfactory. The NADH should be weighed out by laboratory staff and used at 1 mg/ml of pyruvate substrate. When buying both pyruvate substrate and colour reagent allow for student error in pipetting and buy a little extra. The sodium hydroxide used must be accurately standardised to 0·4N and reasonably free of CO_2. This should also be made up just prior to the experiment. Sigma publish a Technical Bulletin, No. 500 for the colorimetric determination of lactic dehydrogenase which is well worth writing away for.

SHIRLEY E. HAWKINS
King's College, London

7 *In vitro* stimulation of erythropoiesis.

Time required
Session 1 : 1½ hr
       18 hr gap
Session 2 : 3 hr
       3 hr gap
Session 3 : 2 hr
       18 hr gap
Session 4 : 2 hr
Session 5 : 4 hr

Assumed knowledge
1 Nature and interdependence of DNA, RNA and protein synthesis
2 Sequence of events during erythroid cell maturation
3 Simple sterile technique.

Theoretical background
Many of the factors which regulate haemoglobin synthesis in the whole animal take effect via erythropoietin, a glycoprotein hormone synthesised by the kidney. *In vivo*, erythropoietin may act on an unidentified erythropoietin-sensitive stem cell, itself already committed to the erythroid pathway, and cause it to differentiate via the recognisable erythroblast series and the reticulocyte to a mature erythrocyte. In cultures of embryonic liver (the main site of erythropoiesis in the middle stages of mammalian embryonic development) erythropoietin appears to act first by promoting RNA and DNA synthesis and subsequently cell division and haemoglobin synthesis. The stimulation of RNA, DNA and protein synthesis are prerequisites for the increase in haemoglobin synthesis, though cell division may not be. The first detectable response to erythropoietin occurs in the proerythroblasts and after 24 hr abnormally large haemoglobinised cells appear in treated cultures.

Hepatic erythropoiesis reaches a maximum at 15 days in the mouse embryo and at this point the liver becomes insensitive to erythropoietin in culture. In practice 13 days embryonic liver has been found most suitable for stimulation in culture.

This experiment is based on Cole and Paul.[2]

Necessary equipment

Sterile items	Per group	Sessions
13 day mouse embryos	4	1
Waymouth's medium (Flow) + 10% foetal bovine serum + 20mM HEPES, 50 units/ml penicillin, pH 7·2	40 ml in 4 oz (100 ml) stock bottle	2
0·25% trypsin (Difco 1:250) in isotonic citrate/saline. Paul[1 2] + 0·3% CMC	1 ml in 5in x $\frac{5}{8}$ in test tube	1
Hanks balanced salt solution (BSS) (Flow)	50 ml / 400 ml	1 / 2, 3, 4
Petri dishes for dissection (Sterilin)	3	1
Fine forceps (Horwell)	2 pairs	1
Scalpels, No. 11 blades	2	1
Graduated pipettes;		
Pasteur	1 can	2
1 ml	1 can	2
10 ml	1 can	1, 2
Universal container (Sterilin)	1	2
Test tubes	36	2
1 ml tuberculin syringes (Gillette Surgical)	6	2, 3, 4

Non-sterile items	Per group	Sessions
Drabkins solution ('Diagen' reagent. . . Diagnostic Reagents Ltd.)	10 ml	2, 4
1N HCl	2 ml	5
Analar butanone (BDH) (ice cold)	20 ml	5
0·2N perchloric acid (ice cold)	100 ml	5
2N perchloric acid	20 ml	5
70°C water-bath	1 only	5

Non-sterile items	Per group	Sessions
Triton X; toluene based scintillator (Koch-Light)	40 ml	5
Scintillation counter		5
Whatman GF/B glass fibre discs 2·5 cm	10	5
80°C oven		5
Isotopes		
^{59}Fe, in mouse serum 10 μCi/μg iron (Radiochemical Centre, Amersham)	40 μCi (in 2 ml)	2, 3, 4
^{3}H-Thymidine (TdR) (20Ci/mM) (Radiochemical Centre, Amersham)	25 μCi (in 1·25 ml)	2, 3, 4
^{3}H-Uridine (5Ci/mM) (UR) (Radiochemical Centre, Amersham)	25 μCi (in 1·25 ml)	2, 3, 4
Scintillation vials	30	5
Dissection microscope or magnifier (10–20X sufficient)		1
Erythropoietin, (EP) human urinary	10 units (in 1·6 ml)	2
Vortex mixer Cytocentrifuge–Shandon (optional) Centrifuge, preferably refrigerated		

Type of group
Groups of 3 or 4 (total around 20). Each individual may handle one isotope.

Instruction to the student
Outline
A trypsinised suspension is prepared from 13 day embryonic mouse liver and cultured in the presence and absence of erythropoietin. The rate of haemoglobin synthesis is determined at the start and after 4 and 24 hr by measuring the incorporation of ^{59}Fe into haem, extracted from the cells into acid buta-

none. The rates of DNA and RNA synthesis are measured at the start and after 4 hr and 24 hr by measuring the incorporation of ^{3}H-thymidine and ^{3}H-uridine respectively. The cytology of the cells is also examined at the start and at the end of 24 hr.

Protocol
Day 1. p.m. Session 1 *Dissection of foetal livers* (sterile):
1. Dissect out 4 livers from 13 day old mouse embryos and place directly into 1 ml 0·25% trypsin in citrate containing 0·3% carboxymethyl cellulose, and sitting on ice.
2. Place in cold room overnight.

Day 2. a.m. Session 2 *Disaggregation of liver and setting up cultures* (sterile):
1. Remove trypsin from livers, taking care not to disrupt the tissue.
2. Place livers at 37°C for 8–10 min (no longer) in hot room.
3. Add 10 ml Waymouth's supplemented medium from stock bottle to a disposable plastic universal container (Sterilin) and place at 37°C.
4. After (2) above take 2 ml from universal container and disaggregate livers gently in test tube by pipetting up and down *10 times*. Try to avoid frothing.
5. Add 2 ml suspension from (4) to universal container, mix, and allow to settle for about 2 min. Carefully collect supernatant free of clumps from the universal container without disturbing the sediment which may have formed, and transfer to medium remaining in stock bottle. Mix.
6. Dispense 1 ml from stock bottle into 38 test tubes per group of 4 and add:

A. 0·15 ml ^{59}Fe transferrin to 2 tubes (see Hints for the Teacher)
B. 0·1 ml ^{3}H-TDR to 2 tubes
C. 0·1 ml ^{3}H-UR to 2 tubes
D. No addition of isotope to 2 tubes
0·1 ml erythropoietin 6u/ml to one each of A,B,C,D
0·1 ml BSS to one each of A,B,C,D.

At this stage it may be convenient for each member of the group to take responsibility for a separate isotope, and one further person to look after the cytology. Thus, A takes 10 tubes: B, 10 tubes: C, 10 tubes and D, 8 tubes plus the residual stock cell suspension.

7　Seal tubes and place in hot room.

8　D. Count residue of cell suspension remaining in 4 oz bottle (on Coulter counter model D,4/40: 0·4 ml suspension in 20 ml counting fluid; or on haemocytometer). Take 5 drops of cell suspension and spin on cytocentrifuge. Dry off rapidly in air, fix in *fresh* methanol (10 min) dry again, and stain—May Grunwald and Giemsa or Leishmann (Gurr)

Staining: See 'Hints for the teacher'

Sampling (non-sterile)

9　A,B,C: after 1 hour remove T_0 samples from hot room and place on ice. Fill tube with cold BSS and spin 5 min 800 g. Discard supernatant *(Radioactive! pour out in radioactive sink)*. Resuspend cells in 10 ml ice cold BSS, and spin again. Discard supernatant, and repeat wash.

10　Discard 2nd BSS wash and dry off inside of tube—be careful not to dislodge pellet!

11　A. Add 1 ml Drabkin's solution to ^{59}Fe—treated samples, resuspend pellet on Vortex mixer, and freeze at $-20°$C.

12　B and C. Freeze ^{3}H samples directly after washing and discarding BSS.

Day 2. p.m. Session 3 *Four hour samples* (non-sterile):

1　4 hr after start,

A. add 0·15 ml ^{59}Fe to 5 tubes (2 with EP and 2 without EP)

B. add 0·1 ml ^{3}H-TdR to 4 tubes (2 with EP and 2 without EP)

C. add 0·1 ml ^{3}H-UR to 4 tubes (2 with EP and 2 without EP)

2　Incubate for 1 hr at $37°$C and collect as in 9—12 above.

3　D. count tubes on Coulter or haemocytometer and make cytopreps.

Day 3. p.m. Session 4 *24 hour samples* (non-sterile): After 24 hr repeat 1, 2 and 3 of Session 3.

Day 3. p.m. Session 5 *Haem extraction and counting* (A):

1　Freeze and thaw tubes a total of 3x (including freeze storage).

2　Add 0·1 ml 1N HCl to each tube, followed by 1·2 ml ice-cold butanone. (Keep tubes at 0—4°C throughout butanone extraction.)

Mix and spin at 1500 rev/min for 10 min at 4°C.*

*If refrigerated centrifuge not available, place tubes on ice for 15 min after centrifugation before collection of supernatant.

3　Collect 0·2 ml of supernatant (be very careful not to disturb interphase where the concentration of inorganic ^{59}Fe is very high) and add to scintillation vial containing a GF/B glass-fibre disc.

4　Dry off vials in oven at $80°$C for 1 hr.

5　Cool vials and add 5 ml toluene-based scintillator.

6　Count in scintillation counter using wide spectral range.

DNA and RNA extraction and counting (B and C)

1　Add 10 ml 0·2N perchloric acid (ice-cold) to ^{3}H-UR and ^{3}H-TdR-labelled tubes and mix.

2　Spin 10 min at 1000g.

3　Discard supernatant (radioactive).

4　Repeat (1—3).

5　Resuspend pellets in 1 ml 2N perchloric acid, rinse, and incubate at $70°$C for 20 min.

6　Cool, spin at 1000g.

7　Decant supernatant into scintillation vial, and add 10 ml triton X/toluene-based scintillator.

8　Count on scintillation counter.

Recording results and Disucssion

1　Graph isotopic incorporation per 10^6 cells against time and determine induction by erythropoietin.

2　Examine cytological preparation and see if you can distinguish haemoglobinised cells (red or brown cytoplasm, nucleus condensed or absent). Does the ratio of these change in your experiment? If so, how? Are these haemoglobinised cells present at the end of the experiment any different from those present at the beginning? Can you suggest why they might be?

Make up a table from your cytological observations, including, if you can, an estimate of the number of primitive cells (large cells with blue cytoplasm and uncondensed nucleus symmetrically placed).

Note any change in cell number and explain.

	Zero time	24 hr	
		−EP	+EP
Percentage of haemoglobinised cells			
Percentage of primitive cells			

3 Can you prepare a model for EP action based on (a) biochemical data and (b) cytological data?

Extensions and applications

This system provides a valuable model for the examination of differentiation since the use of one specific trigger (erythropoietin) will promote a sequence of biochemical and cytological events producing a mature cell making largely one protein. The blockage of induction with antimetabolites such as FUdR, actinomycin D, and puromycin, show how DNA, RNA and protein synthesis are interrelated and necessary for erythroid differentiation. Since erythroid maturation is accompanied by a reduction in cell size, the different populations of immature and mature cells can be separated by their sedimentation velocity and the sequence of molecular events accompanying differentiation can be studied.

Further reading

1 Chui, D.H.K., Djaldetti, M., Marks, P.A. and Rifkind, R.A.
   1971 J. Cell Biol., 51, 585
2 Cole, R.J. and Paul, J.
   1966 J. Embryol. Exp. Morph., 15, 245
3 Goldwasser, E.
   1966 'Current Topics in Developmental Biology', eds. Moscona and Monroy, 173-211. Academic Press, New York
4 Goldwasser, E. and Gross, M.
   1969 In 'Hemic Cells in vitro' 4. Williams and Wilkins, Baltimore, Md
5 Gross, M. and Goldwasser, E.
   1970 J. Biol. Chem., 245, 1632
6 Gross, M. and Goldwasser, E.
   1971 J. Biol. Chem., 246, 2480
7 Hunter, J.A. and Paul, J.
   1969 J. Embryol. exp. Morph., 21, 361
8 Krantz, S.B. Gallien-Lartigue, O. & Goldwasser, E.
   1963 J. Biol. Chem., 238, 4085
9 Marks, P.A. and Kovach, J.S.
   1966 'Current Topics in Developmental Biology'. Eds. Moscona and Monroy, p. 213-52, Academic Press, New York
10 Murphy, M.J., Bertles, J.F. and Gordon, A.S.
   1971 J. Cell Sci., 9, 23
11 McDonald, G.A., Dodds, T.C. & Cruickshank, B.
   1965 'Atlas of Haematology', E. & S. Livingstone, Edinburgh and London
12 Paul, J. and Hunter, J.A.
   1969 Nature, 219, 1362

Hints for the teacher

Mice

The detection of a vaginal plug on the day after mating is taken as day zero. The embryos should be collected by sterile dissection from the uterus into sterile BSS on day 13, immediately before the class.

Mouse serum

This can be obtained commercially from Biocult, although it is very expensive. It can be prepared as follows:

1 Stun mice* and bleed into clean beaker by slitting throat.
   Do not completely sever head or they will not bleed.
2 After bleeding complete, make sure mouse is dead by breaking the neck.
3 Place blood at $4°C$ overnight to clot, and for clot to retract.
4 Decant supernatant serum and dilute 50:50 with balanced salt solution.
5 Clear by centrifugation at 10 000g
6 Sterilise by membrane filtration through three filters (Millipore)—
   (a) Glass fibre prefilter, (b) $0.45\mu m$ or $1.0\mu m$ membrane filter (c) $0.22\mu m$ membrane filter. All these may be packed together in one holder, or used in sequence in three holders or in three operations.
   The yield from each mouse is about 1 ml BSS/serum mixture.

Erythropoietin

This is rather difficult to obtain. Requests should be sent to Erythropoietin Committee, National Heart and Lung Institute† with an explanation of the purpose and amount required. Alternatively, Connaught Medical Research Laboratories, Toronto are able to provide it. Unfortunately, the author cannot undertake to supply this. If sufficient demand were made a joint application could be put to the committee; in this case the author would act as a collecting point for requests and pass these to the person most appropriate to make the application.

^{59}Fe

This is available from The Radiochemical Centre in

*A licence is required.
†Dr J.M. Stengle,
National Heart and Lung Institute,
National Institutes of Health, Bethesda, Md. 20014, USA

Amersham. It would be preferable to make up the serum-^{59}Fe mixture before the class to avoid students handling the high-activity isotope. Add 0·4 ml ^{59}Fe Cl$_3$, 100 μCi/ml, to 2 ml mouse serum, 0·1 ml penicillin, 10^4 units/ml, and 0·5 ml BSS. Place in lead pot at 37°C overnight to allow ^{59}Fe to bind to mouse transferrin.

^{59}Fe-citrate may be used without serum transferrin and will yield comparable results (contact Author)

Counting

The method of counting radioactivity described can be replaced by gas-flow counting if necessary. It is recommended that ^{14}C is substituted for ^{3}H in this case.

Trypsin

This is as detailed by Paul and Hunter.[12] It contains 0·5% carboxymethyl cellulose (CMC) in addition to improve survival (this may be a viscosity effect or a membrane interaction; it is not known which). CMC is obtained from ICI as Edifas B10. It is best prepared as a 3% solution in BSS by heating up to 100°C, mixing, and allowing to cool. Sterilise by steaming as autoclaving may cause breaks in the polymer and a resultant reduction of viscosity.

Drabkins reagent

This is most easily obtained as the haematological reagent from Diagnostic Reagents, Ltd.,
It can be made up, however, as follows:
Potassium ferricyanide   20 mg
Potassium cyanide   50 mg
Water (distilled or deionised) to 1 litre
(Dorcie, J.V. and Lewis, S.M., 'Practical Haematology', Churchill, Edinburgh (1968)

General protocol

As described the experiment is very full. It could be contracted (1) by sticking to one isotope, (2) by omitting isotopes and relying on cytology alone, (3) by using TdR or UR alone for 4 hr.

It is important to have the cells diluted and placed in culture with the appropriate reagents (Session II) as rapidly as possible. Any delay here can reduce induction considerably. Students should know what each is going to do and all preparations made beforehand, even to the extent of placing isotope and EP in appropriate tubes, so that when the livers are disaggregated they can be diluted and inoculated into tubes and immediately transferred to 37°C. It is not necessary to count the cells before this stage is complete as four 13-day livers diluted to 40 ml give

approximately 10^6 cells/ml when Porton White Swiss mice are used. Most other outbred mice should be similar. Once the cultures are set up and at 37°C, the cells should be counted for confirmation. After 24 hr the cell concentration may have increased about 10%.

It is very important that the pH of the medium does not rise above 7·3–7·4 or incorporation of the ^{59}Fe into induced catalase may create a high T_0 value.

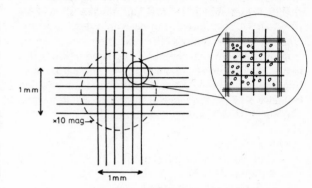

Figure 1 Haemocytometer (improved Neubauer)
1 Count all, or a representative sample, of the cells in the centre mm grid of 25 squares (each square bounded by 3 parallel lines).
2 Cells on lines: count cells on right and lower lines only.
3 Use small squares (bounded by single lines) as a counting aid only.
4 Calculations:

$$\text{Conc. of cells/ml} = n \times \frac{25}{5} \times 10^4 \times D$$

Where n = visual count

25 = Total number of squares in square area of field (1 mm^2)

S = Number of squares counted (each square bounded by 3 lines)

10^4 = factor to correct to 1 ml

D = dilution

(For no dilution, and counting all 25 squares) = $n \times 10^4$

Cytology

The erythroblast series is illustrated in a chart in the book by McDonald, Dodds and Cruickshank.[11] Emphasis should be placed on (1) cell size, (2) nuclear condensation, (3) cytoplasmic haemoglobinisation. The student need not recognise all of the cell types but should attempt to distinguish proerythroblasts and early basophilic erythroblasts from orthochromatic erythroblasts and reticulocytes. It should be possible to detect a large cell with haemoglobin in the cytoplasm but an uncondensed or only partially condensed nucleus in erythropoietin-treated samples after 24 hr.

The Shandon cytocentrifuge is preferable for cytological preparations, but a smear may be used if this is not available. Smearing is best done gently with a camel hair brush, and not with a slide as for conventional blood smears. Dry off rapidly and treat as for cytocentrifuge preps.

Cell counting

This is best done on a Coulter model D set at A.C.4 and threshold 50 (these settings should be confirmed by doing a plateau if possible, although the plateau will be found to be very narrow). In the absence of a Coulter counter, a haemocytometer may be used (see Fig. 1).

Staining foetal liver cells (fixed in anhydrous methanol)

1 May-Grunwald (Gurr) in methanol, diluted 1+2 with 0·06M PO_4 buffer, pH 7·0, 1·5 min
2 Wash in buffer.
3 Stain in 7% Giemsa ($\sim$ 1 drop conc. Giemsa/ml) in pH 7·0 PO_4 buffer 0·06M for 5 min
4 Drain, air dry and mount in DePeX (Gurr).

R. I. FRESHNEY,
Glasgow.

8 Induction of haemoglobin synthesis in a continuous cell strain

Time required
1½ hr on first day and 3 to 4 hr four days later.

Assumed knowledge
Culture of cells in suspension. Outline of erythropoiesis. Elementary biology of mouse RNA tumour viruses.

Theoretical background
The infection of certain strains of mice with the Friend RNA virus produces a tumour of erythroid precursor cells which may be propagated by serial transfer in the same strain of mice or as a cell culture. Different cell strains show a varying degree of spontaneous maturation and induction of haemoglobin synthesis. The strain used in this experiment is capable of induction by dimethylsulphoxide (DMSO), such that more than 60% of the cells become haemoglobinised. Presumably, viral transformation allows these cells to survive indefinitely in culture, unlike normal erythroid precursor cells which cannot be cultured continuously. The rationale for the action of DMSO is not fully understood. It may influence membrane permeability to precursors or other inducers (e.g. hormones in the serum) or it may alter the interaction of the viral and host genomes.

Necessary equipment

Item	Option 1 or 2	Amount	Day
Friend cells—strain 707—Beatson Institute, Glasgow. (Grown to 10^6/ml in log phase)	1	5×10^6 cells each	1
Culture bottles (4 oz medical flats; or No. 50 from Flow Labs., or hexagonal baby bottles). (Dry-heat sterilised or autoclaved) Screw caps and silver foil to cover caps	1	4 each	1

245

Item	Option 1 or 2	Amount	Day
No.120 flasks (Flow Labs), or Roux flasks (sterile)	2	2 each	1
Medium—RPMI 1640 + 20mM HEPES + 15% horse serum. (Flow Labs.,) or Ham's F12 Supplemented with Eagle's MEM amino acids, 20mM HEPES, 15% horse serum (Flow Labs) (Sterile)	1 / 2	50 ml each / 100 ml each	1
DMSO—Analar or spectroscopic grade (BDH) (Self-sterilizing but should be transferred to sterile glass container)	1,2	1 ml each	1
Incubator or hot room at 36°C	1,2	1 (shared)	1
Centrifuge (preferably refrigerated)	1,2	1 (shared)	1,6
Tubes— 15 ml conical centrifuge	1,2	4 ea./2 each	6
50 ml conical centrifuge	2	2 each	
^{59}Fe Cil$_3$, 10 Ci/ml in medium (Radiochemical Centre, Amersham)	1	5 ml/50 μCi each	1
Scintillation vials (Koch-Light)	1	8 each	6
Scintillation fluid, toluene based (Koch-Light)	1	40 ml	6
Scintillation counter	1	1 (shared)	6
Alternative counting Planchette and carriers	1	8 each	6
Gasflow counter		1 (shared)	6
Gamma counter;	1	1 (shared)	6

Item	Option 1 or 2	Amount	Day
Vials		8 each	6
Coulter counter (Model D or other)	1,2	1 (shared)	1 and 6
or Haemocytometer slides and cover slips (Gallenkamp)	1,2	1 each	1 and 6
1N NaOH (Sterilised by autoclaving)	1,2	10 ml	1
1M HEPES (Sterilised by autoclaving)	1,2	10 ml	1-6
Drabkins solution ("Diagen Reagent"—Diagnostic Reagents)	1	10 ml	6
Phosphate buffered saline (PBS) (Wellcome Reagents, Beckenham, Kent)	1 / 2	100 ml / 20 ml	6
Dry ice and alcohol bath.	1	1 (shared)	6
Haemoglobin standard (Diagnostic Reagents)	2	5 ml	6
Butanone, ice cold (BDH)	1	10 ml	6
N HC1	1	5 ml	6
Glass fibre discs Whatman 2·5 cm (in scintillation vials)	1	4 each	6
Spectrophotometer (Scanning 200 nm-600 nm if possible, otherwise 415 nm	2	1 (shared)	6
Cuvettes, 4 ml (quartz if scanning)	2	1 set each size (shared)	6
10 ml test tubes	2	2 each	6

Type of group

Pairs, up to maximum of 10. Alternatively, 4 groups of 3.

Instructions to student

Option 1. ^{59}Fe incorporation.

Option 2. OD_{415} and/or absorption spectrum of cell lysate.

Outline

The cell suspension is diluted in growth medium and two parallel cultures are initiated. Dimethyl sulphoxide (DMSO) is added to one.

Option 1. ^{59}Fe Cl$_3$ is added at the start and the incorporation measured after 6 days in culture.

Option 2. After 6 days the cells are washed, lysed and the OD$_{415}$ measured (or the spectrum analysed if possible).

PROTOCOL
Option 1
Day 0

1 Examine cell suspension provided. Count cells and dilute to 5×10^4/ml in 50 ml growth medium.
2 Add 10 ml to each of 4 culture bottles.
3 Add 0·15 ml DMSO to 2 bottles
4 Add 1 ml ^{59}Fe, 10 μCi/ml, to all four bottles. Retain surplus ^{59}Fe to prepare standards on Day 6.
5 Place bottles in incubator with screw caps loose and cap and neck covered in aluminium foil.
6 Check pH of cultures after 2 hr and adjust with sterile 1N NaOH, or 1M HEPES, if necessary (less than 0·1 ml will be required).

Day 6:

1 Count aliquot of cells (0·5 ml) from each bottle. (*Care! radioactive*).
2 Transfer remaining cells to conical centrifuge tubes, and spin for 5 min at 500g, 4°C.
3 Discard supernatant from (2) above—*Care! This is radioactive and should be disposed of accordingly*. Resuspend pellet in ice-cold PBS and spin again. Repeat wash once more.
4 Add 1 ml Drabkins solution.
5 Freeze and thaw 3 times in dry ice/alcohol, to lyse cells.
6 Acidify with 0·1 ml 1N HCl to cleave haem from globin.
7 Extract haem by adding 1·2 ml butanone, ice cold, to acidified Drabkins lysate on ice. Mix thoroughly, cap and spin at 2000g for 10 min at 4°C to separate phases. If it has been necessary to spin at room temperature, reduce spin to 5 min and then tubes should be placed on ice for at least 10 min before any attempt is made to collect butanone layer.
8 Collect 0·2 ml of butanone layer (upper) containing ^{59}Fe-haem. Do not disturb interphase as it may contain inorganic ^{59}Fe.
9 Add to glass-fibre discs in scintillation vials and transfer to 80°C oven to dry off butanone for 1 hr. Add butanone alone to two vials to act as blanks; prepare two standards using 0·1 ml of a 1:100 dilution of ^{59}Fe chloride solution used at start of incubation. These blanks and standards should be placed in the oven to dry for an hour along with sample vials.
10 Cool and add 5 ml toluene-based scintillator. Count 10 min per vial on the scintillation counter set at a wide spectral range.
    Alternatively, add butanone to aluminium planchettes, dry off and count on gas flow counter; or count butanone samples directly in gamma counter.

Note: ^{59}Fe has several β emission energies over a wide spectral range, and also emits γ-radiation. The ratio of $\beta:\gamma$ is approximately 1:1.

Option 2
Day 1 (start)

1 Examine and count cell suspension provided. Dilute to 5×10^4/ml in 100 ml growth medium.
2 Add 50 ml cell suspension to each of two culture bottles.
3 Add 0·75 ml DMSO to one bottle.
4 Place both bottles in incubator with screw caps loose and covered in aluminium foil (to allow free gas exchange but reduce contamination).
5 After about 2 hr, check pH and correct to 7·4 with 1N NaOH or 1M HEPES if necessary. (If time available this should be repeated each day.)

Day 6

1 Count aliquot (0·5 ml) from each bottle.
2 Transfer remainder to 50 ml centrifuge tubes and spin at 500g for 5 min.
3 Resuspend in 10 ml phosphate buffered saline (no phenol red) and transfer to 15 ml centrifuge tubes.
4 Spin again, 500g, 5 min.
5 Remove as much saline as possible and discard.
6 Resuspend each pellet in 5 ml distilled water and pipette up and down to disaggregate pellet and lyse cells.
7 Spin 2000g to clarify. Collect clear supernatant and transfer to fresh tubes.
8 Read optical density at 415 nm against water. (It may be possible to scan at this stage for classical Hb absorption spectrum.) Read absorption of standard against same blank.

Recording results

1 Tabulate counts per minute with and without DMSO. Convert to picomoles Fe^{+++} incorporated.

2 Calculate haemoglobin present per cell with and without DMSO.

Discussion
1 Do the cells proliferate and differentiate?
2 Is this a high level of haemoglobin synthesis compared to normal erythropoiesis?
3 Have you any suggestion for the effect of DMSO?
4 How accurate is your estimation of iron incorporation into haem?
5 What errors is it subject to?

Extensions and applications
The chief benefit of this system lies in the ability to produce differentiation at will in cell culture, under well-defined conditions, on a given stimulus. Thus many of the molecular events regulating differentiation can be explored by, for example, assessing the effect of DMSO on globin messenger production, or by looking for changes in chromatin which would relate to the induction of differentiation. The interaction of the viral genome with the host cell can also be examined. In some cases, this tumour produces polycythaemia while in others it produces leukaemia. The viral genome can react with the host genome in different ways such that the expression of differentiation and the susceptibility to inducer varies from one case to another, and with different substrains of virus. Since cell strains can be obtained in culture with different capacities for haemoglobin synthesis and its induction, they provide interesting tools for examining the genetics of regulation of mammalian differentiation using cell-fusion techniques.

Further reading
Cole, R.J. and Paul, J.
1966 'The Effects of Erythropoietin on Haem Synthesis in Mouse Yolk Sac and Cultured Foetal Liver Cells', J. Embryol. Exp. Morph., 15, 245–60
Rossi, G.B. and Friend, C.
1967 'Erythrocytic Maturation of (Friend) Virus-induced Leukemic Cells in Spleen Clones', Proc. Nat. Acad. Sci. US, 58, 1373–80
Patuleia, C. and Friend, C.
1967 'Tissue Culture Studies on Murine Virus-induced Leukemia Cells: Isolation of Single Cells in Agar-liquid Medium'. Cancer Res., 27, 726–30
Scher, W., Holland, J.G. and Friend, C.
1971 'Hemoglobin Synthesis in Murine Virus-induced Leukemic Cells in vitro. 1. Partial Purification and Identification of Hemoglobins', Blood, 37, 428–37
Friend, C., Scher, W., Holland, J.G. and Sato, T.
1971 'Hemoglobin Synthesis in Murine Virus-induced Leukemic Cells in vitro. 2. Stimulation of Erythroid Differentiation by Dimethyl Sulfoxide', Proc. Nat. Acad. Sci. US, 68, 378–82

Hints for the teacher
1 Since this cell strain contains virus it is recommended that all glassware used in culture is soaked in Chloros (disinfectant) after use or autoclaved. It is unknown for Friend virus to infect other than DBA or Balb strains of mice; it does not seem to survive long in tissue-culture media; and it does not appear to be infective in vitro. However, better safe than sorry! It is good practice for students to learn to treat viral contaminated glassware appropriately.
2 These cells are now in use in several laboratories but samples may be obtained from the author. They grow rapidly in several media but RPMI 1640 is probably best in this case. Since it is essential not to let the pH rise above 7·4 at the start, and since these cells produce a lot of acid, HEPES buffer is recommended. It is also recommended that the pH of the culture be checked an hour or two after the start to make sure it does not rise above 7·4. The cells should be made available in log phase at about 10^6/ml. They will grow from 5×10^4/ml to more than $2 \cdot 5 \times 10^6$/ml in 6 days with a doubling time of 12–18 hr.
3 Radioactivity precautions will be required for handling ^{59}Fe. This isotope has a half-life of about 40 days and therefore should be obtained within a few weeks of the class. It is a mixed β and γ emitter with a wide energy range and is potentially very dangerous as it will be concentrated in the bone marrow.

R.I. FRESHNEY
Glasgow.

9 Specificity of secondary inductive interaction during organogenesis

Time required
Day 1: 3 hr. Day 2–5: 15 min each. Day 6: 1 hr.

Assumed knowledge
Mammalian embryology, function of primitive germ layers, embryonic induction, organogenesis. Organ culture, elementary tissue-culture technique.

Theoretical background
During organogenesis in vertebrate embryos cell interaction between ectoderm and mesoderm, or between endoderm and mesoderm, is an essential feature of the differentiation of these tissues in mutual association. Thus, somitic mesenchyme will only differentiate into cartilage when it is allowed to associate with spinal cord or notochord. Similarly, primordial kidney mesenchyme will differentiate in association with either kidney epithelium or dorsal spinal cord, and ureteric bud will differentiate into collecting tubules in association with kidney mesenchyme. A commonly expressed pattern is the differentiation of endodermal rudiments, derived as diverticula of the primitive gut, to form many organs such as salivary gland, lungs, liver and pancreas, by interacting with the appropriate adjacent mesenchyme.

In this exercise submandibular salivary gland and lung have been chosen as two systems which have been studied extensively and which are not unduly difficult to dissect. Salivary mesenchyme will induce proliferation and differentiation (formation of acini and amylase production) in salivary epithelium and will promote some proliferation of lung epithelium. Lung mesenchyme will promote proliferation and differentiation of lung epithelium into alveoli but will only promote proliferation and branching in salivary epithelium.

Necessary equipment

	Per pair	Day
13 day mouse embryos (aseptically collected)	4	1
in sterile balanced salt solution, Hanks'	20 ml	1

	Per pair	Day
Instruments (Sterile):- Beaver chuck handles with No. 61 Blades (Gallenkamp)	2	1
No. 3 watchmaker's forceps	1 pr.	1
Dissecting dishes (sterile): 5 cm glass or plastic Petris (Sterilin)	4	1
Organ culture dishes (Falcon Plastics)	4	1
Culture dishes (Sterile): As above—organ culture dishes (Falcon)	6	1
Either perspex supports (see 'Hints for the teacher')	6	1
or: Stainless steel grids (Falcon Plastics) or fine stainless steel grid (The Expanded Metal Co.) Millipore supports (sterilise by autoclaving): 1 cm squares Millipore Cat. No. THWP102F0 (cut from 10 cm squares)	6	1
Fine-drawn Pasteur pipettes (Dry-heat sterilised) Graded sizes from $300 \mu m$-1 mm diameter	12	1
Microtitre trays (Sterile)	1	1
Sandwich boxes (plastic) (8in x 5in x 2in but size not critical)	1	1
50/50: 50% Foetal bovine serum (Flow/Biocult) (sterile) in Hanks' balanced salt solution (Flow/Biocult)	50 ml	1
Trypsin: 0·05% in Ca^{++} and Mg^{++} free saline (See CMF. Hints for the teacher) (Trypsin powder 1:250, Difco (sterilize by filtration.)	10 ml	1

	Per pair	*Day*
Medium:		
Eagle's MEM + 20 mM		
HEPES		
(Flow/Biocult)	10 ml	1
20% foetal bovine serum		
(Flow/Biocult)		
3% chick embryo extract		
(Flow/Biocult)	10 ml	1
+ 200iu/ml penicillin and		
100 iu/ml streptomycin		
70% ethanol for swabbing	250 ml	1
Swabs or cotton wool		1
Dissection microscopes		
(50x-100x)	1	1-5
Camera/camera lucida/graticule		1-5
Rubber teats or mouthpiece		1

Type of group
This exercise is probably best performed in pairs up to about 12 students.

Instructions to the student
Outline
Submandibular salivary glands and lungs are removed from the embryo. Mesenchyme is removed by the combined action of mechanical dissection and mild tryptic digestion. The mesenchyme is then re-combined with each epithelium separately on a Millipore membrane substrate. The combinations are then maintained for 5 days at 37°C with frequent observation.

Dissection of submandibular salivary gland
1 Prepare four, 5 cm Petri dishes and about 12 microtitre wells with 50/50.
2 Take three 13 day mouse embryos and dissect off head cutting through neck and place in 50/50 in one dish.
3 Remove lower jaw by cutting from front of head to back through the angle of the jaw; this should free the whole lower jaw and tongue with a section of the neck. Remove neck section but be careful not to cut jaw or tongue. Cut the jaw in half by a cut down the centre of the tongue.

Take each half in turn and tease the tongue away from the jaw. In the angle so formed the salivary gland will appear like a hand in a mitten. The 'hand' is the epithelium and the 'mitten' the mesenchyme.
4 Place in 50/50 in fresh dish (an organ culture dish is best for this stage) and dissect off as much of mesenchyme as possible and place in microtitre tray.
N.B. Prepare a table with the grid references of the microtitre tray and use this to remind you which fragments are in which wells.
5 Place epithelial rudiments in 0·05% trypsin at 37°C for 5–10 min in another well of the microtitre tray. When the mesenchyme disperses off the epithelium with gentle pipetting in a fine-drawn Pasteur pipette, the epithelium should be transferred free of mesenchyme into 50/50 in another fresh well. If mesenchyme does not come off, transfer to fresh trypsin (serum traces may be inhibiting tryptic activity) and repeat incubation

Dissection of lung
1 Take three 13 day embryos, remove the head, cut through the body, transversely, posterior to the liver. If the anterior trunk fragment is slit on the ventral surface, the heart, liver, gut and lungs may be removed in one cluster. If the lungs do not come out with the heart and liver and gut they will be ovious as the only remaining organs in the thoracic cavity.
2 Dissect off as much of the mesenchyme as possible from the lungs, keeping the pieces as large as possible, and place the mesenchyme in microtitre wells in 50/50.
3 Place epithelial rudiments in fresh trypsin as for salivary glands to remove remaining mesenchyme.
4 When free of mesenchyme transfer epithelial rudiments to 50/50.

Setting up cultures
1 Prepare 6 dishes with 1·1 ml medium in each.
2 Add epithelial and mesenchymal fragments to Millipore supports to make the following combinations. Arrange the mesenchyme around the epithelium in excess. Try to have approximately the same amount of tissue in each culture:-
(a) SE/SM: Salivary epithelium + salivary mesenchyme
(b) SE/SE: Salivary epithelium + salivary epithelium

250

(c) SE/LM: Salivary epithelium + lung mesenchyme
(d) LE/LM: Lung epithelium + lung mesenchyme
(e) LE/LE: Lung epithelium + lung epithelium
(f) LE/SM: Lung epithelium + salivary mesenchyme

A rough sketch should be made of each culture (or photo if available).

3 Make sure all the pieces are in close association and then transfer dishes to polystyrene sandwich boxes (previously swabbed with ethanol but now dry) and place at 37°C.

4 Check daily for growth and morphogenesis and make sure medium is not evaporating. Change medium after three days and terminate after 5.

Termination of cultures

1 Examine cultures by transmitted light for signs of epithelial proliferation, branching, and morphogenesis, and compare with sketch/photo of start.

2 Remove Millipore support complete with explants and place on a slide. Mount a cover slip in saline and examine under microscope, taking care to replenish saline if there are any signs of drying out. A whole mount stain, such as carmine, may be used at this stage if desired, and the specimen cleared in xylene and a permanent preparation made.

Recording results

Sketches or photographs should be used to compare the stages and various combinations. Keep the same scale between drawings, using an eyepiece graticule. Note length of tubules, number of branchings and area covered by epithelial network. Look for any signs of differentiation in epithelium—acini or alveoli.

Discussion

Is mesenchyme necessary for (1) survival, (2) proliferation, (3) morphogenesis? Are the two mesenchymes interchangeable either partially or completely?

What specific and non-specific function does the mesenchyme perform? Can you suggest a mechanism? Can you suggest further experiments which would help to explain the process?

Extension and application

The interposition of membranes of varying porosities between the interacting tissues has shown that the process occurs via a diffusible intermediate, but which will not cross dialysis membrane. Precursor studies have so far failed to reveal the nature of the inducer substance. Some evidence has suggested that part of the interaction depends on the diffusion of precursors of collagen synthesis. More recently there are some results implicating a specific protein inducer substance as yet unidentified.

Cross-reactivity studies with inter-organ and interspecific combinations suggests a higher degree of specificity between organs than between species even of different classes, e.g. bird and mammal.

Further reading

General reviews
Sengel, P.
1970 'Study of Organogenesis by Dissociation and Reassociation of Embryonic Rudiments *in vitro*', in 'Organ Culture', ed. Thomas, J.A., Academic Press, New York, Vol. 9, 379
Wessels, N.K. and Rutter, W.J.
1969 'Phases in Cell Differentiation', Sci. Amer., 220, 36
Lung and salivary gland
Grobstein, C.
1953 'Epithelio-mesenchymal Specificity in the Morphogenesis of Mouse Sub-mandibular Rudiments *in vitro*', J. Exp. Zool., 124, 383
Taderera, J.V.
1967 'Control of Lung Differentiation *in vitro*', Dev. Biol., 16, 489
Other systems
Cooper, G.W.
1965 'Induction of Somite Chondrogenesis by Cartilage and Notochord: a Correlation between Inductive Activity and Specific Stages of Cytodifferentiation', Dev. Biol., 12, 185
Grobstein, C.
1964 'Cytodifferentiation and its Controls', Science, 143, 643
Grobstein, C.
1967 'Mechanisms of Organogenetic Tissue Interactions', Nat. Cancer Inst., Mono. 26, 279
Kratochwil, K.
1969 'Organ Specificity in Mesenchymal Induction Demonstrated in the Embryonic Development of the Mammary Gland of the Mouse', Dev. Biol., 20, 46
Rutter, W.J., Wessels, N.K. and Grobstein, C.
1964 'Control of Specific Synthesis in the Developing Pancreas', J. Nat. Cancer Inst., Mono., 13, 51
Unsworth, B. and Grobstein, C.
1970 'Induction of Kidney Tubules in Mouse

'Metanephric Mesenchyme by Various Embryonic Mesenchymal Tissues', Dev. Biol., 21, 547

Hints for the teacher

Embryos

Embryos should be collected aseptically from mice and placed in sterile balanced salt solution immediately before they are required. The stage of development may be determined by checking mated females for the presence of a vaginal 'plug' first thing in the morning after mating. The presence of such a 'plug' is evidence of insemination and the female should be separated and the date noted as 'day zero'. If outbred mice or hybrids are used, 13 days is appropriate. Pure bred mice may require to be older by a day or so, but will need checking beforehand. If males and females are placed together (male in cage first then females a day or two later) and checked daily for 'plugs', the highest incidence can be detected after 3—4 days when oestrous is induced.

If necessary it is possible to use the same embryos for dissection of lung and salivary gland. If the students work in pairs this can be done without leaving the embryos lying around for too long.

The average litter size is 6—7 and about 10% of the females mated can be expected to provide embryos at the right stage (assuming that the maximum number of pregnancies occur between the 3rd and 4th day after putting animals together).

Millipore filters

These are most easily supported on steel grids as illustrated by Falcon in their catalogue. Similar grids can be prepared from stainless steel about 1 mm mesh. Alternatively, and experimentally rather better, supports can be prepared from perspex strips about 1 in long x ¼ in wide x $^1/_{32}$ in thick with a central $\frac{1}{8}$ in hole. A disc of Millipore (THWP102F0) is stuck over one side of the hole (using glue available from Millipore) to make a shallow well. The strip bridges the well in the centre of the organ-culture dish, and the explants are located in the filter well of the perspex strip.

Transfilter experiments

It is possible to do these induction experiments with the epithelium in the filter well and the mesenchyme adhering to the opposite of the filter disc, embedded in a plasma clot. The clot is made from 1 part chick plasma: 1 part embryo extract: 2 parts culture medium and should be placed over the mesenchyme with the perspex strip upside down and no medium in the culture dish. When the plasma clots the strip is turned back over, well uppermost, and the epithelium positioned directly over the mesenchyme. Growth medium is then added.

Fine-drawn Pasteur pipettes

These can be drawn from standard 'long-form' Pasteur pipettes and should be graded with an eyepiece micrometer. 400—600µm diameter will be found most useful, but some larger and smaller should be provided.

Recording specimens

Ideally a camera with a Polaroid attachment is best but rather expensive to operate. Alternatively, the specimen may be drawn and for this a camera-lucida attachment is very useful. Failing this an eyepiece graticule makes accurate scale drawings easier for the less artistic.

Trypsin

Trypsin may be purchased sterile from Flow or Biocult and diluted for use with Ca^{++}- and Mg^{++}-free saline (CMF). Alternatively, it may be made up from Difco powder (1:250 grade) in CMF and sterilised by filtration. The solution should be prepared one day, pH adjusted to 7·8 and left at 4°C overnight. It is then filtered first through a glass-fibre prefilter and then through a 0·22µm membrane such as Millipore.

CMF

NaCl	8·00 g	
KCl	0·20 g	
$NaH_2PO_4 \cdot H_2O$	0.005 g	1 litre with double
*$NaHCO_3$	1·00 g	distilled water
Glucose	2·00 g	
Phenol red	0·02 g	

*Sterilise by filtration.

If sterilised by autoclaving, glucose should be autoclaved separately as a 20% solution and added later. May be buffered with HEPES, 10mM final, neutralised by 1N NaOH, in which case the bicarbonate should be reduced to 0·1 g/litre. If 1·00 g/litre bicarbonate is used, it must be kept under 5% CO_2 in air. The HEPES system is to be preferred since no gassing is necessary.

R.I. FRESHNEY
Glasgow.

Section 14

Immunology

1 Blood grouping

Time required
30 min

Assumed knowledge
None

Theoretical background
The red blood cells of humans can be classified into four groups depending on whether an individual possesses A or B antigens. The groups are A, B, AB or O (if neither antigen is present). Antibodies against these antigens occur in the sera of individuals who do not possess the particular antigen. In people with A antigens, B antibodies are present in the serum; in those with B antigens, A antibodies are present; in those with AB antigens no antibodies are present; and in those with no antigens, antibodies against both A and B are present. Red cells agglutinate in the presence of the appropriate antiserum and thus the blood group of an individual can be determined by adding his blood to standard anti-A and anti-B sera and noting whether agglutination occurs in one, both or neither of these. This reaction is the basis of the ABO blood-grouping system, which is only one of a number of less well-known systems. Among Caucasians in England approximately 46% belong to Group A, 9% to B, 3% to AB and 42% to O, whereas in Scotland the figures are 35%, 10%, 3% and 52%.

Necessary equipment
Each student will need the following.
Anti-human A serum ⎫ Medical Research Council,
Anti-human B serum ⎬ Blood Group Reference
⎩ Laboratory, Gatliffe Road,
⎭ London SW1.

1 slide, 3in x 1in
Wax pencil
5 ml 0·9% saline
Alcohol (for cleaning finger)
Paper tissues
Sterile disposable blood lancet (Hoslab, Horwell)
1 Pasteur pipette, with teat.

Type of group
Individual students.

Instructions to the student

1 Draw a line with a wax pencil across a clean 3in x 1in slide. Label one half A and the other B.
2 Place drops of anti-A serum and anti-B serum in appropriate positions on the slide.
3 Take a drop of your blood into a larger drop of saline, add one drop of the sample to each of the sera and mix. To obtain blood, clean your left thumb with alcohol and wipe with a paper tissue. Shake your hand vigorously and stab the ball of your thumb with the lancet. Squeeze the thumb and discard the first drop of blood. Take the second into a Pasteur pipette containing about 0·1 ml saline.

   N.B. Under no circumstances use a lancet which has been used by someone else.

4 Read the results after 5 min. If there is a reaction the cells will clump together. If there are signs of clumping but it is difficult to decide whether the result is positive or negative, leave the slide for a further 5 min but do not shake it.

Interpret your results as follows

Agglutination with anti-A serum	blood group A
Agglutination with anti-B serum	blood group B
Agglutination with both anti-A and anti-B	blood group AB
Agglutination with neither anti-A or anti-B	blood group O

Recording results

Record your results as above and note the degree of agglutination. Record the results from the class as a whole and compare them with the figures for Caucasians given above.

Discussion

In your discussion you should compare the results obtained from your class with those listed above and any possible reasons for differences should be considered. The natural antibodies, which you have not determined but can infer, are acquired after birth and it has been suggested that they result from immunisation with A and B antigens from the bacterial flora of the gut. A and B antigens are common in nature and are shared between a number of micro-organisms and man. For example, the ubiquitous bacterium *Escherichia coli* possesses B antigens. Discuss the reasons why the relative proportions of A, B, AB and O individuals vary from population to population and how you could determine whether blood group antigens are inherited.

Extensions and applications

The method described is a class one and results obtained should not be interpreted as meaning that an individual actually does possess the antigens indicated by the test. In most cases the results will be correct but blood donation and reception are matters of life and death and blood grouping should be carried out by a qualified medical laboratory technician. In fact sub-groups of the major groups, particularly A, should be determined. Agglutinogens (the A and B antigens) may be determined in the way described, but if the antisera (agglutinins) are to be determined then known samples of A and B blood must be added to the unknown serum. Rhesus groups may also be determined by agglutination techniques.

Further reading

Weir, D.M.
   1971 'Immunology for Undergraduates', 2nd edn., Livingstone, Edinburgh, Chapter 10
Humphrey, J.H. and White, R.G.
   1970 'Immunology for Students of Medicine', 3rd edn, Blackwell, Oxford, pp. 220-6

Hints for the teacher

This is a simple exercise but it is essential to stress that a fresh sterile lancet should be used every time because of the dangers of infectious hepatitis. Used lancets should be collected in a jar of strong disinfectant and disposed of carefully. An easy method of blood grouping is the use of Laboratory Class Eldoncards (Harris Biological Supplies Ltd.). These are supplied in packs containing everything necessary except lancets. In addition to ABO groups, Rhesus groups can be determined and each pack contains a useful illustrated pamphlet.

F.E.G. COX
King's College
London.

2 Viral haemagglutination

Time required
2 hr (30 min, 1 hr gap and a further 30 min).

Assumed knowledge
None

Theoretical background
Red blood cells tend to agglutinate in the presence of antisera against them and this agglutination is seen in its simplest form in the tests for the ABO blood groups of man (see Section 14.1). Red blood cells can easily be coated with a number of antigens and agglutination occurs in the presence of antisera against these antigens; the red cells act as passive carriers of the antigens and aid the recognition of the antibody-antigen reaction. The technique using antigen-coated red blood cells in this way is called passive haemagglutination. Haemagglutination techniques are used for the identification of antisera against disease organisms, for example the Salmonellae causing enteric fever, and for monitoring antibody responses to antigens in experimental systems. In brief, the red cells are coated with a substance such as tannic acid which causes an antigen to adhere to them, then coated with the antigen. Various dilutions of the antiserum are placed in wells in a perspex tray and the coated red cells are added to each well. Where there is an antibody-antigen reaction, the cells agglutinate and the titre of the serum is the greatest dilution which gives this reaction. In practice, red cells tend to agglutinate spontaneously, so an alternative method in which the antibody is allowed to react with the antigen before the coated cells are added is used. In this situation there is no antibody available to agglutinate the red cells and the reaction is inhibited. This technique is called the passive haemagglutination inhibition technique.

Another kind of haemagglutination is brought about by the action of viruses. Certain viruses, such as those responsible for mumps or influenza, cause red blood cells to agglutinate spontaneously, and this agglutination is inhibited in the presence of specific anti-viral antibodies. This reaction forms the basis of a test for detecting antibodies against viruses which affect red blood cells in this way. Viral haemagglutination will be demonstrated in this exercise.

Necessary equipment
Agglutination tray to hold 0·25 ml fluid per well (Biocult Linbro Disposo trays 5-MRC 96 in packs of 100 are ideal)
3 x 1 ml pipettes
50 ml saline
Mumps antigen (Wellcome Reagents VHO4)
Young chick (to be available before class)

Type of group
Pairs

Instructions to the student
You are provided with the following:
   agglutination tray
   1 ml mumps antigen
   2 ml 1% suspension of chicken erythrocytes
   normal saline
Put 0·05 ml saline in wells 2−12 in the first row. Put 0·1 ml antigen in well 1 in the first row. Prepare a series of dilutions of the antigen by taking 0·05 ml of the antigen from well 1 to well 2, mixing it and then repeating the process in well 3 and so on until well 11. Remove 0·05 ml from well 11 and discard. Add 0·1 ml of the red blood cell suspension to wells 1−12. Gently shake the rack to ensure even distribution of the cells, cover and leave at room temperature for 1 hr.
Read the test as follows:
Complete agglutination (+). Cells distributed evenly all over the bottom of the well.
No agglutination (-). Cells compacted in a small red button in the bottom of the well.
Partial agglutination (±). Non-agglutinated cells overlying a thin layer of agglutinated cells.
The titre is the highest dilution of antigen causing complete agglutination.
Well 12 is your negative control.

Recording results
Simply record the results you observe and compare the titre you obtained with those obtained by others in the class.

Discussion
Discuss the reasons why viruses cause red blood cells to agglutinate and the possible nature of diseases caused by such viruses. Discuss the possible errors which are inherent in this technique.

Extensions and applications
The viral haemagglutination technique can be used both for the identification of viruses if the antiserum

is known and for the qualification and quantification of the antiserum if the virus is available.

Further reading
Weir, D.M.
1971 'Immunology for undergraduates', 2nd ed., Livingstone, Edinburgh, pp. 140-1

Hints for the teacher
This is a simple non-immunological technique which is useful for introducing students to haemagglutination and the handling of immunological equipment and results. Ideally the gap of an hour in the middle of the exercise could be used for the human ABO blood grouping technique (see Section 14.1) Normal passive haemagglutination tests take too long to prepare and perform and the results tend to be disappointing in inexperienced hands.

Procedure
Bleed the chicken into 0·9% saline, 1 ml blood to 4 ml saline. Use a *freshly* killed chicken. Open it up rapidly and withdraw the blood from the heart using a hypodermic syringe fitted with a wide bore needle. Alternatively cut the brachial vein and bleed into a beaker. Wash the red cells in saline by centrifugation at 900 g for 10 min three times. After each centrifugation remove the 'buffy coat' of white cells on the top of the packed red cells. Finally resuspend to give a suspension of 1% (v/v) of packed cells in saline. Each group will require about 2 ml so 1 ml originally should be ample for about 25 groups. Distribute the blood in labelled tubes.

Add 1 ml distilled water to the antigen and add to 9 ml saline. Distribute in 1 ml amounts in labelled tubes. 1 ml will supply 10 groups with 10% antigen.

F.E.G. COX
King's College
London.

3 Immunodiffusion

Time required
1½ hr on day of exercise. ½ hr to interpret plates, 1, 4 or 7 days later. ½ hr to stain plates, but preparation takes about 4 days.

Assumed knowledge
None.

Theoretical background
When certain soluble antigens combine with antibodies a precipitate is formed. This can be demonstrated by allowing the two liquids to come into contact in a capillary tube when a ring of precipitation will occur at the interface between them. The two liquids may be separated by a layer of agar, in which case the precipitate will occur in the agar because of the diffusion of the antibody and antigen towards each other. The most widely used and satisfactory immunodiffusion technique is to introduce the antibody and antigen into wells in a layer of agar in a Petri dish. When this is done, the molecules diffuse out from the wells and form arcs of precipitate where the reacting substances come in contact with one another. Because different antigens diffuse into the agar at different rates, separate arcs are formed for each antibody-antigen reaction. If a number of wells are cut in the agar and filled with antigen and the antibody is introduced into another well equidistant from them, lines of precipitate will form. If the antigens are the same then the line of precipitate will be continuous; if not there will either be no reaction or the lines of precipitate will intersect one another. This technique, which is called double diffusion in two dimensions or the Ouchterlony technique, can be used for the detection and identification of antigens if the antibody is known or to detect the presence of antibodies if the antigens are known.

Necessary equipment
For each group of students:
3 x 4in Petri dishes
15g Ionagar (Oxoid)
150 ml 1% sodium azide in distilled water
2 x 500 ml conical flasks plugged with cotton wool + weights to hold down
Container big enough to act as boiling water bath for the two flasks

Tripod
Bunsen burner
1 x 25 ml graduated pipette
1 x 0·5 cm cork borer
1 needle with flat edge
Repelecote-silicon (Hopkin and Williams Ltd.)
1 ml rabbit anti-horse serum (Burroughs Wellcome)
1 ml horse serum (homologous serum) (Burroughs Wellcome)
1 ml calf serum (heterologous serum) (Burroughs Wellcome)
(any other precipitating sera will do)
3 x 1 ml immunological pipettes—disposable (Sterilin)
10 ml saline (0·9%)
3 Pasteur pipettes with teats
9 test tubes, 1in
1 test-tube rack
Wax pencil or pentel

For staining
100 ml 0·2% acid fuchsin (BDH) in 50-10-40 methanol-acetic acid-distilled water
1 litre 50-10-40 methanol-acetic acid-distilled water
500 ml beaker
3 x 3 x 2in slides
3 pieces of filter paper 3in x 2in.

Type of group
Best done in pairs.

Instructions to the student
You are provided with an antiserum and two antigens, one homologous with the antiserum, the other heterologous. Prepare three Ouchterlony plates as follows.

1 Make up a 1% solution of ionagar (w/v) in 1% sodium azide in a conical flask (*sodium azide is poisonous, as are its fumes—be very careful*). Allow to soak for a few minutes and then place in a pan of boiling water until the agar dissolves and the solution clarifies (about 20 min). Make up 150 ml.

2 At the same time place some distilled water in a conical flask in the water bath. Rinse a 25 ml pipette in the hot water and leave in the flask.

3 While this is happening, drop some silicone into a 4in Petri dish and allow it to spread until it coats the whole of the bottom.

4 Using the warm pipette and *working quickly*, place 20 ml agar into the Petri dish. Avoid bubbles. Keep the dish as level as possible. Allow

to gel (about 20 min). Replace the pipette in the boiling water. Prepare three such plates.

5 Using a cork borer, punch a hole approximately 0·5 cm diameter in the centre of the plate. Punch four other holes equally spaced with their edges 0·5 cm from the centre hole so that the final appearance is like this, $\circ^\circ_\circ$. Remove the centres with a flat needle inserted diagonally into each well.

6 Dilute your antigens and antiserum with saline to give dilutions of ½, ¼ and ⅛.

7 Carefully fill the centre well of one of your plates with the ½ antiserum. Fill two adjacent wells with one of your ½ antigens and the other two with the other ½ antigen. Prepare two other plates, one with ¼ reagents, the other with ⅛. Mark the bottom of the dish with a wax pencil or pentel so that you can identify your wells.

8 Place in an incubator at 38°C and examine after 24 hr, on a bench and examine after 4 days or in a refrigerator and examine after 7 days.

When you have prepared your plates and are waiting for them to cool down, test your antigens for reactivity with the antiserum using a precipitation technique in a capillary tube. Perform a separate test for each dilution of reagent.

1 Take a clean capillary tube and place it into the anti-serum at such an angle that about 1 inch flows into the tube. Run the liquid to the other end of the tube and draw up about 1 inch of serum. Hold your finger over the top of the tube and place the tube upright in a piece of plasticine.
(*N.B.* When performing this test the heaviest liquid must always be at the bottom, and this will determine whether the anti-serum or antigen is to be drawn up first).

2 Examine the tubes under a light against a dark background.

3 Record your results in terms of time taken for the reaction, intensity and appearance.

Permanent preparations of Ouchterlony plates

When the precipitation arcs are clearly visible, wash the plates thoroughly for 1−2 days in several changes of saline. Carefully remove the agar from the Petri dishes and place on a 3in x 2in microscope slide and trim off the surplus agar. Fill each hole with distilled water and cover the agar with a piece of wet filter paper. Leave at room temperature until the agar has dried completely (1−2 days). Stain the dried slides by placing them, with the filter paper in position, in a trough of 0·2% acid fuchsin in a 50-10-40 mixture of methanol-acetic acid-distilled water for 5 min. Rinse in tap water and de-stain in a methanol-acetic acid-water mixture until the gel is clear and the lines of precipitation stand out clearly. Finally rinse in distilled water and leave to dry in a vertical position.

Other exercises

When the basic techniques have been mastered you can carry out a variety of different exercises using Ouchterlony plates. For example, the antiserum can be fractionated by precipitation with ammonium sulphate to give albumins and globulins and the fractions tested against antigen in the central well. Alternatively, if the antigens are sera these too can be fractionated and tested against whole antisera or fractions of it.

Recording results

Hold the unstained plate against a dark background in such a way that it is illuminated from the side, and observe the lines of precipitation. Record your results by drawing the lines on a piece of unlined paper. Make sure that all the lines have been drawn and that their thickness and intensity have been correctly interpreted. Look particularly for 'lines of identity'—lines which join up with those of adjacent wells, and 'lines of non-identity'—those which intersect. The unstained plates can be recorded photographically by placing them on unexposed photographic plates and illuminating them from above. No rules can be set for this as only trial and error can produce satisfactory results. Stained plates can be drawn or traced. In all cases where a complex of experiments have been carried out an attempt should be made to interpret the meaning of each line.

Discussion

This exercise is basically one designed to teach a useful technique. The discussion must be based on ancillary experiments carried out.

Extensions and applications

The Ouchterlony technique is widely used for the identification of anti-sera in diseases, for example amoebiasis and experimental malaria. A single well is filled with antigen and the test sera together with positive and negative controls are placed in peripheral wells. In this way large numbers of sera can be tested and only a small amount of antigen used. This technique is also widely used to monitor immune responses to antigens administered experimentally.

The reaction can be examined over a period of time by taking serum from the experimental animal, diluting it and observing the greatest dilution at which the reaction is evident. The greater the dilution that gives a satisfactory result, the 'stronger' the antibody.

The Ouchterlony plate method can also be used in a wide range of cell biological investigations. For example, if a particular kind of cell is used as an antigen to immunize a rabbit the antibody will react with the original cell extract. The antibody will also react in different ways with fractionated components of the cell if these are placed in peripheral wells around a central well filled with antiserum. In this way pure cell fractions can be checked immunologically and impurities noted.

Further reading

Ouchterlony, O.
1973 'Immunodiffusion and Immunoelectrophoresis', in Weir, D.M. (ed.) 'Handbook of Experimental Immunology', Blackwell, Oxford

Hints for the teacher

Agar should normally be made up in a buffer solution. 1% azide provides a suitable ionic strength for diffusion and also acts as a preservative which means that the plates do not become overgrown with micro-organisms.

Prepared antisera are expensive and suitable materials are often available from immunologists. Check that the reagents do produce a precipitate, though.

F.E.G. COX
King's College
London.

4 Cunningham plaque assay for antibody-secreting cells

Time required
5 hr, approximately, including 1 hr for incubation when students may be absent.

Assumed knowledge
1 Standard pipetting, including use of 0·1 ml pipettes. Ability to manufacture dropping micro-pipettes
2 Use of haemocytometer
3 Meaning of 'antigen' and 'antibody', 'immune response'.

Theoretical background
This practical demonstrates antibody production by single cells, using the Cunningham modification of the Jerne plaque assay, and shows the development of an immune response following injection of antigen.

A plaque assay for identifying and counting individual antibody secreting cells was first described by Jerne, Nordin and Henry.[6] Principally it has been used to study immune responses to erythrocyte antigens because the assay reveals antibody-secreting cells by localised haemolysis, but by binding other antigenic molecules onto erythrocytes it is possible to obtain lysis by antibody directed against the other antigenic specificities. Such antigens include haptens like DNP (dinitrophenyl), proteins like bovine serum albumin, bacterial antigens such as pneumococcal polysaccharide and so on.

The assay is simple and its most popular application has been the straightforward measurement of the immune response, e.g. following various treatments of the cells involved in the inductive stages of antibody formation. It is also versatile, and can be readily modified to investigate many properties of antibody secreting cells. Thus, it is possible to study the class of immunoglobulin made by the cells by using appropriate 'developing' antisera (see later). Tritium autoradiography can be performed on plaque-forming cells to provide detailed information on the kinetics of the proliferation of their precursors (Tannenberg and Malaviya,[10] Koros, Mazur and Mowery[7]). Techniques have also been devised for isolating individual antibody secreting cells by micromanipulation following their identification by the plaque assay (Nossal, Cunningham, Mitchell and Miller,[8] Nossal and Lewis[9]).

In the original version of the assay, (see Fig. 1, direct path), a mixture of two cell suspensions is immobilised in semi-solid agarose over a solid agar base: one suspension consists of the lymphoid cells whose content of antibody-forming cells is to be

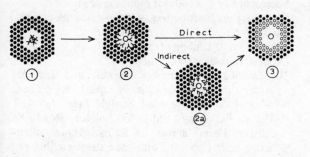

Figure 1 Principle of plaque formation.

    SRBC = sheep erythrocytes

    Ab = anti-SRBC antibody

DIRECT	INDIRECT
(Measures IgM)	(Measures IgG)
1 Plaque-forming cells secrete Ab	1 Plaque-forming cells secrete Ab
↓	↓
2 Ab coats nearest SRBC	2 Ab coats nearest SRBC
↓	↓
3 Complement lyses Ab–SRBC	Add developing serum
	↓
	2a Anti-Ab fixes to Ab
	↓
	3 Complement lyses complex

assayed, the other of sheep erythrocytes if these were the immunising antigen. If some other antigen had been used for immunisation, the second cell suspension would be erythrocytes 'sensitised' by previously coupling the immunising antigen chemically to the erythrocyte membrane. After incubation for about an hour at 37°C, during which the released antibody binds to erythrocytes immediately adjacent to the antibody secreting cell, guinea-pig serum is added as a source of complement to complete the lysis of erythrocytes by binding to the red cell-antibody complex. The result is to create a localised zone of haemolysis surrounding the antibody-releasing cell, which may be easily seen at a low magnification as a clear area with the cell at its centre. Since lysed erythrocytes (erythrocyte 'ghosts') are invisible by direct illumination, the clear area will appear empty except for the antibody-secreting cell. Occasionally more than one leucocyte or a number of erythrocytes may be seen inside the clear area. This usually occurs when the lymphoid cell suspension has been used at a high cell concentration (appropriate when the proportion of antibody forming cells is expected to be low, for example early in the response). These cells which 'contaminate' a plaque are rat cells derived from the original suspension, and this is why the erythrocytes are not lysed. (Incidentally, rat erythrocytes are noticeably larger than sheep erythrocytes). Plaque formation requires the active metabolism of the secreting cells and can be inhibited by poisons such as dinitrophenol, puromycin and actinomycin D (Ingraham and Bussard[5]).

This procedure detects antibodies of *high* haemolytic efficiency, chiefly IgM, one molecule of which is sufficient to fix complement and set in motion the chain of events leading to the lysis of one erythrocyte. It may be extended to detect antibody of low haemolytic efficiency (mainly IgG, of which several thousand molecules may be needed: some IgG subclasses cannot fix complement at all and can only cause agglutination) if a 'developing' anti-globulin serum is added after the first incubation and before the addition of complement (Dresser and Wortis[3]) (see Fig. 1, indirect path); such a serum (raised by immunising animals with immunoglobulins of the species whose lymphoid cells are being examined—in this case, rat immunoglobulins) latches on to the IgG antibody molecules stuck to the surface of the target erythrocytes, and the whole complex can then fix complement to lyse the red cell. Plaques obtained in the *absence* of developing serum (indicating IgM-secreting cells) are called *direct* plaques: the extra plaques developed using anti-globulin serum (indica-

ting mainly IgG-secreting cells) are known as *indirect* plaques. (You should note that the action of developing serum is actually much more complex than this simplified account would suggest: its effects are crucially dependent upon its concentration—high concentrations *inhibit* plaque formation—(Why?).

A streamlined version of the original Jerne assay due to Cunningham is described here (Cunningham, Smith and Mercer,[1] Cunningham and Szenberg,[2]). All the reagents are incubated with the cell suspensions simultaneously, not sequentially; and the use of agar as a supporting medium is avoided (eliminating the need for operating at 45°C to keep the agar molten) by employing cell suspensions as monolayers. The Cunningham method is also cheaper and more sensitive (yielding about 2–3 times the number of plaque-forming cells).

Necessary equipment
Each pair of students will need the following
2 rats, appropriately immunised*
1 dissecting board and 6 pins
2 pairs scissors (coarse and fine)
2 pairs forceps (e.g. dagger forceps)
2 pairs fine forceps (e.g. watchmaker's forceps)
2 Petri dishes (e.g. 5 cm diameter)
70% alcohol
Absorbent cotton wool
2 small filter funnels (e.g. 5·5 cm diameter) with small sterile cotton wool plug
2 x 15 ml pointed centrifuge tubes
4 specimen tubes (e.g. 75 x 9 mm)
Pasteur pipettes and teats
1 haemocytometer (e.g. Neubauer-ruled; Section 13.7)
Red magic marker or grease pencil
Microscope (e.g. x100, for counting cells)
Dissecting microscope (e.g. x14 for counting plaques)
Ice bucket, ice
10 ml, 1 ml, 0·1 ml pipettes
Dropper to deliver 0·025 ml
10 Cunningham chambers: manufacture each by laying three strips of adhesive tape, sticky on both sides, (¼in double-sided Scotch Tape, No. 410: Herts. Packaging Ltd., 53 London Road, St. Albans, Herts) across an alcohol-cleaned microscope slide (7·5 x 2·5 cm)—one strip at either end and one in the middle; gently press a second slide on top to form chamber (Fig. 2)

*Wash sheep blood (Wellcome) three times in phosphate-buffered saline. Inject intravenously 1 ml 1% (v/v) washed erythrocytes, at varying times before the class.

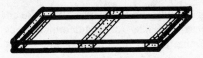

Figure 2 Cunningham chamber. Stippled strips represent double-sided sticky tape.

Sheep blood (0·4 ml of 25% suspension in handling buffer), washed thrice in phosphate-buffered saline freshly on the day of the practical
100 ml handling buffer, sterile: (Dulbecco 'A' + 'B' (Oxoid) containing 10% (v/v) foetal calf serum or a serum (e.g. horse) checked to be non-toxic before the class).
1 ml guinea-pig serum (see 'Hints for the teacher')
0·4 ml developing serum (dilution in handling buffer of rabbit anti-rat γ globulin serum: see 'Hints for the teacher')
White blood cell counting fluid (1% acetic acid in distilled water, with 0·15% w/v toluidine blue).

The following is shared equipment for whole class.
37°C incubators
Milligram balances
Centrifuges to run at 4°C (MSE Mistral)
Wax baths in heated sand trays (Gallenkamp) (mix Vaseline and paraffin wax in equal proportions).

Type of group
Work in pairs, dealing with two rats per pair. Each rat should be taken at a different time after immunisation, so that between them the whole class can draw out the time course of the immune response.

Instructions to the student
Prepare cell suspension
You are provided with two recently dead corpses of rats killed at various times after an intravenous injection (via tail vein) of 1 ml 1% (v/v) sheep erythrocytes (SRBC). This dose contains about 2×10^8 red cells. There will also be a group of non-immunised rats. The following instructions apply to each rat: carry them out with one animal while your partner works with the other.

Pin out the rat, belly up, and swab the ventral skin with 70% alcohol. This stops hair getting everywhere. Remove the spleen with the forceps and scissors provided (*do not crush the spleen:* hold it by the hilar membrane) and transfer it to the weighed Petri dish. Re-weigh and note the spleen weight. Add 5–10 ml of the handling medium, chop the spleen into about 8

large chunks and tease it apart with the two pairs of fine forceps until all large lumps have been disaggregated.

Filter the suspension through cotton wool into the conical 15 ml centrifuge tube and wash through the cells remaining on the filter with handling medium using a Pasteur pipette to a final volume of 15 ml. Centrifuge the cells for 10 min at 450 g and 4°C, remove the supernatant and resuspend in 10 ml of handling medium using a Pasteur pipette, centrifuge and resuspend to 10 ml again. After mixing the cells well, remove a 0·1 ml sample and dilute in 0·9 ml white blood cell counting fluid (1% acetic acid in distilled water with 0·15% toluidine blue). Count the suspension in a haemocytometer. On the basis of this count calculate:

1 The concentration of leucocytes in the final suspension
2 The total cell yield obtained from the spleen
3 The cell yield per mg spleen. This is a check on the preparative technique
4 An appropriate dilution to assay. This depends on the time since immunisation and will be explained in greater detail.

Elementary precautions to be observed when handling cells.
To keep cells in good shape you must *treat them gently*. In particular, *don't* centrifuge unnecessarily vigorously: 450 g is quite adequate for these cells and the 15 ml tubes you will use. *Don't* blow air bubbles into the suspension when resuspending cell pellets: the bubble is itself traumatic, and cells trapped in foam at the surface are subjected to excessive shearing forces because of surface tension. *Don't* subject cells to unnecessary temperature changes, which interfere with their metabolism. Work at 0–4°C all the time until you are ready for the assay. *Always* remember to mix a suspension thoroughly before sampling for any reason (e.g. counting or for assay). Cells settle surprisingly quickly.

Prepare an appropriate cell dilution to assay
Assay
(Description for one suspension only: repeat in exactly the same way for the other). Arrange 4 sample tubes in the ice bucket (duplicate tubes for incubations with or without 'developing' antiglobulin serum). Mix, in the order stated in the table, reading down the list.

Tube	*A*	*B*	*C*	*D*
Handling medium			1 drop	1 drop
Developing serum	1 drop	1 drop	-	-
Complement, 2 drops	All tubes			
25% SRBC, 1 drop	All tubes			
Spleen cell suspension, 0·1 ml	All tubes			

Notes:
1 'Developing serum' is a dilution in handling medium of an anti-(rat heavy chain) serum raised in a rabbit.
2 Complement is recently thawed neat guinea-pig serum, absorbed with SRBC (why?); keep this on ice right until you need it, since it is labile.
3 25% SRBC is thrice-washed sheep blood, commercially obtained and diluted finally ¼ in handling medium.
4 Use a 0·1 ml pipette to measure the spleen cell suspension; wipe the outside with absorbent paper before expelling the contents.
5 1 drop = 0·025 ml approximately, using the droppers provided (Fildes and Smart[4]).

Filling the chambers
Mix the contents of a sample tube by resuspension in a Pasteur pipette (no frothing!) hold the chamber approximately at an angle to the vertical in one hand, appose the tip of the pipette to the lowermost point of the edge of the bottom half of the chamber, and run the contents in gently using the capillarity of the glass to fill the half-chamber, continue with the remainder for the top half of the chamber. If you try to run it in too fast, it will go all over the floor; but don't be too slow, or the cells will settle differentially within the pipette and an uneven monolayer will result. The volumes of the samples have been so arranged that the contents of one sample tube should nearly fill one chamber: certainly there should be no excess unused.

You are provided with two extra chambers simply to practise filling (without cells).

Seal the chambers along their edges with molten wax. Incubate on a *horizontal* surface for 1 hr at 37°C. Check the sealing after 10 min incubation and re-seal if necessary—there should be no signs of evaporation along the edge.

Counting
Mark each half of the chamber on one side with a 4 x 4 grid using the red marker. With the chamber on the *horizontal* stage of a stereomicroscope, arrange the illumination so that plaques appear dark against a bright background (of unlysed cells): it needs to be slightly incident, not direct. Count all the plaques, using the grid pattern to keep track of where you have counted.

Do not keep the chambers at an angle for any length of time, or the plaques lose their shape.

Calculate
1 The mean total plaques per chamber.
2 The mean direct plaques per chamber.
3 The mean indirect plaques per chamber.
Convert these to plaques per 10^6 cells.

Recording results
See 'Instructions to the student'. Also plot the numbers of direct and indirect plaque-forming cells per 10^6 spleen cells on a log scale against time after immunisation, using data from the whole class.

Discussion
Examine the kinetics of the immune response. Is it likely that the increased numbers of plaque-forming cells arise by division of pre-existing plaque-forming cells, or are more recruited from an outside source?

What experiments might you do to find out whether individual cells switch from IgM to IgG synthesis, or whether the later IgG synthesis is performed by a different population of newly differentiated cells?

Extensions and applications
See 'Theoretical background'

Further reading
1 Cunningham, A.J., Smith, J.B. and Mercer, E.H.
    1966 J. Exp. Med. 124, 701
2 Cunningham, A.J. and Szenberg, A.
    1968 Immunology, 14, 599
3 Dresser, D.W. and Wortis, H.H.
    1965 Nature, Lond., 208, 859
4 Fildes, P. and Smart, W.A.M.
    1926 Brit. J. Exp. Path., 7, 68
5 Ingraham, J.S. and Bussard, A.
    1964 J. Exp. Med., 119, 667
6 Jerne, N.K., Nordin, A.A. and Henry, C.
    1963 'Cell Bound Antibodies', ed. Amos and Koprowski, Wistar Inst. Press, Philadelphia, p. 109
7 Koros, A.M.C., Mazur, J.M. and Mowery, M.J.
    1968 J. Exp. Med., 128, 235

8 Nossal, G.J.V., Cunningham, A.J., Mitchell, G.G. and Miller, J.F.A.P.
       1968 J. Exp. Med., 128, 839
9 Nossal, G.J.V. and Lewis, H.
       1971 Immunology., 20, 739
10 Tannenberg, W.J.K. and Malaviya, A.N.
       1968 J. Exp. Med., 128, 895

Hints for the teacher

1 A simpler version of the experiment, requiring less time, and not involving the preparation of the anti-immunoglobulin reagent, would measure only direct plaques. However an important immunological phenomenon (the IgM–IgG sequence) is thereby missed.

2 Common snags: locating the spleen, resuspending cells (not using Pasteur pipette), use of counting chamber, distinguishing plaques from air bubbles, sealing chambers with wax (chambers immersed too far).

3 Number of cells to assay per chamber: best determined by preliminary experiment using strain of rat that will be used in class.
   As a guide:

Days after immunisation:	0–3	4–5	6 and later
No. of cells to assay:	Use suspension undiluted	2×10^5	10^6

4 Preparation of materials:

   (a) *Rats* should be immunised intravenously on various days before the class. Include two unimmunised rats. (Time: washing of sheep blood—45 min intermittent: injection—5 min). Animal licence and certificate A needed.

   (b) *Complement:* pool fresh guinea-pig serum from at least 2 donors well in advance: absorb with washed SRBC (30′, 4°C; cells: serum approximately 1:5): freeze in 1 ml aliquots. Assay for complement activity (see Kabat, E.A. and Mayer, M.M. 'Experimental Immunochemistry', 2nd edn., Thomas, New York, 1961, p. 149; titre to be greater than 180 units/ml. Store deep-frozen (preferably −70°C, otherwise, −20°C) until day of use. Time: 1½ days.

   (c) *Handling buffer:* important to check that serum (if not foetal calf) is not toxic, by preliminary experiment. Time: ½ day.

   (d) *Cunningham chambers:* as described in 'Necessary equipment'. Time—about 1 hr per 100 chambers, including cleaning of slides.

   (e) *Anti-immunoglobulin antiserum:* raise by immunising intramuscularly at least 2 rabbits with 10 mg ammonium-sulphate-precipitated rat globulins (Kekwick, Biochem. J. 34, 1248 (1934)) in Freund's complete adjuvant (Difco). At least three immunisations, spaced at 10–12 week intervals, should be given. Bleed 14 days after the last immunisation, absorb with SRBC and freeze. Serum must be assayed at several dilutions before the class to determine optimal concentration for development of indirect plaques. As a guide, a dilution of 1/80 has been found best for one serum raised in this laboratory.

5 Cell yield per spleen should be about 0·7 to 1·5 x 10^6 cells/mg net weight.

S.V. HUNT
Oxford.

5 Separation and assessment of lymphocyte function

Time required
Separation of lymphocytes 90 min by method A or B; 30 min by method C. Rosettes 120–150 min, this includes the time of initial separation by method C above. Histocompatibility 60 min after separation and counting of lymphocyte population. 120 min for harvesting.

Assumed knowledge
A working familiarity with laboratory techniques, but more importantly with radioactive handling and preferably tissue culture, is necessary. Third-year students should be able to do this. For consistent results one needs practice as much as anything.

Theoretical background
Lymphocyte is a generic name given to small mono-nuclear cells found circulating in the blood, lymph nodes, thymus, bone marrow and other tissues. These cells are thought to be the basis for both 'cellular' and 'humeral' immunity. Although morphologically similar, i.e. a round cell with a large nucleus and a small amount of cytoplasm, there are probably several different cell lines present within this group. They form a subgroup of blood leucocytes (white cells) which also include polymorphonuclear cells, large mononuclear cells, macrophages, and many undefined cell types. The proportions of these different cell lines varies from organ to organ and in disease states, and also probably in normal people from time to time. The basic function of lymphocytes is to recognise antigens and respond by division and the production of immunoglobulins and other effector molecules.

Representative samples of these cells can be isolated from the peripheral circulation and their function then tested *in vitro*. For practical purposes it is far simpler to recognise lymphocyte response by cell division and DNA synthesis than by production of immunoglobulin or effector molecules. Though it must be realised that since lymphocyte subgroups can respond in a different way to the same antigen (mitogen) then the type of response tested will partially define the lymphocyte population being assayed. Note, mitogens are theoretically substances causing cells to multiply, without implying immunological sensitivity to these substances. However, it is now recognised that mitogens can stimulate lymphocytes to produce immunoglobulin and effector molecules without dividing and further that the lymphocyte spectrum responding to different mitogens varies.

In the following experiments the general methods for isolation of the lymphocytes will be demonstrated. It is possible to get purer populations by combining methods. For instance, following the sedimentation procedure, the harvested lymphocytes can then be put on a density gradient. Phagocytic cells can be removed following ingestion of iron filings by gently applying a magnet, which will pull the phagocytosing cells down leaving the lymphocytes etc. in the supernatant. Adherent cells such as macrophages can be removed by leaving the separated white cells in contact with a glass surface, when the macrophages will adhere to the glass. For most practical purposes, these refinements are not necessary. The following methods depend on either the tendency of red cells to form rouleaux (aggregates) (this is helped by gelatin or dextran) or the slight dehydration of the red cells (by Ficoll) thus increasing their density and causing these cells to pass through the Triosil density gradient.

After separation of the lymphocytes there are usually three questions to be answered. Firstly, what proportion of the lymphocytes will respond to a non-specific mitogen such as PHA (phytohaemagglutinin), pokeweed or Concanavalin A; this represents viable T (thymus-derived) cells, i.e. cells concerned with cell-mediated immunity, and it is a method of testing the patient's immune competence, and potentially of following immunosuppressive treatment. Secondly, are the parient's lymphocytes sensitive to a particular antigen such as candida or tuberculin and if so to what extent; this again can be used to help localise defective immune responses. Thirdly, a variation of the first two questions which is a means of testing for histocompatibility where the histocompatibility antigens on the lymphocyte are used as activators of added lymphocytes from another individual. This is done by the mixed lymphocyte reaction, or its much better refinement the 'one-way reaction', in which lymphocytes from two different individuals are brought together. The first two tests can be augmented by testing for rosette formation, which is used non-specifically as described here to identify T cells. A slightly different test can be performed using red cells coated with an antigen;

this is a much more specific test for the particular lymphocytes responding to that antigen.

Not described here are methods for looking at the immunoglobulins on the surface of the lymphocytes, which can be shown by immunofluorescence or isotope techniques and which pick up B (bursa-derived) cells or immunoglobulin-producing cells. And lastly, but of increasing vogue, are the tests for humeral effectors (lymphokines) produced by activated lymphocytes. These are substances of varying molecular weight which are liberated by lymphocytes and produce their effect on other cells, some of these effects being target-cell specific, others not. Such effects have been described on the macrophage whose metabolism may be increased but mobility reduced; other changes include the ability to produce chemotaxis and erythematous skin reactions, cellular division and also cytotoxic effects on target cells. These are all rather specialised tests still being evaluated and need corresponding laboratory facilities.

Experiment I Separation of Lymphocytes
This is the initial step before all other procedures.

Necessary equipment
All materials used for tissue culture must be bought ready prepared or washed in detergent such as 5% solution Decon 90 (Gallenkamp) and then 2% hydrochloric acid, followed by several washings in water before being sterilised.

1 Glass or plastic syringes (e.g. 10–20 ml) with No. 1 needles.
2a Preservative-free heparin (1000 iu/ml) (BDH)
   or
2b A method of defibrination, e.g. glass beads in a 100 ml conical flask or wooden applicator sticks and 25 ml universal bottles.
3 Sterile 25 ml glass or plastic screw-cap universal bottles (Sterilin)
   Sterile Pasteur pipettes plugged with cotton wool.
   Sterile glass pipettes 1, 2, 5, 10 ml plugged with cotton wool.
   Media: Eagles Minimal Essential, TC 199 can be used. For class work HEPES modification (Flow Labs) is advisable to avoid excessive alkalisation from loss of CO_2
4a Triosil 440 (Vestric Ltd.)
   Ficoll (Pharmacia)
   Dilute Triosil to 34% i.e. 20 ml Triosil 440 + 24 ml water
   Dissolve Ficoll in water to 9% by weight.

Reagent for use: 40 ml diluted Triosil and 96 ml Ficoll Solution. Mix and sterilise by autoclaving in 4 oz bottles.
   or
4b 3% gelatin solution in normal saline or phosphate-buffered saline, sterilised by filtering through very hot 80°C Sietz filter (using pressure from gas cylinder). Solution is not stable and should be used within 24 hr.
   or
4c Plasmagel (Laboratoire Roger Bellon, 159 Avenue du Poule, Neuilly, France). (This is a stable gelatin solution that is already prepared and sterilised).
5 Haemocytometer (improved Neubauer). White cell counting fluid, 3% acetic acid with drop or two of background stain, e.g. Giemsa, added.
6 Microscope—preferably with phase-contrast attachments, and fluorescence, if fluorescent staining is employed (see below).

Type of group
Students working in pairs.

Instructions to the student
1 10–20 ml of venous blood is drawn from either human or rabbit (see 'Hints for the teacher').
   To separate lymphocytes the blood must be prevented from coagulating. This is most easily done by using heparin at 20 units/ml. However, fibrinogen is not removed by this method and fresh heparin must be used in the subsequent washing procedures to prevent fibrin being formed. A cleaner but more time-consuming method is to use a 100 ml flask with glass beads in it to agitate the blood until the clot has formed on the beads, or to stir the blood with a sterile wooden stick until the clot has been removed (rabbit ear vein bleeding without a syringe must be done into heparinised containers).
2 Lymphocytes can now be separated from blood in three ways (Fig. 1).
   (a) Gravity sedimentation: The heparinized blood is allowed to settle in the container which should be about 3 times as long as wide and set at a slight angle. The use of the original syringe is very useful for this with the nipple and needle pointing upwards. After about 1½ hr there will have been clear separation into opalescent supernatant and dark red lower layer. Occasional taps on the syringe barrel will help to dislodge the red

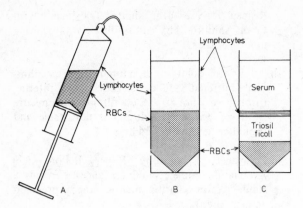

Figure 1 Lymphocyte stimulation: (A) Gravity sedimentation; (B) Gelatin or dextran sedimentation; (C) Density gradient. (RBCs = red blood cells)

cells adhering to the syringe wall. The needle is replaced and bent over and the supernatant ejected into a fresh container.

(b) Gelatin or plasmagel sedimentation: Spin defibrinated blood at 1000 g for 5–10 min until clear serum supernatant is present. Mark upper layer of serum on bottle. Aspirate (suck off by vacuum) $\frac{2}{3}$ or ¾ of serum and save. Add gelatin of volume equal to red cells and remaining serum. Recap and invert several times and leave for 5 min. Now increase volume by 50% with media and again invert several times avoiding bubbles. Transfer to fresh container (preferably plastic) avoiding contact with cap (so that red cells will not be retained on cap). Leave for about 1 hr. Once separation has started an occasional tap or twist of the bottle will dislodge red cells on wall.

(c) Density gradient: If heparinised whole blood is to be used then it should be diluted with 3 parts heparinised media and the density gradient should have heparin added.

Alternatively, defibrinated blood is spun down (1000 g for 5 mins) and 2/3–¾ of the serum removed and saved. If this serum is to be used further, it should be recentrifuged at 2000 rev/min to remove any remaining cells. The original volume is restored with media. In either case an equal volume of the reconstituted blood is carefully layered upon an equal volume of the Ficoll-Triosil mixture. The bottle is then centrifuged at 2000–3000 g for 10 min as soon as possible. The bottle is removed and by using transmitted light it will be seen that it consists of four layers (Fig. 1) a lower packed red cell layer over which is an opalescent layer surmounted by a narrow dense band (which contains the lymphocytes) which in turn is covered by a layer of serum.

Using a Pasteur pipette, the narrow dense band is carefully aspirated until it disappears, a certain amount of the adjacent layers will also be removed. Dilute this aspirate with an equal volume of media and centrifuge. Remove most of the supernatant and gently agitate the remaining cell pellet, which may look disappointingly red! Now dilute with 5 ml of media and thoroughly mix. Withdraw 0·1 ml and add to 0·9 ml of the white-cell counting fluid. Count lymphocytes in the haemocytometer. The concentration and thus total number of lymphocytes are now known. As a working rule it will be found that 1 ml of blood provides 10^6 lymphocytes with (i) from 5% (Method (c)) to 30% (Methods (a) and (b)) polymorphonuclear leukocytes and (ii) one red cell to each leukocyte.

Viability of lymphocytes

Before using the lymphocytes it is useful to test the viability of the cells. The easiest method is to add 1 drop of 1% eosin (Gurr) to 7 drops of cell suspension, wait for 2 min and count the percentage of pinkish cells. These should be less than 5%. A more accurate and sensitive method is to use Fluorescein diacetate (Koch-Light) 5 mg/ml acetone stock solution kept at $-20°C$, diluted 1/1000 just before use. Add 0·1 ml to 0·9 ml cell suspension, warm to 37°C for 2 min and then examine under fluorescent light and count cells that fluoresce. These are cells that are damaged and have allowed penetration of fluorescein diacetate, which is then converted to the fluorescent compound. Thus fluorescent cells are damaged cells.

Experiment II Lymphocyte culture with mitogens

Necessary equipment

1 3in x ½in glass tubes with corresponding loose-fitting 'Oxoid' aluminium caps. Place tubes in metal trays (Luckham Ltd.) with cap on, put in

plain metal tin or paper bag, and sterilise by autoclaving.
2 *Wire* racks to hold above-mentioned tubes
3 Large bell dessicator or airtight container for culturing above-mentioned tubes in racks in 5% CO_2 in air
4 Source of 5% CO_2 in air
5 $37°C$ incubator
6 Phytohaemagglutinin (Wellcome), Concanavalin A (Miles Seravac). Candida antigen is prepared by freezing and thawing cultured organisms. then sonicating them and sterilising extract by Seitz filter (Astell Labs). Foetal calf serum (may be used instead of human serum).

Instructions to the student

The lymphocytes obtained above are diluted with media to 10^6 cells per 0·7 ml (if insufficient cells have been obtained then a lesser concentration above $5 \times 10^5 / 0·7$ ml may be used).

The cultures are now set up in 3in x ½in tubes in quadruplicate as follows, *adding the constituents in the order shown:*

	A	B		
Tubes	1—4	5—8	9—12	13—16
Serum	0·2 ml	0·2 ml	Controls and	
Cell suspension	0·7 ml	0·7 ml	different cells	
Mitogen		0·1 ml		
Media	0·1 ml			

The culture tubes in their racks are placed in a desiccator or other gas-tight box and 5% CO_2 in air is blown in with the lid of the desiccator off centre. The gas is turned off and the lid rapidly replaced. Incubate at $37°C$; re-gas after every manoeuvre. Tubes 1—4 form the background control for unstimulated cells. Tubes 5—8 show the result of stimulation. It is usual in this type of work to compare different cells or the effect of different culture conditions (serum, ' se of mitogen, time of culture). For this tubes 9—12, 13—16 etc. can be used. Remember that cells from different donors need their own control (unstimulated) tubes.

The degree of activation of the lymphocytes is represented by $B-A$ or BA. Using this basic scheme, the following experiments can be designed and must be done before the system is used practically.

1 What is the optimum time for measuring the lymphocyte response?
2 What is the optimum concentration of antigen and how does the response curve vary with different concentrations of antigen?
3 How long must antigen be present for optimum response?

Measurement of activation

The easiest method is to add $1 \mu Ci$ ^3H thymidine (Radiochemical Centre) to each culture 4 hr before harvesting cells.
The cells can be harvested in their culture tubes as follows (*non-sterile equipment*):
1 Centrifuge at 900 rev/min (MSE Super minor).
2 Aspirate or pipette off supernatant.*
3 Add 2 ml phosphate buffered saline (PBS Dulbecco A, Oxoid).
4 Centrifuge.
5 Remove supernatant.
6 Resuspend in 1 ml PBS.
7 Add 1 ml 10% TCA, mix, wait 15 min (TCA kept at $4°C$).
8 Centrifuge.
9 Remove supernatant.
10 Add 1 ml methanol.
11 Centrifuge and remove methanol.
12 Resuspend pellet in 1 ml methanol.
13 Wash into scintillation vial with two washes of scintillation fluid.

Zipette type dispensers (Jencons) are ideal for the PBS and TCA washes since they wash material from the sides of the tubes. The final 1 ml of methanol must be accurately dispensed.

Scintillation fluid

Either hyamine hydroxide, 1OX (Koch-Light) and toluene, or Triton X100 (1 part) and xylene (2 parts) with scintillator added in either case.

Morphological assessment

After spinning the cultures down, a drop of plasma is added to each tube and the cells thoroughly resuspended:
1 A smear of these cells is made and rapidly dried.
2 Methanol wash for 3 min.
3 May-Greenwald (Gurr) stain 2 min.
4 Giemsa—diluted 1/10 for 10 min.
5 Wash twice in saline
6 Dry and mount.
7 Count large cells with multiple nucleoli. This can

*With practice, careful decanting of the supernatant can be undertaken and the final drop of fluid removed by touching the test-tube lip on tissue paper.

be fully appreciated only by experience and comparison with controls.

Autoradiography

Activation may be assessed by using autoradiography to determine how many cells have undergone DNA synthesis during the period of addition of ^{3}H thymidine. This method takes 1−2 weeks before a result is obtained but does provide additional data, and can be used in conjunction with staining techniques.

Experiment III Histocompatability testing

The principle is the same as when mitogens are used, but the activation is now due to histocompatibility antigens on cells from different donors using lymphocytes as the source of antigen. Thus cells from two individuals are mixed in equal quantities. However, since both sets of cells could react and summate to the response, one set of cells is rendered incapable of responding by blocking DNA synthesis, either by irradiation or the use of drugs. In practice it is important to know whether the host will reject the donor and thus the donor cells are blocked. By this means several potential donors can be compared. The least activation indicates the best donor.

Material: Mitomycin C, Kyowa Hakko Kogyo Co. Ltd., Ohtemachi Building, Ohtemachi, Chiyoda-Ku, Tokyo.

Instructions to the student

Separate both sets of cells as above.

Donor cells are irradiated with 1200 rad or incubated for 30 min with 25 μg/ml mitomycin C and then washed once with media.

Set up cultures as before using 5×10^5 cells from each donor.

	A	B	C
Tubes	1−4	5−8	9−12
Serum	0·2 ml	0·2 ml	0·2 ml
Cells			
suspension A	0·4 ml	−	0·4 ml
Cells			
suspension B	−	0·4 ml	0·4 ml
Media	0·4 ml	0·4 ml	−

Degree of incompatibility $C-(A + B)$. The maximum response occurs at 6 days.

Experiment IV Rosette formation

Instructions to the student

1 Separate lymphocytes as above, ideally by method C.
2 Wash three times in Hanks' media.
3 Bring concentration of lymphocytes to approx. 3×10^6 ml in well dispersed state.
4 Wash sheep red cells 3 times in Hanks' solution, then dilute sheep red cells to 1% in Hanks' solution (the sheep red cells in serum will keep for 5−7 days at 4°C) (Burroughs Wellcome).
5 Add 0·2 ml of lymphocyte suspension to 0·2 ml sheep red cell solution.
6 Gas with 5% CO_2 in air and spin at 500 rev/min for 7 min.
7 Incubate for 60 min at 4°C, gently agitating occasionally.
8 Dilute the incubated cells with an equal volume of media, place a drop on a haemocytometer, and count immediately the number of rosettes formed (a rosette has more than 5 red cells around the lymphocyte).

Since final concentration of lymphocyte and the number of rosettes formed is known, the proportion of lymphocytes forming rosettes can be worked out.

Further reading

Bach, F.H. and Voynow, N.K.
   1966 'One way Stimulation in Mixed Leukocyte Cultures', Science, 153, 545
Böyum, A.
   1968 'Leukocyte Separation', Scand. J. Clin. Lab. Invest 21, Suppl. 97
Ling, N.R.
   1968 'Lymphocyte Stimulation', North-Holland, Amsterdam

Hints for the teacher

1 A Home Office licence is required for the withdrawal of blood from living animals, and a medical colleague should supervise if human blood is used.
2 The isolated lymphocytes, if kept in 20% serum at room temperature, can be used the following day but should preferably be used within 4 hr. Heparinised whole blood at room temperature must be used within 2 hr.
3 Phytohaemagglutinin−dilute ampoule to 50 ml with sterile normal saline and use 0·1 ml of this to 1 ml final culture volume. Range about $\frac{1}{10}$−10 times this concentration. Concanavalin A, final concentration 10 μg per ml, range 1−25 μg.

4 Metal trays can be obtained from Luckham Ltd.
5 Radioactive materials and contaminated glassware must be disposed of according to the
   regulations in force in the College. Consult the
   Radiation Protection Officer.

P.J.L. HOLT
Medical School
Manchester.

Book list

Data reference books

'Difco Manual of Dehydrated Culture Media and Reagents for Microbiological and Clinical Laboratory Procedures', Difco Laboratories Inc., Detroit, USA

Dawson, R.M.C., *et al*, 'Biological Laboratory Data', Methuen, London (1965)

Hale, L.J., 'Biological Laboratory Data', Methuen, London (1965)

Stecker, P.G., ed. 'The Merck Index', 8th edn., Merck & Co., New Jersey (1968)

Sober, H.A., ed. 'Handbook of Biochemistry', The Chemical Rubber Co., Ohio (1968)

Weast, R.C., ed. 'Handbook of Chemistry and Physics', 48th edn., The Chemical Rubber Co., Ohio (1967)

'The Extra Pharmacopeia', 24th edn., The Pharmaceutical Press, London (1958)

'The U.S. Pharmacopeia', 17th edn., US Pharmacopeial Convention, New York (1965)

'*Analar* Standards for Laboratory Chemicals', 6th edn., Analar Standards Ltd., London (1967)

Technique books

Humason, G.L., 'Animal Tissue Techniques', 3rd edn., W.H. Freeman & Co., San Francisco (1972)

'The UFAW Handbook on Care and Management of Laboratory Animals', 4th edn., Churchill Livingstone, Edinburgh (1972)

Ross, K.F.A., 'Phase Contrast and Interference Microscopy for Cell Biologists', Edward Arnold, London (1967)

Chayen, J., and Denby, F.F., 'Biophysical Techniques as Applied to Cell Biology', Methuen, London (1968)

Norris, J.R., and Robbins, D.W., 'Methods in Microbiology', Academic Press, New York (1972)

Bracegirdle, B., 'Photography for Books and Reports', David and Charles, Newton Abbot (1970)

Pearse, A.G.E., 'Histochemistry: Theoretical and Applied', Vol. I and II, 3rd edn., Churchill, London (1968 and 1972)

Sharma, A.K., and Sharma, A., 'Chromosome Techniques', Butterworths, London (1965)

Colowick, S.P., and Kaplan, N.O., 'Methods in Enzymology', Vols. 1–23, Academic Press, New York (1955–72)

Gibbs, B.M., and Skinner, F.A., 'Identification Methods for Microbiologists', Academic Press, London (1966)

Kay, D., 'Techniques for Electron Microscopy', Blackwell, Oxford (1965)

Prescott, D.M., ed. 'Methods in Cell Physiology', Vols. 1–5, Academic Press, New York (1964–72)

Stock, R., and Rice, C.B.F., 'Chromatographic Methods', 2nd edn., Chapman and Hall, London (1967)

Bergmeyer, H.U., 'Methods of Enzymatic Analysis', Academic Press, New York (1965)

Altman, P.L., and Dittmer, D.S., eds., 'Metabolism; a Biological Handbook', FASEB, Bethesda, Maryland (1968)

Work, T.S., and Work, E., 'Laboratory Techniques in Biochemistry and Molecular Biology', Vols. 1 and 2, North-Holland, Amsterdam (1970)

Plummer, D.T., 'An Introduction to Practical Biochemistry', McGraw-Hill, London (1971)

General cell biology books

Brachet, J., and Mirsky, A., eds., 'The Cell', Vols. 1–6, Academic Press, New York (1960–4)

Lima De Faria, A., ed., 'Handbook of Molecular Cytology', North-Holland, Amsterdam (1969)

Roodyn, D.B., ed., 'Enzyme Cytology', Academic Press, London (1967)

Packer, L., 'Experiments in Cell Physiology', Academic Press, New York (1967)

Bottle, R.T., and Wyatt, H.V., 'The Use of Biological Literature', Butterworths, London (1972)

Thornburn, C.C., 'Isotopes and Radiation in Biology', Butterworths, London (1972)

Marriott, F.H.C., 'Basic Mathematics for the Biological and Social Sciences', Pergamon Press, Oxford (1970)

Journals

Journal of Cell Biology
Journal of Cell Science
Experimental Cell Research
Journal of Molecular Biology
Journal of Ultrastructure Research
Z. für Zellforschung
J. de Microscopie
Quarterly Journal of Microscopical Science
Journal of the Royal Microscopical Society
Histochemie
Chromosoma
Nature
Science
International Review of Cytology
Advances in Cell and Molecular Biology
Current Topics in Cellular Regulation

Addresses of contributors

Dr R.L.P. Adams Department of Biochemistry, University of Glasgow, G11 6NV

Dr T. ap Rees Department of Botany, University of Cambridge, Cambridge

Dr J.M. Ashworth Department of Biology, University of Essex, Wivenhoe Park, Colchester, CO4 3SQ

Dr M. Balls School of Biological Sciences, University of East Anglia, Norwich, NOR 88C

Dr C.R. Bantock Department of Biology and Geology, Polytechnic of North London, Holloway, London N7 8DB

Dr W. Barry Department of Physiology, University College, Cardiff, CF1 1XL

Dr R.J. Berry Department of Radiotherapy, Churchill Hospital Research Institute, Headington, Oxford, OX3 7LJ

Dr N.P. Bishun Research Department, Marie Curie Memorial Foundation, The Chart, Oxted, Surrey

Dr J. Boss Department of Physiology, School of Veterinary Science, University of Bristol, Park Row, Bristol, BS1 5LS

Dr S. Bradbury Department of Human Anatomy, University of Oxford, Oxford

Dr R. Cammack Department of Plant Sciences, King's College, London, SE24 9JF

Dr W. Cockburn Botanical Laboratories, School of Biological Sciences, University of Leicester, Leicester, LE1 7RH

Mr F.G. Cowie Paterson Laboratories, Christie Hospital and Holt Radium Institute, Manchester, M20 9BX

Dr F.E.G. Cox Department of Zoology, King's College, Strand, London, WC2R 2LS

Dr J. Creanor Department of Zoology, University of Edinburgh, Edinburgh

Prof. A.S.G. Curtis Department of Cell Biology, University of Glasgow, Glasgow, G11 6NV

Dr H. Davies Department of Biophysics, King's College, London, WC2R 2LS

Dr Mary Dawson Department of Pharmaceutical Technology, University of Strathclyde, Glasgow

Dr P. Doyle Research Department, Marie Curie Memorial Foundation, The Chart, Oxted, Surrey

Dr J.G. Edwards Department of Cell Biology, The University, Glasgow, G11 6NV

Dr M.C.W. Evans Department of Botany, University College, London, WC1

Dr M.L. Fenwick Sir William Dunn School of Pathology, Oxford

Dr R.I. Freshney The Beatson Institute for Cancer Research, Royal Beatson Memorial Hospital, 132 Hill Street, Glasgow, G3 6UD

Dr P.B. Gahan Department of Biology, Queen Elizabeth College, London, W.8.

Dr P.M. Godsell School of Biological Sciences, University of East Anglia, Norwich, NOR 88C

Mrs M.V. Haigh Paterson Laboratories, Christie Hospital and Holt Radium Institute, Manchester, M20 9BX

Dr D.O. Hall School of Biological Sciences, King's College, London, SE24 9JF

Prof. E. Jean Hanson† School of Biological Sciences, King's College, London, WC2R 2LS

Dr D.M. Hawcroft Department of Biological Studies, Lanchester Polytechnic, Coventry, CV1 5FB

Dr Shirley E. Hawkins Department of Zoology, King's College, London, WC2R 2LS

Dr P.J.L. Holt Medical School, Royal Infirmary, Manchester M13 9WL

Dr Alma Howard Paterson Laboratories, Christie Hospital and Holt Radium Institute, Manchester, M20 9BX

Dr S.V. Hunt Sir William Dunn School of Pathology, University of Oxford, Oxford, OX1 3RE

Dr J. Jacob Institute of Animal Genetics, West Mains Road, Edinburgh, EH9 3JN

Dr D. Kay Sir William Dunn School of Pathology, University of Oxford, Oxford, OX1 3RE

Dr V.A. Knight Department of Physiology, University College, Cardiff, CF1 1XL

Dr N. Maclean Department of Zoology, University of Southampton, Southampton

Dr A.H. Maddy Department of Zoology, University of Edinburgh, Edinburgh

Dr M.A. Mayo Scottish Horticultural Research Institute, Invergowrie, Dundee, DD2 5DA

Dr C.W.F. McClare Department of Biophysics, King's College, London, WC2R 2LS

Dr J.M. Palmer Department of Botany, Imperial College, London, SW7

Miss Mary Plant Churchill Hospital Research Institute, Headington, Oxford, OX3 7LJ

Dr T.R. Ricketts Department of Botany, School of Biological Sciences, The University, Nottingham, Notts

Dr E. Sidebottom Sir William Dunn School of Pathology, University of Oxford, Oxford, OX1 3RE

Dr J.D. Simnett Developmental Biology Laboratory,

†deceased

University of Newcastle-upon-Tyne, Newcastle-upon-Tyne, NE2 4AB

Prof. H.E. Street Botanical Laboratories, School of Biological Sciences, University of Leicester, Leicester, LE1 7RH

Dr A.M. Whitaker Virology Department, Wellcome Research Laboratories, Langley Court, Beckenham, BR3 3BS

Dr G. Wiernik Churchill Hospital Research Institute, Headington, Oxford, OX3 7LJ

Dr D.C. Williams Research Department, Marie Curie Memorial Foundation, The Chart, Oxted, Surrey

Dr R.S. Worley School of Biological Sciences, University of East Anglia, Norwich, NOR 88C

Addresses of suppliers

Astell Laboratory Service Co. (Hearson Co.) 172 Brownhill Road, Catford, SE6 2DL. (01-698-4814/5217)

Baird & Tatlock, Freshwater Road, Chadwell Heath, Essex. (01-590-7700)

BDH Chemicals Ltd. Poole, Dorset, BH12 4NN. (0202-5520)

Beckman–RIIC Ltd., Sunley House, 4 Bedford Park, Croydon CR9 3LG. (01-686-7435)

Becton, Dickinson U.K. Ltd. (Falcon Plastics, etc) 74 Astmoor Industrial Estate, Runcorn, Cheshire. (09285-6622)

Beecham Research Laboratories, Beecham House, Great West Road, Brentford, Middlesex. (01-560-5151)

Biocult Laboratories, Sandyford Industrial Estate, Paisley, Scotland. (041-887-6111)

BOC Cryoproducts, Deer park Rd., Morden, London, SW19 (01-540-7468)

Boehringer Corp. (London) Ltd. Bilton House, 54–8 Uxbridge Road, Ealing, London W5. (01-579-6944)

Boro' Laboratories (Plastics) 1 Station Buildings, Catford Bridge, London SE6. (01-690-2901)

Burroughs & Wellcome–see Wellcome Reagents Ltd.

Calbiochem Ltd. 10 Wyndham Place, London W1. (01-402-6361)

Cambrian Chemicals Ltd. (Biochemicals) Beddington Farm Road, Croydon. CRO 4XB. (01-689-0681)

The Cambridge Culture Collection of Algae and Protozoa, 36 Storey's Way, Cambridge CB3 ODT (0223-61378)

Chance & Propper, P.O. Box 53, Spon Lane, Smethwick, Worley, Worcs. (021-553-5551)

Ciba-Geigy (UK) Ltd. 42 Berkeley Square, London W1. (01-499-6621)

Coulter Electronics Ltd. High Street, South Dunstable, Beds. (0582-66211)

Dawe Instruments Ltd. Western Avenue, Acton, London NW3. (01-992-5026)

Decon, Ellen St., Portslade, Brighton BN4 1EQ

Degenhardt & Co. Ltd. Carl Zeiss House, 31/36 Foley Street, London W1 (01-636-8050)

Diagnostic Reagents Ltd., Thame, Oxon. (084421-2426)

Difco Laboratories, P.O. Box 14B, Central Avenue,

East Molesey, Surrey. (01-979-9951)

Diversey Ltd., Cockfosters, Barnet, Herts

Down Brothers & Mayer & Phelps Ltd. Church Path, Mitcham, Surrey CR4 3UE. (01-468-6291)

Du Pont Co. (UK) Ltd. Wilbury House, Wilbury Way, Hitchin, Herts. (0462-52671)

East Anglia Chemicals Ltd. (Reeve Angel Scientific), Weir Road, London SW12 (01-673-4434)

Esco (Rubber) Ltd. Walsingham House, 35 Seething Lane, London EC3. (01-480-6140)

Evans Electroselenium Ltd. (EEL), St. Andrew's Works, Halstead, Essex. (078-742461)

Expanded Metal Co., Burwood House, Caxton St., London SW1. (01-222-2766)

Fisons Scientific Apparatus Ltd. Bishop Meadow Road, Loughborough, Leics. (050-935781)

Flow Laboratories Ltd. Heatherhouse Road, Irvine, Ayrshire, Scotland. (029-472833)

Gallenkamp P.O. Box 290, Technico House, Christopher Street, London EC2. (01-247-3211)

Gerrard & Haig Ltd. Gerrard House, Worthing Road, East Preston, Near Littlehampton, Sussex. (09062-4151)

Gillette Surgical Industries Ltd. Great West Road, Isleworth, Middlesex. (01-560-1234)

Glaxo Laboratories Ltd.: see Vestric Ltd.

Grant Instruments (Cambridge) Ltd. Barrington, Cambridge, CB2 5QZ. (0223-870528)

Griffin and George Ltd. Ealing Road., Alperton, Wembley, Middlesex. (01-997-3344)

Gurr, Edward, Ltd. Cressex Industrial Estate, High Wycombe, Bucks. (0494-32761)

Gurr, George T. Lane End Road, High Wycombe, Bucks. (0494-21124)

Hamilton Co. Inc.—see V.A. Howe & Co. Ltd.

Harris Biological Supplies Ltd. Oldmixon, Weston-super-Mare, Somerset. (0934-27534)

Hopkin & Williams, P.O. Box 1, Romford, Essex. RM1 1HA. (01-590-7700)

Horwell, A.R. 2 Grange Way, Kilburn High Road, London NW6. (01-328-1551)

Hoslab, 12 Charterhouse Square, London EC1. (01-253-4840)

Howe, V.A. 88 Petersborough Road, London SW6. (01-736-8394)

Ilford Ltd. Ilford, Essex. (01-478-3000)

Imperial Chemical Industries Ltd. Imperial Chemical House, Millbank, London SW1. (01-834-4444)

Imperial Chemical Industries Ltd. Industrial Chemical Department, Templar House, 81-7 High Holborn, London WC1. (01-242-9711)

Jencons (Scientific) Ltd. Mark Road, Hemel Hempstead, Herts. (0442-4641)

Johnson & Jorgensen Ltd. Merringham Road, London SE7 (01-858-6141)

K & K Laboratories Inc. (Rare/Fine Chemicals), 121 Express Street, Plainview, N.Y. 11803. U.S.A.

Koch-Light Laboratories Ltd. Colnbrook, Bucks. SL3 OB2 (2262-9645)

Kodak Ltd. Box 33, Swallowdale Lane, Hemel Hempstead, Herts. (0442-58621)

Kodak-Eastman Organic Chemicals, Kodak Ltd. Kirkby, Liverpool. (051-546-2101)

LKB Instruments Ltd. 232 Addington Road, South Croydon, Surrey CR2 8YD. (01-657-0286)

Luckham Ltd. Labrow Works, Victoria Gardens, Burges Hill, Sussex. (0444-65348)

May & Baker Ltd. Laboratory Sales Division, Dagenham, Essex. (01-592-3060)

MicroBio Laboratories, 46 Pembridge Rd., London, W11. (01-229-6634)

Miles-Seravac (Pty) Ltd. Stoke Court, Stoke Poges, Slough, Bucks. (369-2151)

Millipore (UK) Ltd. Millipore House, Abbey Road, London NW10 7SP. (01-965-9611)

MSE—Measuring & Scientific Equipment Ltd. Manor Royal, Crawley, Sussex. (0293-31100)

National Collection of Industrial Bacteria, Ministry of Technology, Torry Research Station, P.O. Box 31, 135 Abbey Road, Aberdeen, Scotland. (0224-23131)

National Collection of Yeast Cultures, Brewing Industry Research Foundation, Nutfield, Surrey.

Northern Media Supply Ltd. Crosslands Lane, Newport Road, Brough, Yorkshire HU152 PG. (0430-23131)

Oxoid Division (Oxo Ltd). Southwark Bridge Road, London SE1. (01-928-4515)

Palmer, C.F. (London) Ltd. Lane End Road, High Wycombe, Bucks. (0494-21124)

Parke and Davis Co. Staines Road, Hounslow, Middlesex. (01-570-2361)

Payne, C.E. & Sons Ltd. 6 Iveley Road, Clapham, London SW4 OHS. (01-720-5801)

Pharmacia (GB) Ltd. Paramount House, 75 Uxbridge Road, London W5 5SS. (01-579-0102)

Philip Harris Scientific, Ludgate Hill, Birmingham 3. (021-236-4041)

Portex Ltd. Hythe, Kent. (0303-66863)

Pye Unicam Ltd. York Street, Cambridge, CB1 2PY. (0223-58866)

Radiochemical Centre Ltd. Amersham, Bucks. (02404-4444)

Ralph N. Emanuel Ltd. (Aldrich Organic Chemicals) 264 Water Road, Wembley, Middlesex. HAO 1PY. (01-998-4414)

Rank Brothers, Bottisham, Cambridge. (022-028369)

Rank Strand Electric (Cinemoid) 250 Kennington Road, London SE11 5RD. (01-735-7811)

Scientific Glass Engineering Pty. Ltd. (SGE) 657 North Circular Road, London NW2 7AY. (01-452-6244)

Scientific Supplies Ltd. (Jobling etc) Scientific House, Vine Hill, Clerkenwell Road, London EC1. (01-837-7765)

Searle, G.D. & Co. Ltd., Lane End Road, High Wycombe, Bucks (0494-21124)

Shandon Scientific Co. Ltd. 65 Pound Lane, Willesden, London NW10. (01-459-8671)

Sigma London Chemical Co. Ltd. Norbiton Station Yard, Kingston-upon-Thames, Surrey KT2 7BH. (01-549-3171)

Squibb Chemicals Ltd. E.R. Squibb & Sons Ltd. Regal House, Twickenham, Middlesex. (01-892-0164)

Sterilin Ltd. 12-14 Hill Rise, Richmond, Surrey. (01-940-9982)

TAAB, 52 Kidmore End Rd., Emmer Green, Reading. (0734-475388)

Union Carbide (UK) Ltd. Redworth Way, Aycliff Industrial Estate, Nr. Darlington, Co. Durham. (032571-2581)

Universal Containers Ltd. Kingston Rd., Staines, Mddx. (see Gallenkamp)

Vestric Ltd. P.O. Box 21, Lock Field Avenue, Enfield, Middlesex. (01-804-7222)

Vicsons Ltd. 148 Pinner Road, Harrow, Middlesex. (01-427-0706)

Wellcome Reagents Ltd. Wellcome Research Laboratories, Beckenham, Kent. BR3 3BS. (01-658-2211)

Wesley Coe (Glassware) Ltd. 32 Scotland Road, Cambridge. (0223-63252)

Whatman Biochemicals Ltd. Springfield Mill, Maidstone, Kent. (0622-61688)

Wright Scientific Ltd. (Chromatography Columns) Lower Road, Kenley, Surrey CR2 5NH. (01-668-4611)

Xlon Products Ltd. (Plastics) Xlon House, Glyn Street, London SE11. (01-735-8551)

List of American Suppliers

1. Laboratory Equipment

American Optical Corporation, Scientific Instrument Division, Buffalo, NY 14215. (617-765-9711)

American Instrument Co. (Aminco) 8030 Georgia Avenue, Silver Spring, MD 20910. (301-589-1727)

Bausch & Lomb, Scientific Optical Products Division, 22973 Bausch Street, Rochester, NY 14602. (716-232-6000)

Beckman Instruments, Inc. Spinco Division, 1117 California Avenue, Palo Alto, CA 94304. (415-326-1970)

Bel-Art Products, Pequannock, NJ 07440. (201-694-0500)

Buchler Instruments, (Division of Searle Analytic Inc.) 1327 Sixteenth Street, Fort Lee, NJ 07024. (201-224-3333)

Brinkmann Instruments, Inc. Cantiague Road, Westbury, NY 11590. (516-334-7500)

Corning Scientific Instruments, Medfield, MA 02052. (617-359-2341)

Du Pont Company, Instrument Products Division, Sorvall Operations, Newtown, CT 06470. (203-426-5811)

Fisher Scientific Co. 711 Forbes Ave., Pittsburgh, PA 15219. (412-562-8300)

Gilson Medical Electronics, Box 27, Middleton, WI 53562. (608-836-1551)

Greiner Scientific, Corp. 20 N Moore Street, New York, NY 10013. (212-966-4700)

Hamilton Company, P.O. Box 17500, Reno, NV 89510. (702-786-7077)

Kimble Products, Box 1035, Toledo, OH 43666. (419-242-6543)

Kontes Glass Co. Vineland, NJ 08360. (609-692-8500)

LKB Instruments Inc. 12221 Parklawn Drive, Rockville, MD 20852. (301-881-2510)

Labindustries, 1802M Second Street, Berkeley, CA 94710. (415-843-0220)

LaPine Scientific Co. 6001 S Knox Ave., Chicago, IL 60629. (312-735-4700)

Mettler Instrument Corporation, Box 100, Princeton, NJ 08540. (609-448-3000)

Nalge, (Nalgene Labware) Dept. 2311, Nalgene Labware Div., P.O. Box 365, Rochester, NY 14602. (716-586-8800)

New Brunswick Scientific Co. Inc., 1130 Somerset Street, P.O. Box 606, New Brunswick, NJ 08903. (201-846-4600)

Hewlett-Packard, Palo Alto, CA 94304. (415-493-1501)

ISCO, Lincoln, NE 68505. (402-434-0231)

The London Company, 811 Sharon Drive, Cleveland, OH 44145. (216-871-8900)

Quickfit Div., Instrutec Corp., 7 Just Road, Fairfield, NJ 07006. (201-227-4050)

Sargent-Welch Scientific Company, 7300 N. Linder Avenue, Skokie, IL 60076. (312-677-0600)

Arthur H. Thomas Company, Vine Street at Third, Philadelphia, PA 19105. (215-627-5600)

VWR Scientific, Box 3200, San Francisco, CA 94119. (415-467-2600)

Zeiss, 444 Fifth Avenue, New York, NY 10018. (212-736-6070)

2. Chemicals

Aldrich Chemical, 940 W St. Paul Ave., Milwaukee, WI 53233. (414-273-3850)

J.T. Baker Chemical Company, Phillipsburg, NJ 08865. (201-859-2151)

Eastman Organic Chemicals, 343 State St., Rochester, NY 14650. (716-458-4080)

Fisher Scientific Co. 711 Forbes Ave., Pittsburgh, PA 15219. (412-562-8300)

K & K Laboratories, 121 Express Street, Plainview, NY 11803. (516-433-6262)

Mallinckrodt, Science Products Division, St. Louis, MO 63147. (314-231-8980)

Pierce Chemical, Box 117, Rockford, IL 61105. (815-968-0747)

3. Biochemicals and Enzymes

Boehringer Mannheim Corporation, 219 E. 44th St., New York, NY 10017. (212-682-5656)

Calbiochem. Box 12087, San Diego, CA 92112. (714-453-7331)

Miles Laboratories, Box 272, Kankakee, IL 60901. (815-939-4417)

P-L Biochemicals, 1037 E McKinley, Ave., Milwaukee, WI 53205. (414-271-5040)

Pharmacia Fine Chem, 800 Centennial, Piscataway, NJ 08854. (201-469-1222)

Schwarz-Mann, Mountain View Ave., Orangeburg, NY 10962. (914-359-2700)

Sigma Chemical Co. P.O. Box 14508, St. Louis, MO 63178. (314-771-5750)

Worthington Biochemical Corp. Freehold, NJ 07728. (201-462-3838)

4. Isotopes

Amersham/Searle, 2636S Clearbrook Drive, Arlington Heights, IL 60005. (312-593-6300)

ICN, 26201 Miles Road, Cleveland, OH 44128. (216-831-3000)

New England Nuclear, 575 Albany Street, Boston, MA 02118. (617-426-7311)

5. Bacteriological, Cell Cultures

American Type Culture Collection, 12301 Parklawn Drive, Rockville, MD 20852. (301-881-2600)

Clay Adams, Div. of Becton, Dickinson & Co. 299 Webro Road, Parsippany, NJ 07054. (201-887-4800)

Bio-Rad Laboratories, 32nd & Griffin Avenue, Richmond, CA 94804. (415-234-4130)

Difco Laboratories, Box 1058A, Detroit, MI 48232. (313-961-0800)

Millipore Corp., Ashby Road, Bedford, MA 07130. (617-275-9200)

New Brunswick Scientific Co. Inc., 1130 Somerset Street, P.O. Box 606, New Brunswick, NJ 08903. (201-846-4600)

Searle Analytic Inc., 2000 Nuclear Drive, Des Plaines, IL 60018. (312-289-6600)

Scientific Film Catalogues

Scientific Film Information—compiled by Dr. Gisele Hodges, Cancer Research Fund, Lincoln's Inn Fields, London, WC2.

U.K.

1 The British National Film Catalogue. Published by the British Film Institute, 81 Dean Street, London, W.1.

> Most nearly comprehensive list of non-fiction and short films made in the U.K. since 1963. Gives a synopsis. Catalogue in form of four quarterly issues and one annual hardback cumulative volume published in spring of following year.

2 Science Film Catalogue. Published by the British Film Institute, 81 Dean Street, London, W.1.

> Contains data and synopses of some 500 films in the B.F.I. library. Films selected with the assistance of the British Universities Film Council. Acquires and distributes on behalf of the B.U.F.C. films recommended for university use which are not available in this country through any other channel of distribution; e.g. films made abroad in course of university research.
> Note: now expanded into Higher Education Film Library—see below.

3 Higher Education Film Library Catalogue—Part One: Films on Biological and Medical Sciences. Published by the British Universities Film Council, 72 Dean Street, London, W.1.

> This library was set out at end of 1971 and incorporates the Science Film Library of the British Film Institute.

4 The Central Film Library, Government Building, London, W.1.

> Distributor of all films produced by the Central Office of Information and by various sponsors in U.K. and abroad. Has approximately 1500 titles and is largest lending library of non-fiction films in U.K. Published two catalogues. Main catalogue lists general subjects but also a great number of films of a technological and scientific nature.

5 The British Association for the Advancement of Science, 23 Savil Row, London, W.1.

> General catalogue of films including biological subjects.

6 The National Audio-Visual Aid Library. Educational Foundation for Visual Aids (E.F.V.A.), Paxton Place, London SE27. "Visual Aid Catalogue", in eight parts.

> Part 6 Section I of Catalogue lists films dealing with palaeontology, biology, botany, zoology.
> Part 6 Section II of Catalogue lists films dealing with human biology, hygiene, health.

7 Royal Institute of Chemistry, The Chemical Society, Burlington House, London, W.1.

> "Index of Chemistry Films (1970)". A comprehensive list of films, film strips and film loops on chemistry and related topics.

8 The Institute of Physics, 47 Belgrave Square, London, S.W.1.

> "Physics Films". Supplement to "Physics Education July 1971". A selection of films with reviews on physics and allied subjects which are considered suitable for university level use.

9 The British Medical Association, Department of Audio Visual Communication, B.M.A. House, Tavistock Square, London, W.C.1.

> Catalogue lists under subject headings some 500 medical films. Includes teaching films which are part of the Encyclopaedia Cinematographica.

10 The Open University, Film Library, 25 The Burroughs, Hendon, London, N.W.4 4A.T.

> Audio tapes and films on Biology & Environmental Sciences. For hire or for sale.

11 "Medical Films available in Great Britain (1971)". Compiled and published by the British Industrial and Scientific Film Association, Watergate House, 1 Watergate, Blackfriars, London, E.C.4.

> A comprehensive catalogue of films on medicine and allied topics classified by subject. (Prepared for the British Life Assurance Trust for Health Education with the British Medical Association).

12 "Medical Film 1973". Department of Audio-visual Communication, B.M.A. House, Tavistock Square, London, W.C.1.

> Contains complete list of films in the B.M.A. Film Library and those which have received the BLAT Certificate of Educational Commendation.

13 "The Microcirculation Film Catalogue (1972)". Compiled by Stephen D. Carlill; published by British Medical Association Publications Department, B.M.A. House, Tavistock Square, London, W.C.1.

A catalogue of films available internationally on microcirculation. A directory of workers who have used or are using cine techniques for research in the field.

14 "A Catalogue of Medical and Biological Videotapes and Films (1973)". Published by the University of London Audio-Visual Centre, 11 Bedford Square, London, W.C.1.

A complete list of all productions in the medical and biological sciences made by the Audio-Visual Centre which are available for hire or sale. Supplements appear in June and December each year.

15 "Audio-Visual Materials for Higher Education (1973)". Compiled and published by the British Universities Film Council, Royalty House, 72 Dean Street, London, W.1.

List of over 2000 films, videotapes, sound recordings and tape slides selected as being useful for some aspect of university work. Films classified by subject according to the U.D.C. system.

16 "Sources of Biological Films". Compiled by the Institute of Biology. Published in Journal of Biological Education, 1971, 5, 93.–98.

List of film libraries holding films of interest to biologists. (to be republished in amended form).

17 Federation of European Biochemical Societies (Secretary-General, Prof. H.R.V. Arnstein, Dept. of Biochemistry, King's College, Strand, London, W.C.2.)

List of films of biochemical interest suitable for teaching purposes which can be made available in Europe. Synopsis given.

18 "Films on Animal Development". Compiled by J.R. Hinchliffe. Published in Journal of Biological Education, 1972, 6, 119–123.

List and evaluation of films on animal development.

19 "Films relating to Microbiology". Compiled for the Society of General Microbiology. Published in Journal of Biological Education, 1968, 2, 173–189.

List of films useful for teaching of microbiology. Assessment of films by committee members of the Teaching Group of the Society of General Microbiology.

Foreign

1 French Scientific Films Library (Contemporary Films Ltd), 55 Greek Street, London, W.1.

"French Films on Science and Technology"

2 Institut für den Wissenschaftlichen film, 34 Göttingen, Nonnenstieg 72, Germany.

International collection of research film documents held by the Institute for Scientific Film at Göttingen. Over 2000 films available for both teaching and research purposes. Films listed in the "Encyclopaedia Cinematographica". Major subject areas covered are zoology, botany, microbiology, ethnology, technical sciences, and history of technology. Majority of films are silent; booklet provided with each film. Films are on 16mm; some available on super 8mm.

3 Canada House Film Library, Canada House, London, S.W.1.

"Films on the Biological Sciences".

Lists 700 titles and is produced by the Canadian National Scientific Film Library.

4 Tissue Culture Association, Film Librarian, Pasadena Foundation for Medical Research, 99 North El Molino Avenue, Pasadena, California 91101, U.S.A.

Catalogues of a) TCA Motion picture film rental collection. April 1970

and b) TCA film collection (Pasadena) 1970.

Scientific Film Associations

Scientific Film Associations from "Cinematographic Institutions" 1973. (UNESCO, Department of Mass Communication)

International Scientific Film Association (ISFA): National Branches

ISFA is a non-profit-making and non-governmental organisation, which groups the national associations representative of the scientific film movement of various countries. ISFA has stimulated the forming of national scientific film associations in a number of countries and developed practical procedures for furthering its main functions, which are: "The freest, widest and most efficient exchange of: information about production, the use and the effect of all types of scientific films; films themselves and cinematic material; the personal experience, skills and ideas of workers in scientific cinematography".

Each year, it organizes in a different country an international congress and festival where selected films are presented and specialized papers are read. In addition, the specialized sections (research, higher education, popularization of science) hold meetings in the course of the year. Headquarters: 38 Avenue des Ternes, 75–Paris 17. The following information is in the form supplied by ISFA.

1 ARGENTINA
   Investigaciones Cinematograficas de la Universidad de Buenos Aires, Peru 222, Buenos Aires. Director: Mr. Aldo-Luis Persano.

2 AUSTRALIA
   Commonwealth Scientific and Industrial Research Organization, 314 Albert Street, P.O. Box 89, East Melbourne (Victoria)–3002. Secretary (Administration): Mr. L.G. Wilson, Officer in Charge; Film Unit: Mr. Stanley T. Evans.

3 AUSTRIA
   Bundesstaatliche Haupstelle für Lichtbild und Bildungsfilm, Abteilung Wissenschaftlicher Film, 5 Schonbrunnerstrasse 56, A–1060 Vienna. Director: Dr. Dankward G. Burkert.

4 BELGIUM
   Institut National de Cinematographic Scientifique, 31 rue Vautier, 1040–Bruxelles. Director: Mr. Alan Quintart.

5 BRAZIL
   Institute Nacional de Cinema, Praca da Republica, 141–A–2e andar, Rio de Janeiro. President: Mr. Ricardo Crave Albin.

6 BULGARIA
   Popular Science Films Studio, 9 Boulevard Biruzov, Sofia. Director; Mr. Tontcho Tchoukovsky.

7 CANADA
   Canadian Science Film Association, c/o Mrs. J. Winestone, Canadian Education Association, 252 Bloor Street West, Toronto–5–Ontario.

8 CZECHOSLOVAKIA
   Czechoslovak Scientific Film Association, at Czechoslovak Academy of Sciences, Zahradnikova 28, Brno.

9 FRANCE
   Institute de Cinematographic Scientifique, 38 Avenue des Ternes, 75–Paris 17. President: Dr. Bernard Vallancien. Director: Mr. Jean Painleve.

10 FEDERAL REPUBLIC OF GERMANY
   Institut für den Wissenschaftlichen Film,

Nonnenstieg 72, 34–Gottingen. Director: Professor Dr. Ing. Gotthard Wolf.

11 DEMOCRATIC REPUBLIC OF GERMANY
Nationale Vereinigung ürden Wissenschaft-lichen Film in der DDR, Alt Newawes 116/118, 1502–Potsdam Babelsberg. President: Professor Wofgang Bethmann.

12 HUNGARY
Magyar Film es Müveszek Szevetsege, Gorkij Faser 38, Budapest. VI. President: Mr. Agoston Kollanyi.

13 ISRAEL
Israel Scientific Film Organization, P.O.B. 7181, Jerusalem. Chairman: Dr. E.L. Huppert.

14 ITALY
Associazione Italiana de Cinematografia Scientifica, Via Alfonso Borelli 50, Rome. President: Professor Alberto Stefanelli.

15 JAPAN
The Japan Science Film Institution, 2–1 Surugadai Kanda, Chiyoda-ku, Tokyo. Head Director: Dr. Sinitrio Tomonaga. Executive Director: Mr. Sakuichiro Kanzawa.

16 PEOPLE'S REPUBLIC OF KOREA
Korean Scientific Film Association, Pyong Yang.

17 NETHERLANDS
Netherlands Scientific Film Association, Hengevoldstraat 29, Utrecht. Secretary: Dr. R.L. Schuursma.

18 PHILIPPINES
The Scientific Film Association of the Philippines, c/o National Science Development Board, P.O. Box 3596, Manila. Executive Secretary: Mr. Mauro L. Gonzales.

19 POLAND
Polish Scientific Film Association, Al. Ujazdowskie 45, Warsaw. President: Professor Jan Jacoby.

20 RUMANIA
Studio Cinematografic Alexandru Sahia, B-dul Aviatoriler 106, Bucharest, I.S.F.A. Delegate: Mr. Ion Bostan.

21 SPAIN
Associacion Espanola de Cine Cientifico, Patronato "Juan de la Cierva", Serrano 150, Madrid 2. President: Mr. Guillermo F. Zuniga.

22 UNION OF SOVIET SOCIALIST REPUBLICS
Association of Filmmakers of the USSR, Vassilieveskaya 13, Moscow. President of the Board: Mr. Lev Kulidzhanov.

23 UNITED KINGDOM
The British Industrial and Scientific Film Association, Watergate House, 1 Watergate, Blackfriars, London, E.C.4. Director: Mr. Keith Bennett.

24 UNITED STATES OF AMERICA
American Science Film Association, 7720 Wisconsin Avenue, Bethesda, Maryland 20014. President: Mr. Donald A. Benjamin.

25 URUGUAY
Associacion Uruguaya de Cine Cientifico, Juan L. Cuestas 1525, Montevideo. President: Mr. Remember Caprio. General Secretary: Mr. Dassori-Barthet.

Corresponding Members

1. CUBA
Professor Nicholas Cossio, Ministerio de Educacion, Direccion Nacional de Extension Cultural 36–4708, Mariano (13), Havana.

2 MEXICO
Mr. Galdino Gomez Gomez, Director de la Cinemateca Mexicana, Instituto Nacional de Antropologia e Historia, Departemento Instituto Nacional de Antropologia e Historia, Departmento de Promocion y Diffusion, Cordoba 45, Mexico 7 D.F.

3 SWITZERLAND
Mr. Rene Schacher, Communaute d'action pour le developpement de l'information audio-visuelle, 10, Avenue d'Epenex, 1024–Ecubiens (Vaud).

4 VENEZUELA
Dr. Marcel Roche, Director of the Venezuelan Institute of Scientific Research, Ministry of Health and Social Assistance, Apartado 1827, Caracas.

Scientific Film Publications

U.K.

1 "Information". Published by the Department of Audio-Visual Communication, British Medical Association, BMA House, Tavistock Square, London, W.C.1.

> Bimonthly bulletin. Contains details of developments in medical education and education technology. Information on new teaching materials, reviews of films, audiotapes and transparencies, abstracts of research articles.

2 "BUFC Newsletter". Published by British Universities Film Council Ltd., Royalty House, 72 Dean Street, London, W.1.

> Newsletter published three times a year. Includes details of new films available, of conferences, festivals and publications relating to the use of film and associated media for university teaching and research.

3 "University Vision". Published by British Universities Film Council Ltd., Royalty House, 72 Dean Street, London, W.1.

> Bulletin published twice a year. Provides forum for discussing the application and selection of films and related media for use in university teaching and research over a broad spectrum of subjects.

Foreign

1 "Science Film". Quarterly publication produced by the International Scientific Film Association, 38 Avenue des Ternes, 75-Paris 17, France.

2 "Research Film-Le Film de Recherche-Forschungsfilm". Joint publication of the Encyclopaedia Cinematographica, Institut für der Wissenschaftlichen film, 34 Göttingen, Nonnenstieg, 72, Germany, and of the Research Film Section of the International Scientific Film Association.

> Bi-annual publication. Comprises original publications from all domaines of the research film, topical news items, reports on the activities of the two sponsoring organisations and lists newly issued films of the Encyclopaedia Cinematographica with indications as to contents.